Cover picture

Young pustule with uredospores of *Puccinia striiformis* (West.) (yellow stripe rust) emerging through the epidermis of a wheat leaf (1.300 x, SEM-Laboratory University of Basel in cooperation with Sandoz Agro AG Basel/Switzerland)

K. Mendgen · D.-E. Lesemann (Eds.)

Electron Microscopy of Plant Pathogens

With 281 Figures

Springer-Verlag Berlin Heidelberg NewYork
London Paris Tokyo HongKong Barcelona

Prof. Dr. rer. nat. Kurt Mendgen

Universität Konstanz
Fakultät für Biologie
Lehrstuhl Phytopathologie
D-7750 Konstanz, FRG

Dr. Dietrich-Eckhardt Lesemann

Biologische Bundesanstalt
für Land- und Forstwirtschaft
Institut für Viruskrankheiten der Planzen
Messeweg 11/12
D-3300 Braunschweig, FRG

ISBN-13: 978-3-642-75820-1 e-ISBN-13: 978-3-642-75818-8
DOI: 10.1007/978-3-642-75818-8

Library of Congress Cataloging-in-Publication Data
Electron microscopy of plant pathogens / K. Mendgen, D.-E. Lesemann, eds. p. cm. Results of the International Symposium Electron Microscopy Applied in Plant Pathology, held at the University of Konstanz in September 1989. Includes bibliographical references.

1. Plant diseases Congresses. 2. Phytopathogenic microorganisms Congresses. 3. Electron microscopy—Congresses. I. Mendgen, K. (Kurt). 1944- . II. Lesemann, D.-E. (Dietrich-Eckhardt), 1939- III. International Symposium Electron Microscopy Applied in Plant Pathology (1989: University of Konstanz)

Softcover reprint of the hardcover 1st edition 1991

2131/3020-543210—Printed on acid-free paper

Contributors

You will find the addresses at the beginning of the respective contribution

Preface

Plants, fungi, and viruses were among the first biological objects studied with an electron microscope. One of the two first instruments built by Siemens was used by Helmut Ruska, a brother of Ernst Ruska, the pioneer in constructing electron microscopes. H. Ruska published numerous papers on different biological objects in 1939. In one of these, the pictures by G. A. Kausche, E. Pfankuch, and H. Ruska of tobacco mosaic virus opened a new age in microscopy. The main problem was then as it still is today, to obtain an appropriate preparation of the specimen for observation in the electron microscope. Beam damage and specimen thickness were the first obstacles to be met. L. Marton in Brussels not only built his own instrument, but also made considerable progress in specimen preparation by introducing the impregnation of samples with heavy metals to obtain useful contrast. His pictures of the bird nest orchid root impregnated with osmium were revolutionary when published in 1934.

It is not the place here to recall the different techniques which were developed in the subsequent years to attain the modern knowledge on the fine structure of plant cells and of different plant pathogens. The tremendous progress obtained with tobacco mosaic virus is reflected in the chapter by M. Wurtz on the fine structure of viruses in this Volume. New cytochemical and immunological techniques considerably surpass the morphological information obtained from the pathogens, especially at the host-parasite interface. Information on nonpathogenic systems, such as on mycorrhizae and lichens, is added, since this may help to elucidate the function of the morphological changes observed during the interaction of two organisms.

The results published in the book were presented to a great part as reviews during the International Symposium Electron Microscopy Applied in Plant Pathology, held at the University of Konstanz in September 1989. Further information has been added where it seemed appropriate in order to give a general account of the fine structure of different plant pathogens. This book is the result of many efforts of collaborators and students in the department of Plant Pathology at the University of Konstanz. The authors of the different chapters also referred the articles of their colleagues. In addition, Bill Bushnell, Wolfram Köller, Gerard de Leeuw, Klaus Rudolph, Roland Rohringer, and Urs Wyss helped with reviewing.

Last but not least, we would like to thank the Deutsche Forschungsgemeinschaft, the Deutsche Phytomedizinische Gesellschaft, and the Freunde and Förderer der Universität Konstanz for financial help in organizing the Symposium and in editing this book.

Konstanz and Braunschweig
Summer 1990

K. Mendgen
D.-E. Lesemann

Contents

Chapter 1 H. C. Hoch

Preservation of Cell Ultrastructure by Freeze-Substitution

1 Introduction

Currently, most cytologists "fix" cells with chemical fixatives for examination with light and electron microscopes. There is little doubt in examining such cells that we have gained enormous information pertaining to cell structure and function. Sadly, however, all but the most critical and astute cytologists are probably unaware that the cells they fix and examine represent little of the way in which the cell existed immediately prior to fixation. Most researchers probably have never examined the cell during the fixation process, and as a result have little appreciation of the violent and dramatic changes that occur over a period of seconds and minutes (frequently up to 30 min) before it is finally stabilized. Not only does the process of chemical fixation take time, but the organization of the cytoplasm is drastically altered during the time required for the fixative to diffuse into the cell (Fig. 1). Various organelles such as nuclei, for example, are frequently described as passing through the septal pore of a fungal hypha when, in fact, they probably were nowhere near the septum prior to fixation. In addition to the alteration in the position of organelles during fixation, considerable membrane blebbing and fusion occurs in the post-fixation process [cf. 23, 24]. Mersey and McCully [38] and others [2, 3, 14, 36, 41, 48] have documented the time-course of fixation as well as the concomitant cytoplasmic distortion for various cells. Most cytologists would gain a great appreciation for the dynamics of the fixation process by reading these and other references on the nature of chemical fixation.

Recently, cryofixation or freeze-substitution of cells has become recognized as a preferred method of cell preparation for cytological studies, primarily because cellular organization is maintained in a state very similar to that immediately prior to freezing. Freeze-substitution is conceptually simple. Specimens are frozen in an appropriate cryogen or against a cold metal surface, then placed in a liquid medium at a temperature that maintains the cell in a frozen state until all cell water has been

Department of Plant Pathology, New York State Agricultural Experiment Station, Cornell University, Geneva, New York 14456-0462, USA

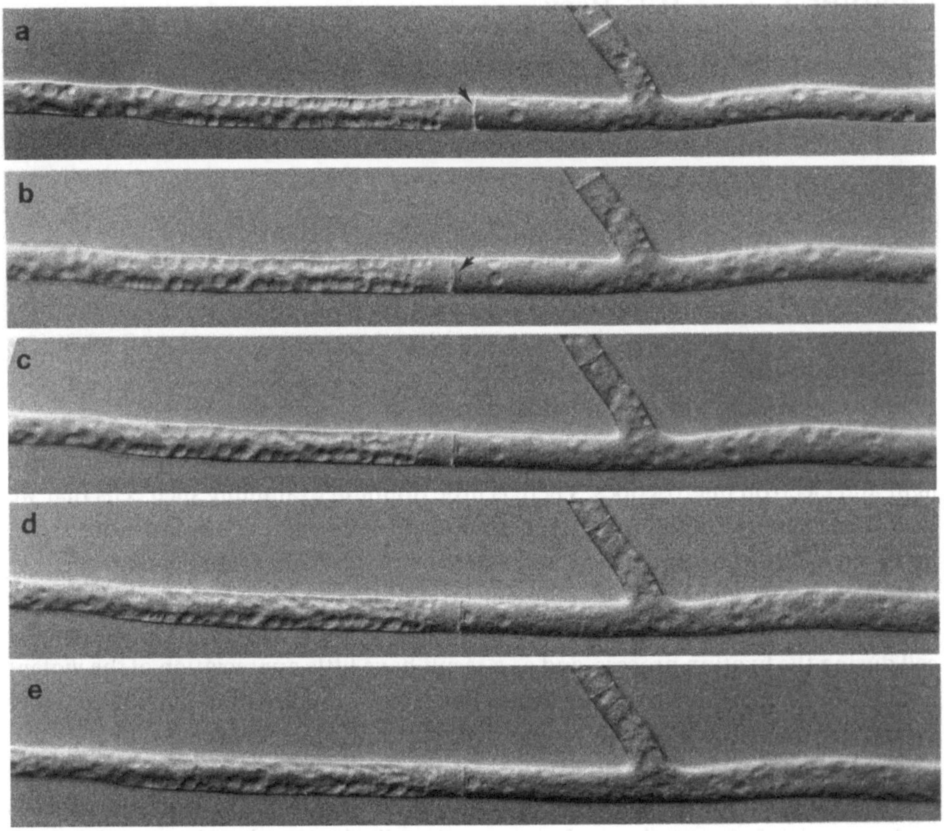

Fig. 1. Light micrographs of a *Rhizoctonia solani* hypha before (**a**) and during (**b**–**e**; 1, 10, 30 s, and 2 min, respectively) fixation with phosphate-buffered glutaraldehyde (2%). Several changes are noted during the fixation process, e.g., septa (*arrows*) and vacuoles become distorted as the cytoplasm surges through the cell (note change in appearance of the vacuoles to the left of the septum). × 500

substituted with the surrounding medium, that usually contains a chemical fixative. Subsequently, the specimens are infiltrated at either sub-zero or room temperatures with resins, and polymerized. Whereas the idea is simple, numerous potential problems have, until recently, discouraged most researchers from pursuing cryotechniques.

Previous reviews and research reports have discussed in considerable detail various methods for specimen preparation and freezing protocols. Some of the more recent and useful reports pertaining to the technique as applied to fungal and plant cells are those by Heath [17, 18], Hoch [23, 24], Hoch and Staples [27], Howard and O'Donnell [30], and Welter and coworkers [58]. In addition to these references, excellent general reviews by Gilkey and Staehelin [15], and Steinbrecht and Müller

[53] provide very useful information on equipment options and techniques concerned with freeze-substitution. Within most of the above cited papers are further references pertaining to safety precautions as well as to handy "tricks of the trade" that should be consulted before one attempts cryofixation for the first time.

2 The Cryofixed Cell

For the period up through 1980, less than six reports had been published in which the ultrastructure of fungi, preserved by freeze-substitution, was documented in detail. Most of those were published in 1979 and 1980. Since then, numerous papers concerned with the ultrastructure of fungi and fungal-plant interactions have been published [e.g., 6–9, 17, 18, 20, 21, 23–27, 29–31, 39, 40, 46, 47, 52, 57, 58]. An important reason for this increase in the reported use of freeze-substitution is the recognition of a greatly enhanced improvement in the preservation of cell structure. Many structural artifacts related to conventional chemical fixation have been eliminated, albeit many remain with freeze-substitution. Just as white is no longer as white when compared to something whiter, fixation artifacts frequently are not recognized until a better preservation method is employed. As biologists we are sometimes "tunnel-visioned" to such artifacts. As an example, the toridial swelling of the dolipore septum, a characteristic morphological feature of many basidiomycete fungi, had been considered (and unfortunately, still is too frequently) well preserved by conventional fixation methods. When similar septa were preserved by freeze-substitution, a very different morphology was observed (Figs. 2, 3) [26]. Critical re-examination using light microscopy of the same septum before and after chemical fixation indicated that, indeed, they were artifactually swollen [26]. Similarly, Golgi in basidiomycete and ascomycete fungi are particularly difficult to preserve using conventional protocols. However, those fixed using freeze-substitution appear quite different morphologically, and probably more closely resemble the organelle in situ, at the instant it was frozen (Figs. 4, 5) [23–25, 27, 29, 39, 40, 47, 52].

A characteristic ultrastructural feature of freeze-substituted cells are membranes (including membrane-bounded organelles and the plasmalemma) having smooth profiles (Figs. 6–8) [6–9, 23, 25, 27, 29, 39, 40, 46, 47]. Double membrane-bounded organelles, such as nuclei and mitochondria, generally exhibit a parallel and smooth profile of the two membranes. Lipid bodies are somewhat electron-opaque, depending on the amount of extraction, fixation, and/or staining. Vacuoles appear to be uniformly filled with a slightly granular content [23, 25]. Other distinct advantages of cryopreserved cells include the apparent preservation of virus-like particles (Fig. 8) [46, 47] and pharmacologically induced cytoplasmic changes [9] not reliably preserved with conventional fixation techniques.

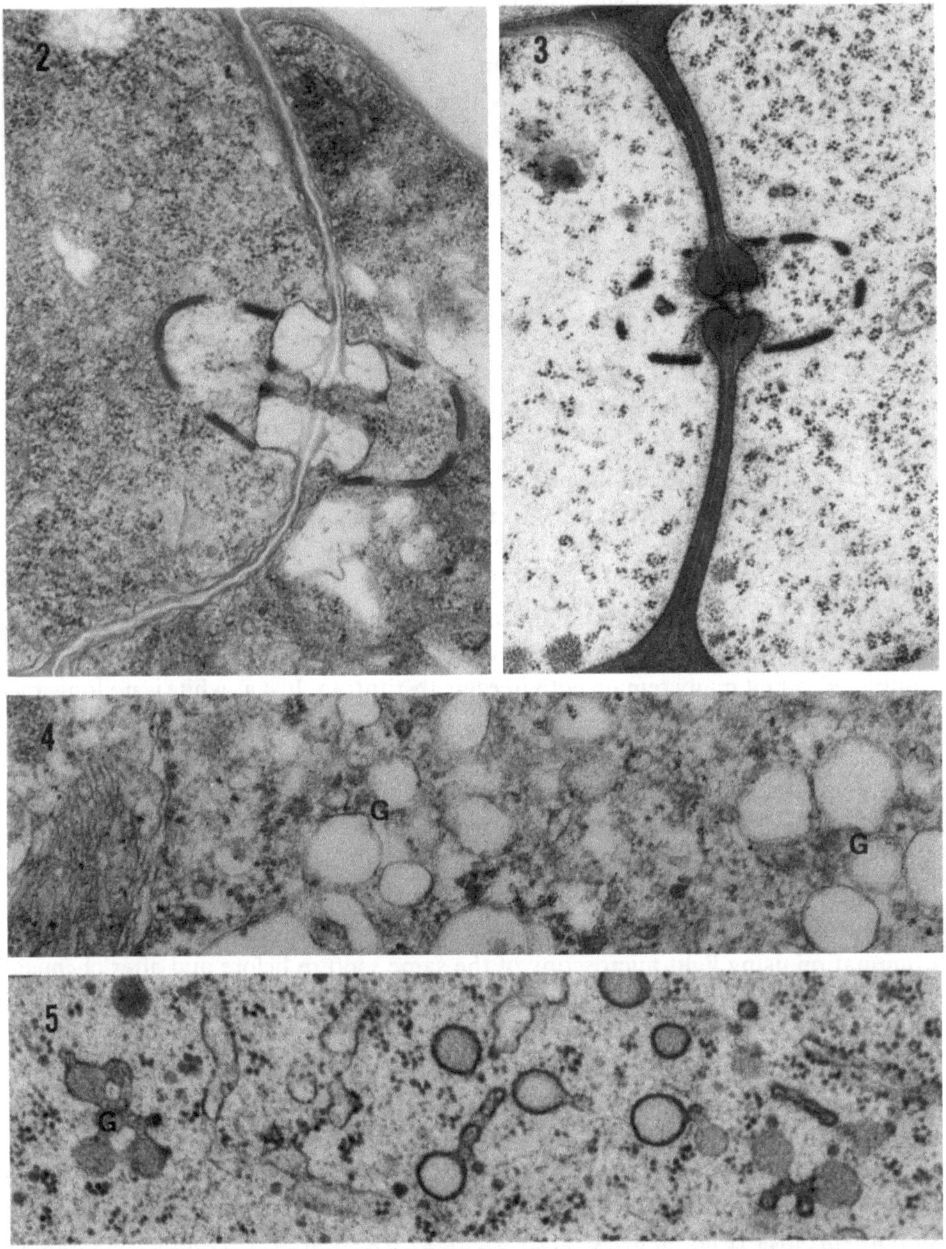

Figs. 2–5. Conventionally fixed and freeze-substituted hyphae of *Laetisaria arvalis* and *Uromyces appendiculatus* depicting dolipore septa and Golgi bodies, respectively. The toroidial swelling of the septum preserved by conventional fixation is swollen and electron-translucent (**2**), whereas a similar septum preserved by freeze-substitution has retained its in vivo size and shape (**3**). Golgi bodies (G) of *U. appendiculatus* fixed conventionally are swollen and not readily recognizable as such organelles (**4**), whereas those fixed cryogenically appear to be intact with the ballooned cisternae interconnected (**5**). Specimens in **2** and **4** were fixed with 2.5% glutaraldehyde and 1% OsO_4 at 20 °C, whereas those represented in **3** and **5** were substituted in 2% glutaraldehyde and 2% OsO_4, followed by 2% OsO_4 in acetone, and 2% OsO_4 and 0.1% U acetate in acetone at − 80 °C, respectively. All were postsection stained with $BaMnO_4$/Pb citrate/U acetate [22]. **2** and **3** both × 30000; **4** and **5** × 50000 and × 60000, respectively. (**3** from Hoch and Howard 1981)

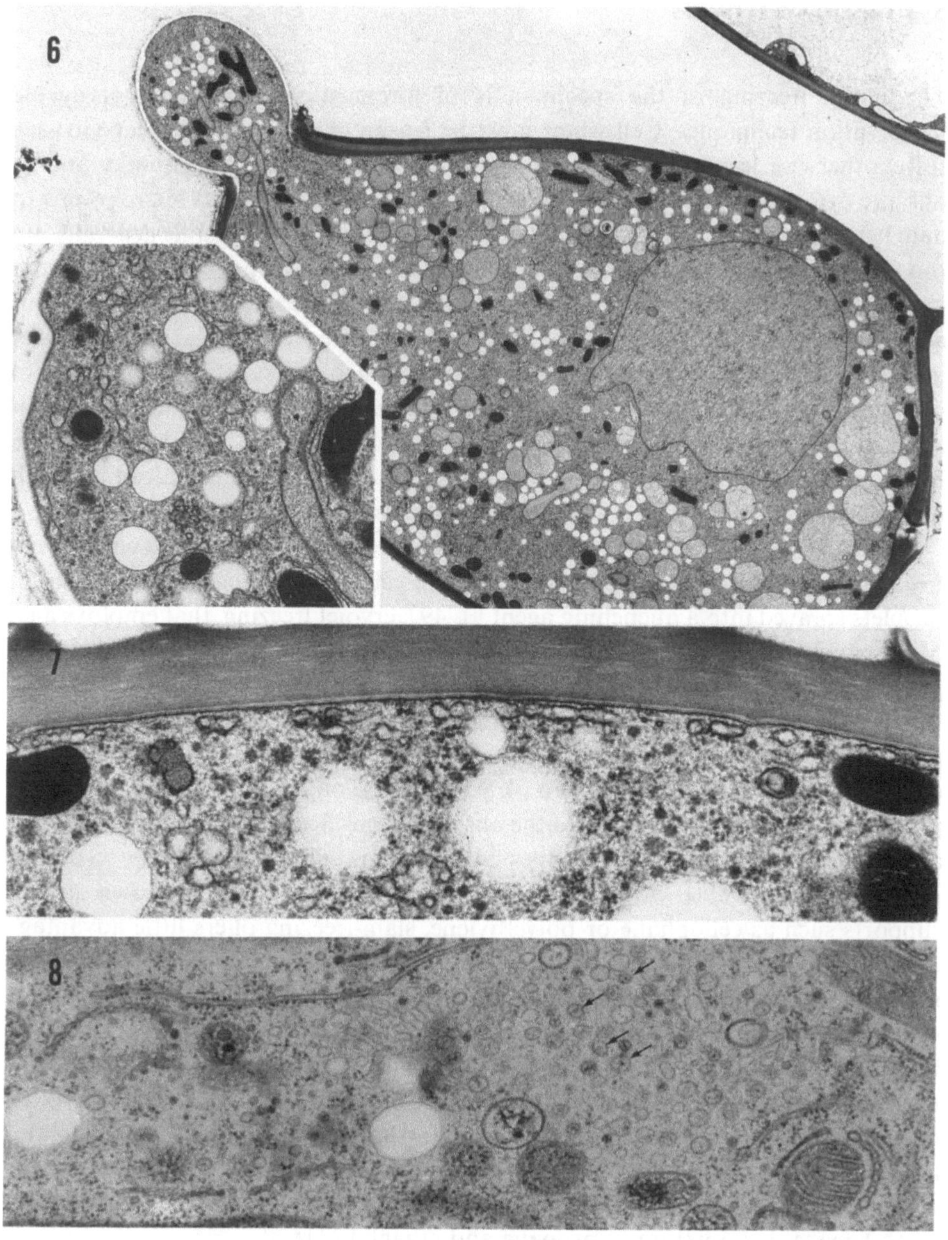

Figs. 6–8. Freeze-substituted cells of *Gymnosporangium clavipes, Authuriomyces peckianus*, and *Endothia parasitica*, respectively. Membranes of various organelles as well as the plasma membrane are preserved with smooth profiles in a germinating teliospore of the rust fungus, *G. clavipes* (**6** and *inset*), and of the aeciospore of *A. peckianus* (**7**). Virus-like particles (*arrows*) present in hypovirulent isolates of *E. parasitica* have only been observed in freeze-substituted cells (**8**). All specimens were substituted in 2% OsO_4 and 0.1% uranyl acetate in acetone and postsection stained with $BaMnO_4$/Pb citrate/U acetate [22]. **6** and *inset* × 3200 and × 15000, respectively **7** × 33000 and **8** × 30000. **6** and **7** from C. Mims and E.A. Richardson unpubl.; Fig. **8** from Newhouse et al. [46]

3 Freezing Methods

The initial freezing of the specimen is of foremost importance in cryogenic preservation techniques. Cell water must be frozen in a vitreous or near-vitreous state, otherwise ice crystals of various sizes can distort cell morphology and/or obscure cytoplasmic detail. Specimens depicted in most published micrographs to date have some degree of ice crystal formation. However, the crystals generally are too small to be easily resolved in the micrograph, and thus do not cause a significant loss of ultrastructural information. Authentic vitrification of cell water, without added cryogenic protectants, probably occurs only under very special circumstances involving extremely small sample sizes [15]. To achieve satisfactory freezing of biological specimens, they must be cooled at extremely fast rates, in the order of approximately 10^4 °C per second, to the final temperature at which they are momentarily maintained, e.g., − 190 °C or lower. They are later placed at much higher temperatures (ca. − 85 °C) at which the water is substituted.

A variety of techniques have been developed to freeze cells. They include spray freezing, whereby the specimens, usually small cells or particles, are suspended in droplets sprayed into a quenching agent [1, 49]; cryojet freezing, that employs a jet of propane sprayed onto the specimen from one or two sides [20, 32, 43, 45, 51]; plunging specimens rapidly into a cryogenic agent such as propane or Freon [11–13, 23, 27, 29, 39, 40]; and "slamming", in which the specimen is rapidly thrust against a polished metal (usually copper) block cooled with liquid helium [8, 10, 12, 17, 19, 56] or alternatively with liquid nitrogen [4, 50]. Because of the excellent heat exchange properties of the metal surface and the absence of gasification of the liquid cryogen around the specimen, some cell types can be slam-frozen satisfactorily to depths greater than is possible with propane. In general, however, for fungi grown on thin supports such as cellophane or polyethylene, slam-freezing offers little advantage over rapid plunging into appropriate cryogens. Furthermore, because of the time needed to warm the plunger, mount the specimen, and clean the cooling block, slammers are not amenable to processing a large number of samples. In addition, liquid helium is expensive. My experience with several commercial slammers has been that fungal cells were either smashed, a problem that could most likely be overcome with equipment modification, or that they exhibited a far greater frequency of cell rupture, whereby the cytoplasm was partiality extruded from the cells. In general, cryopreservation of fungi can be satisfactorily achieved by plunging into propane or a mixture of propane and ethane [23].

None of the aforementioned techniques has lent itself satisfactorily to freezing thicker plant material, such as leaf segments, without extensive ice crystal formation. Recently, considerable success has been reported for these tissues frozen with the newer technique of high pressure freezing [31, 57, 58]. High pressure freezing [42, 44] involves placing the sample, sandwiched between small freezing platelets, in a chamber into which liquid nitrogen is rapidly introduced immediately following an increase in atmospheric pressure in excess of 2000 bar. The technique does not have the limited sample number disadvantage common to slamming, in that samples can be frozen at less than 5-min intervals. However, adaptation of the specimens to the freezing platelets presents somewhat of a problem in handling excised tissues. For best freezing, the material should conform as closely as possible to the interior depression (ca. 1 mm diam) of the platelets. Introduction of 1-hexadecene to the space surrounding the tissues appears to improve freezing [54, 37; see also Mendgen et al., this Vol].

4 Specimen Preparation

The handling of specimens for freezing also warrants careful consideration in order to achieve ice crystal-free preparations. Generally, excellent freezing of cytoplasm can be achieved only to depths of about 10 μm by plunging, cryojet freezing, and slamming techniques. Considerably greater depths of near-vitreous ice formation can be achieved with high-pressure freezing protocols. This is especially important for plant material where tissues are frequently hundreds of microns thick (see Chap. 3). Previously, such material could not be satisfactorily frozen on a routine basis. Because most fungal and bacterial cells are less than 10 μm thick, they can be adequately frozen without too much difficulty. It is important, however, to have the specimens in a physical configuration in which they can be easily manipulated and frozen. Such cells are generally grown in as thin a cell layer as possible, usually on support membranes and not on agar media as is frequently used with conventional chemical fixation protocols. Most filamentous fungi can be grown on membranes permeable to nutrients and water, laid over appropriate agar-based growth media. Membrane supports such as dialysis tubing, cellophane, polycarbonate membranes with pores (e.g., Nucleopore Filters, Nucleopore Filter Corporation, Pleasanton, CA), aluminum foil, polyethylene, and mylar have served as excellent support membranes [18, 23, 25–27, 29, 30, 39, 40, 46, 47]. Further details regarding the preparation and manipulation of these and similar support membranes can be found in recent reviews [23, 30].

Cells grown in liquid culture (e.g., yeasts, zoospores, bacteria, algae) frequently require other types of support. A wire loop (3-mm diam) coated with a thin Formvar film to support the cells may be best for handling these items during freezing [6, 30]. Small pieces of filter paper have also served as carriers of isolated cells, such as fungal zoospores [6, 7]. Specialized specimen holders that sandwich cells between gold or copper discs have been devised for use with plunging or propane jet freezing [8, 12, 16, 20, 45, 51].

Substitution

Cell water is usually substituted by an appropriate organic solvent at temperatures around −85 °C. Recently, molecular distillation at subzero temperatures has been introduced as an alternative, albeit expensive, method of removing frozen cell water without solvents [34]. Most researchers, however, currently employ solvent substitution. The time for substitution to be completed depends on the sample size and substituent, but usually occurs within 1 to 3 days at −85 °C. Various substitution fluids have been used, including acetone, methanol, ethanol, diethylether, glycols, and tetrahydrofuran [cf. 30, 50, 53]. Likewise, various reactive chemicals have been included in the substituents to further stabilize cell components, particularly proteins and lipids. Such additives are necessary for most studies and have included OsO_4, glutaraldehyde, acrolein, uranium salts, $HfCl_4$, tannic acid, oxalic acid, $KMnO_4$, as well as others, singly or in combination [cf. 18, 21, 30, 50]. For the most part, OsO_4 plus uranyl acetate in acetone has most consistently yielded the best cell preservation and as a consequence is used most widely. Three fungi, *Laetisaria arvalis, Rhizoctonia solani*, and *Pythium ultimum*, are illustrated in Figs. 9–22, having been cryopreserved and subsequently substituted in various media. Although substitution with an organic solvent alone preserves considerable cytoplasmic detail, lipids are extracted and electron opacity of retained structures is severely limited even with prolonged post-section staining. Incorporation of OsO_4 at concentrations between 1 and 5%, in acetone, is the most widely used formulation for a substituent. Considerable contrast is obtained for many cellular features including the plasma membrane, mitochondria, ribosomes, and Golgi. However, lipid bodies and membranes of most other organelles (nuclei, endoplasmic reticulum, vacuole, etc.) are poorly preserved, regardless of the concentration of OsO_4 (Figs. 9, 10, 14, 18) [6, 7]. Incorporation of glutaraldehyde or acrolein into the substituent, either alone or with OsO_4, does not generally improve preservation of these latter structures (Fig. 13) [25, 29]. Incorporation of heavy metal salts such as those of uranium or hafnium with OsO_4 or as a postosmication treatment greatly enhance preservation of most membranes and lipids (Figs. 11, 15–17, 19, 20) [23, 30]. Concentrations of uranyl acetate as low as 0.01% in the substituent (with OsO_4) are effective in adding contrast. Concentrations much above 0.2% uranyl acetate are generally unnecessary, since the solution (acetone) becomes saturated in at −85 °C. Potassium permanganate, which was long ago replaced as a chemical fixative for cell ultrastructure, preserves cytoplasmic structure surprising well (Figs. 12, 21, 22) when incorporated into the substituent (acetone) at −85 °C. It is however, not as good as OsO_4 and uranyl acetate.

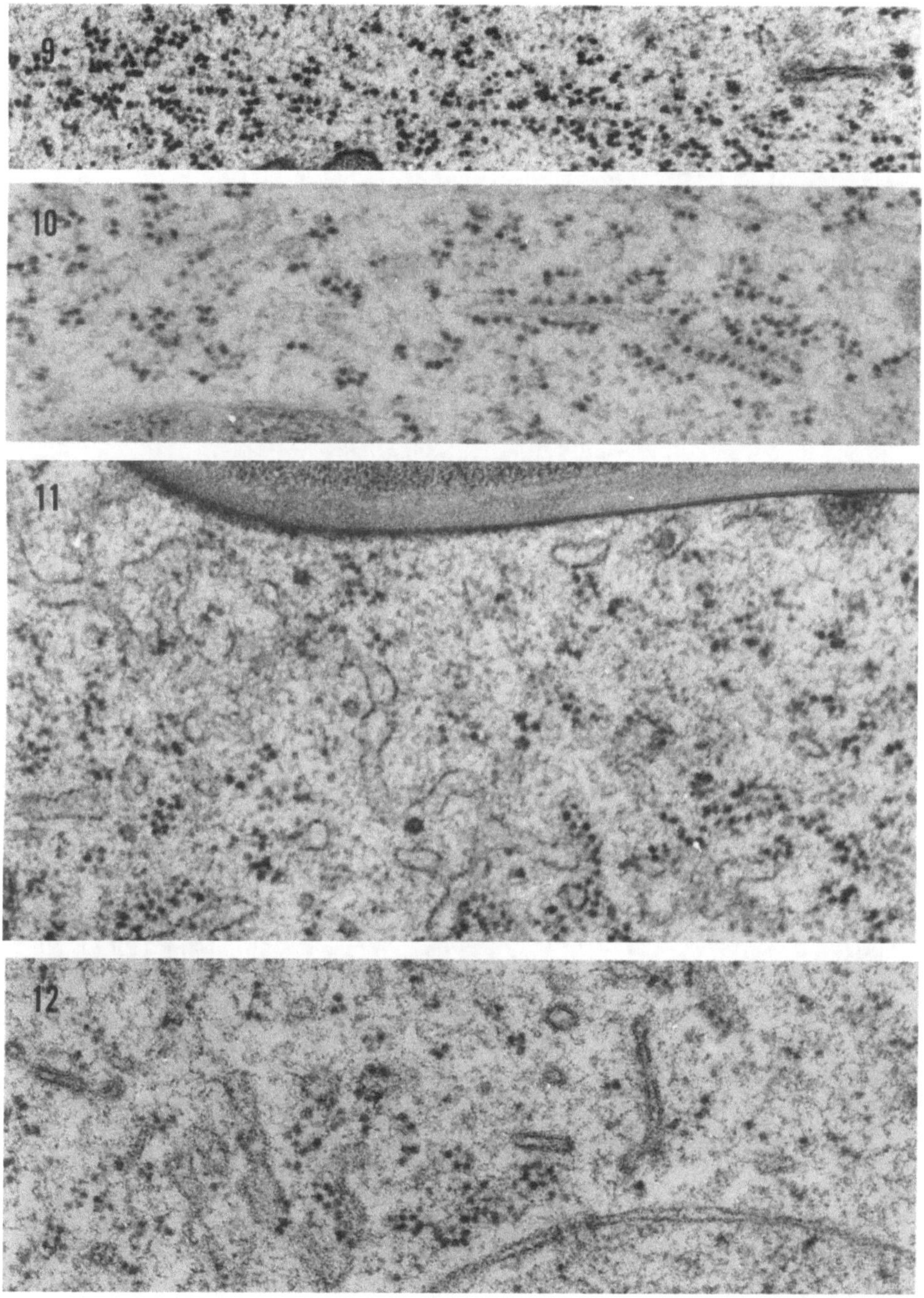

Figs. 9–12. Cytoplasm of *Laetisaria arvalis* preserved by freezing in a mixture of propane/ethane (90%/10%) at −192 °C and substituting in various media. Specimens depicted in **9–12** substituted in 2% OsO_4, 5% OsO_4, 2% OsO_4 plus 0.01% U acetate, and 0.1% $KMnO_4$, respectively, in acetone for 48 h at −80 °C. Endoplasmic reticulum and most other membranes were extracted when substituted in OsO_4 media alone (**9** and **10**). Incorporation of U acetate at concentrations as low as 0.01% preserved cell membranes (**11**). Substitution in $KMnO_4$ while yielding grainy cytoplasmic features, preserved membranes, and ribosomes (**12**). × 50 000, × 60 000, × 70 000, and × 60 000, respectively

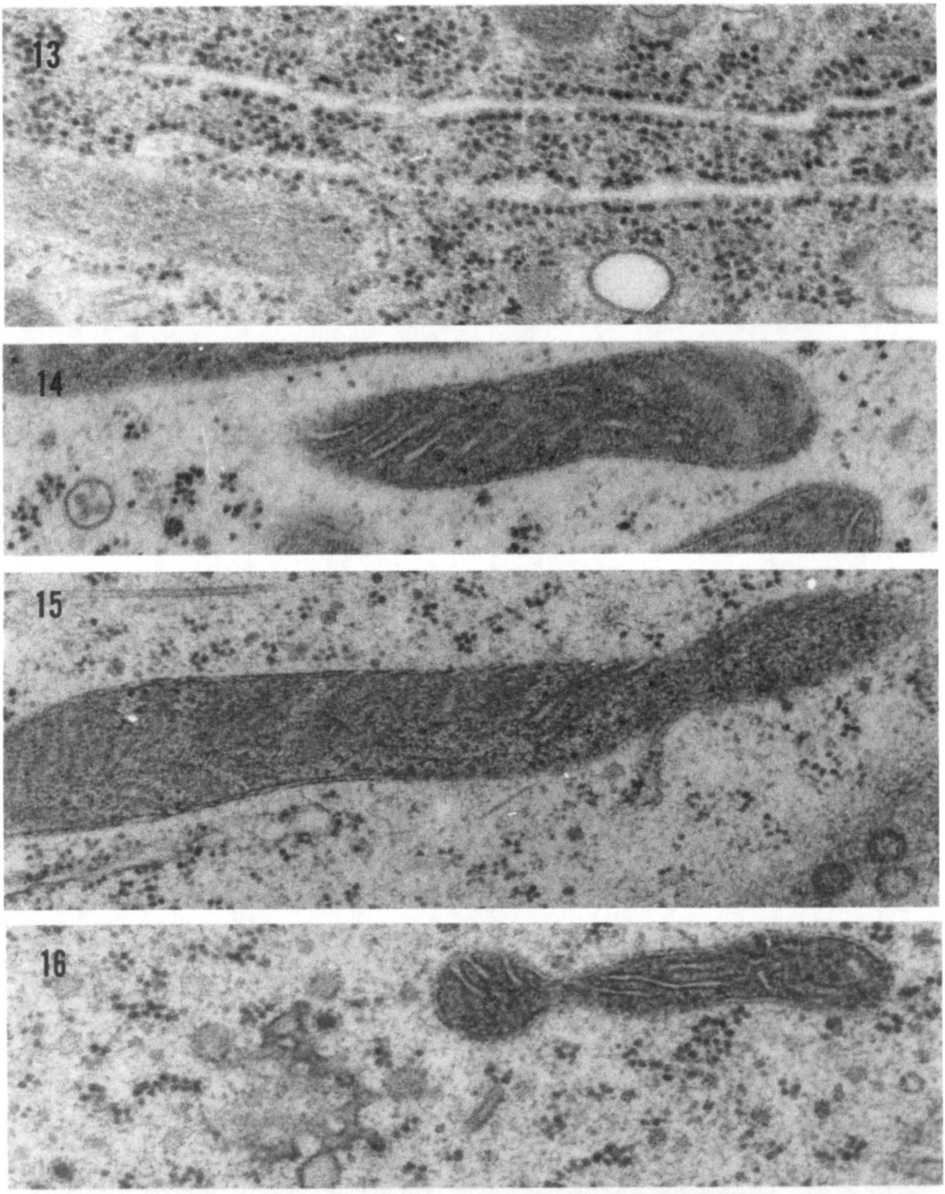

Figs. 13–17. Cytoplasm of *Rhizoctonia solani* preserved by freezing in a mixture of propane/ethane (90%/10%) at $-192\,°C$ and substituting in various media. Specimens depicted in **13–17** were substituted for 48 h at $-80\,°C$ in acetone containing 2% OsO_4 plus 2% glutaraldehyde, 2% OsO_4, 2% OsO_4 (then 1% U acetate in acetone at 20 °C for 1 h), and 2% OsO_4 (then 0.1% $HfCl_4$ in acetone at 20 °C for 1 h), respectively. Most membranes were extracted when substituted in OsO_4 media alone or in media containing OsO_4 plus

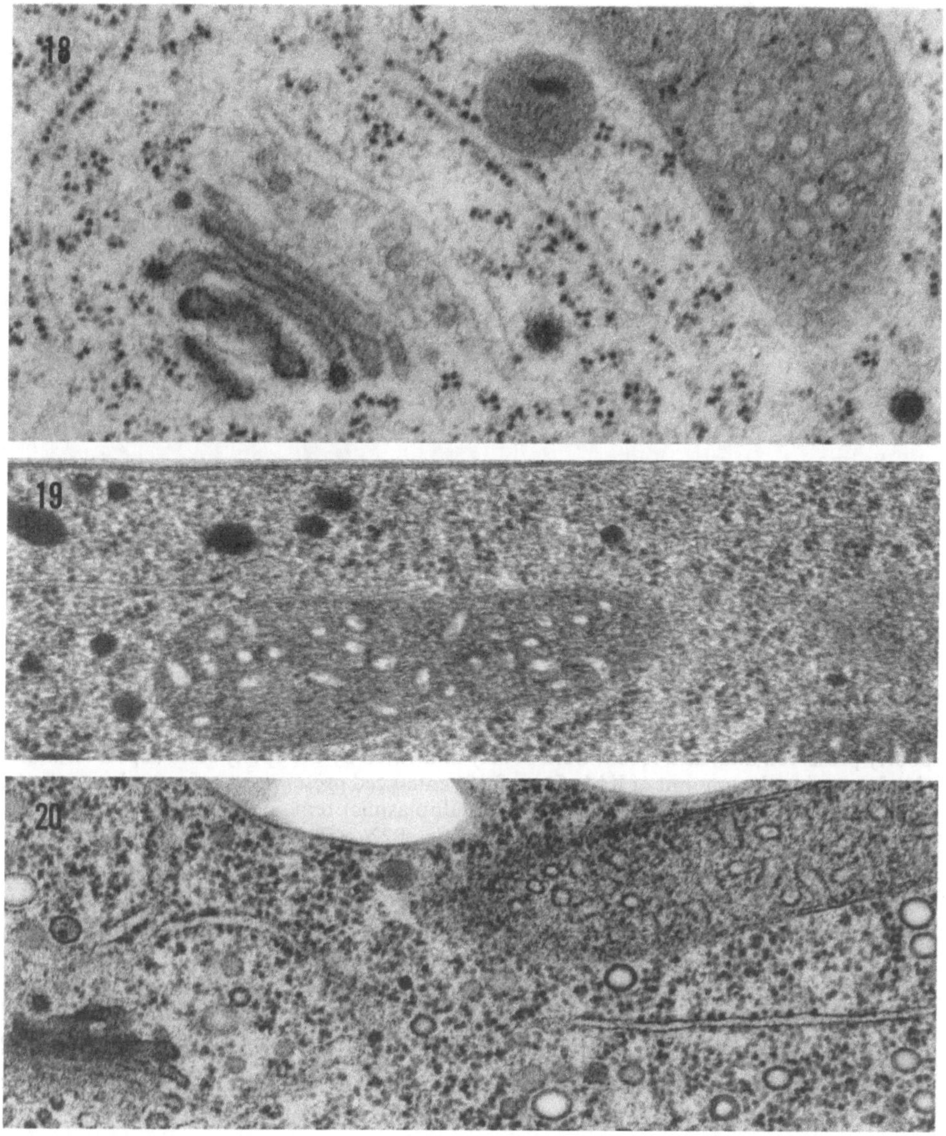

Figs. 18–20. Cytoplasmic features of ***Pythium ultimum*** preserved by freezing in a mixture of propane/ethane (90%/10%) at −192 °C and substituting in various media. Specimens depicted in **18–20** were substituted for 48 h at −80 °C in acetone (except **19**, in ethanol) containing 5% OsO_4, 2% OsO_4 plus 0.1% U acetate, and 2% OsO_4 plus 0.1% U acetate, respectively. Most membranes, except Golgi membranes, were extracted when substituted in OsO_4 media alone (**18**). Substitution in ethanol produced a generally "muddy" appearance of the cytoplasm (**19**). Incorporation of U acetate into the initial substitution medium greatly enhanced membrane contrast (**20**). All × 50 000

◄ ———

glutaraldehyde (**13** and **14**). Substitution in OsO_4 followed by room temperature treatment with U acetate or $HfCl_4$ for 1 h greatly enhanced the contrast of most membranes (**15–17**). All × 60 000

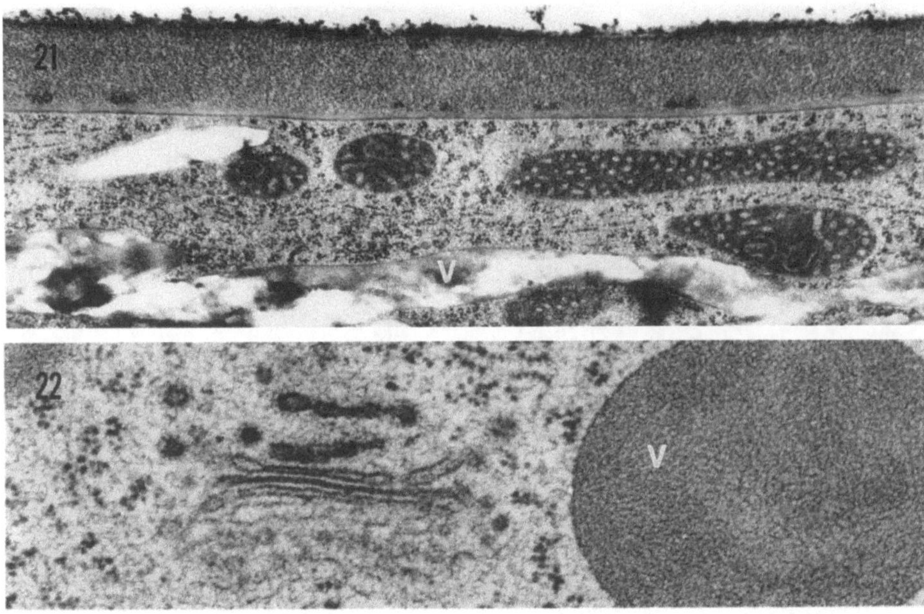

Figs. 21 and 22. Cytoplasmic features of *Pythium ultimum* preserved by freezing in a mixture of propane/ethane (90%/10%) at $-192\,^{\circ}C$ and substitution in acetone containing $KMnO_4$ for 48 h at $-80\,^{\circ}C$. $KMnO_4$ (0.05%) preserved ribosomes, cell wall, mitochondrial matrices, plasma and Golgi membranes well; however, lipid (not shown) and vacuolar contents (v) were poorly preserved and, as a result, were poorly infiltrated by the epoxy resin (Epon-Araldite) (**21**). Increasing the amount of $KMnO_4$ to 0.1% enhanced preservation of vacuolar contents, although many organelle membranes (e.g., endoplasmic) remained poorly preserved (**22**). Note the slight ice crystal formation in the vacuole of **22**. × 20000 and × 48000, respectively

6 Problems and Artifacts Associated with Freeze-Substitution

A problem encountered with nearly all freeze-substituted cells sheathed by a wall (i.e., plants, fungi) is the frequent separation of the cytoplasm from the wall (Figs. 23, 24). Separation occurs at the time of sectioning and usually appears to be located between the plasma membrane and the wall matrix. The reason for such separation is not known, but may be related to the bonding of plastic resins to the membrane-wall interface. The problem is usually tolerated as an annoyance, with the seriousness of separation varying with the cell type. Occasionally entire cells "fall" out of the sections. The most serious problem, however, encountered using freeze-substitution for cell preservation is the occurrence of freeze-damaged cells. The amount of ice crystal formation depends to a large extent on the size of the specimen. Ice crystal damage can be so severe that most cytoplasmic details are obscured [cf. Fig. 21, in Ref. 23]. The outer membrane of mitochondria (Fig. 25) as well as the nucleoplasm [23, 50] appear to be the cell components most sensitive to ice crystal damage. When other components of the cytoplasm appear to be well preserved,

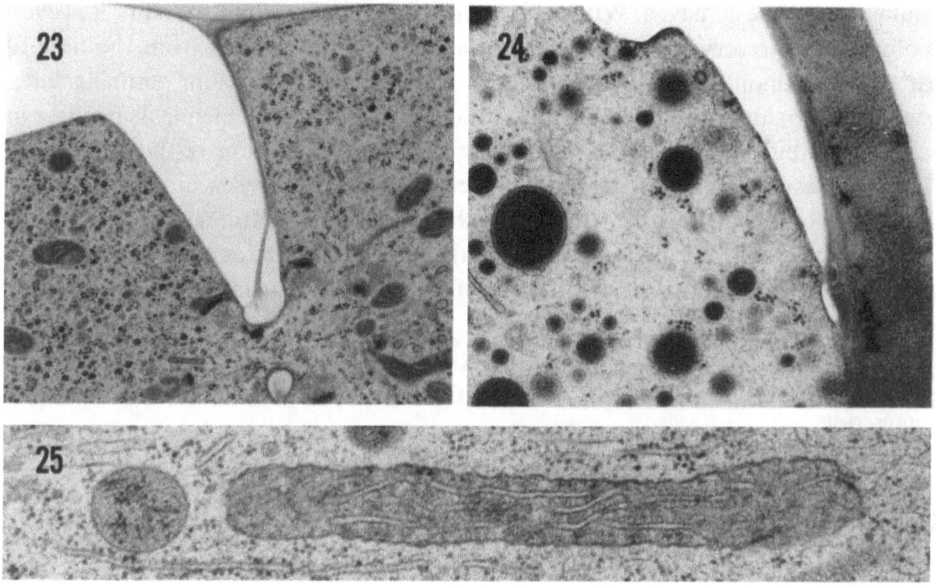

Figs. 23–25. Two serious problems that are frequently encountered in freeze-substituted fungal material are tearing away of specimens from the cell walls during sectioning and ice crystal formation. Cells most often separate at the cytoplasm-wall interface as depicted in *Rhizoctonia solani* and *Phycomyces blakesleeanus* (**23** and **24**). Mitochondria and nucleoplasm (not shown) frequently are the most susceptible cell components to ice crystal damage. A wavy membrane profile is indicative of such damage in mitochondria (**25**). × 18 000, × 30 000, and × 45 000, respectively

these latter are the last structures to be free of damage. Loss of lipids, particularly from membranes, can occur if sufficient osmication has not occurred before the material is warmed to room temperature. Significant osmication does not occur at − 80 °C [56]. The more polar substituents such as methanol and ethanol cause more loss of membranes (lipids) than do acetone or tetrahydrofuran [5]. Collapse of freeze-substituted walled cells and extrusion of small amounts of cytoplasm during the freezing process are occasional problems encountered with cryofixation techniques; however, these can generally be avoided through methods modification [23, 36]. Inadequate preservation of polymeric actin remains a problem common to both conventional and cryofixation protocols. Whereas some actin profiles are observed in freeze-substituted cells, the large arrays of actin cables frequently observed in light microscope studies [28, 55] are only rarely seen in cryopreparations [23, 25, 33].

7 Conclusions

"The purpose of fixation is to stabilize the cell structure so that it does not change with respect to the in vivo condition..." [35]. This has been the goal of cell biologists since the advent of the microscope, but by the very nature of fixation processes, this goal will never be achieved. With each improvement in fixation protocols, however, biologists do preserve cells in a more "natural" state. Progress from the time of early acid-alcohol fixation protocols through the use of potassium permanganate, osmium tetroxide, formaldehyde, glutaraldehyde, and other chemical fixatives in various combinations, has provided better and better renditions of the in vivo cell. In recent years, significant strides have been achieved in preserving cell structure with cryofixation protocols. Although such methods are not new, the quality and reproducibility of freezing has been greatly improved. The current status of the technique is such that most cytologists will be amiss not using cryomethods.

Acknowledgments. Some of the research reviewed here was supported in part by grants from the US National Science Foundation, the USDA-CRGO, and the Whitehall Foundation. I am pleased to acknowledge the help of Dr. Beth Hazen in discussing and reviewing the manuscript.

References

1. Bachmann L, Schmitt WW (1971) Improved cryofixation applicable to freeze-etching. Proc Natl Acad Sci 68: 2149–2152
2. Bajer A, Mole-Bajer J (1969) Fomation of spindle fibers, kinetochore orientation, and behavior of the nuclear envelope during mitosis in endosperm. Fine structural and in vitro studies. Chromosoma 27: 448–484
3. Boyles J, L'Hernault N, Laks H, Palade GE (1981) Evidence for vesicular shuttle in heart capillaries. J Cell Biol 91: 418a
4. Boyne AF (1979) A gentle, bounce-free assembly for quick-freezing tissues for electron microscopy: Application to isolated torpedine ray electrocyte stacks. J Neurosci Methods 1: 353–364
5. Bridgman PC, Reese TC (1984) The structure of cytoplasm in directly frozen cultured cells. I. Filamentous meshworks and the cytoplasmic ground substance. J Cell Biol 99: 1655–1668
6. Cho CW, Fuller MS (1989) Observations of the water expulsion vacoule of *Phytophthora palmivora*. Protoplasma 149: 47–55
7. Cho CW, Fuller MS (1989) Ultrastructural studies of encystment and germination in *Phytophthora palmivora*. Mycologia 81: 539–548
8. Dahmen H, Hobot JA (1986) Ultrastructural analysis of *Erysiphe graminis* haustoria and subcuticular stroma of *Venturia ineaqualis* using cryosubstitution. Protoplasma 131: 92–102
9. Dahman H, Hoch HC, Staub Th (1988) Differential effects of sterol inhibitors on growth, cell membrane permeability and ultrastructure in two target fungi. Phytopathology 78: 1033–1042
10. Dempsey GP, Bullivant S (1976) A copper block method for freezing non-cryoprotected tissues to produce ice-crystal-free regions for electron microscopy. J Microsc 106: 251–270
11. Elder HY, Gray CC, Jardine AG, Chapman JN, Biddlecombe WH (1982) Optimum conditions for cryoquenching of small tissue blocks in liquid coolants. J Microsc 126: 45–61
12. Escaig J (1982) New instruments which facilitate rapid freezing at 83 K and 6 K. J Microsc 126: 221–229
13. Feder N, Sidman RL (1958) Methods and principles of fixation by freeze-substitution. J Biophys Biochem Cytol 4: 593–600

14. Fitzharris TP, Bloodgood RA, McIntosh JR (1972) The effect of fixation on the wave propagation of the protozoan axostyle. Tiss Cell 4: 219–225
15. Gilkey JC, Staehelin LA (1986) Advances in ultrarapid freezing for the preservation of cellular ultrastructure. J Electron Microsc Tech 3: 177–210
16. Handley DA, Alexander JT, Chien S (1981) The design and use of a simple device for rapid quench-freezing of biological samples. J Microsc 121: 273–282
17. Heath IB (1984) A simple and inexpensive liquid helium cooled 'slam freezing' device. J Microsc 135: 75–82
18. Heath IB, Rethoret K, Arsenault AL, Ottensmeyer FP (1985) Improved preservation of the form and contents of wall vesicles and the Golgi apparatus in freeze substituted hyphae of *Saprolegnia*. Protoplasma 128: 81–93
19. Heuser JE, Reese TS, Dennis MJ, Jan Y, Jan L, Evans L (1979) Synaptic vesicle exocytosis captured by quick freezing and correlated with quantal transmitter release. J Cell Biol 81: 275–300
20. Hippe S (1985) Ultrastructure of haustoria of *Erysiphe graminis* f. sp. *hordei* preserved by freeze-substitution. Protoplasma 129: 52–61
21. Hippe S, Hermanns M (1986) Improved structural preservation in freeze-substituted sporidia of *Ustilago avenae*—a comparison with low-temperature embedding. Protoplasma 135: 19–30
22. Hoch HC (1977) Use of permanganate to increase electron opacity of fungal walls. Mycologia 69: 1209–1213
23. Hoch HC (1986) Freeze-substitution in fungi. In: Aldrich HC, Todd WJ (eds) Electron microscopy of microorganisms. Plenum, New York, pp 183–212
24. Hoch HC (1987) Freeze-substitution: A technique that yields improved ultrastructural preservation of fungi. In: Fuller MS, Jaworski A (eds) Zoosporic fungi in teaching and research. Southeastern Publishing, Athens, pp 161–166
25. Hoch HC, Howard RJ (1980) Ultrastructure of freeze-substituted hyphae of the basidiomycete *Laetisaria arvalis*. Protoplasma 103: 281–297
26. Hoch HC, Howard RJ (1981) Conventional chemical fixations induce artifactual swelling of dolipore septa. Exp Mycol 5: 167–172
27. Hoch HC, Staples RC (1983) Ultrastructural organization of the non-differentiated uredospore germling of *Uromyces phaseoli* variety *typica*. Mycologia 75: 795–824
28. Hoch HC, Staples RC (1983) Visualization of actin in situ by rhodamine-conjugated phalloin in the fungus *Uromyces phaseoli*. Eur J Cell Biol 32: 52–58
29. Howard RJ, Aist JR (1979) Hyphal tip cell ultrastructure of the fungus *Fusarium*: Improved preservation by freeze-substitution. J Ultrastruct Res 66: 224–234
30. Howard RJ, O'Donnell KL (1987) Freeze substitution of fungi for cytological analysis. Exp Mycol 11: 250–269
31. Knauf GM, Welter K, Müller M, Mendgen K (1988) The haustorial host-parasite interface in rust-infected bean leaves after high-pressure freezing. Physiol Mol Plant Pathol 34: 519–530
32. Knoll G, Oebel G, Plattner H (1982) A simple sandwich-cryogen-jet procedure with high cooling rates for cryofixation of biological materials in the native state. Protoplasma 111: 161–176
33. Lancelle SA, Cresti M, Hepler PK (1987) Ultrastructure of the cytoskeleton in freeze-substituted pollen tubes of *Nicotiana alata*. Protoplasma 140: 141–150
34. Linner JG, Livesey SA, Harrison DS, Steiner AL (1986) A new technique for removal of amorphous phase tissue water without ice crystal damage: A preparative method for ultrastructural analysis and immunoelectron microscopy. J Histochem Cytochem 34: 1123–1135
35. Martelli P, Russo M (1984) Use of thin sectioning for visualization and identification of plant viruses. In: Maramosch K, Koprowski H (eds) Methods in virology, Vol VIII. Academic Press, New York
36. McCully ME, Canny MJ (1985) The stabilization of labile configurations of plant cytoplasm by freeze-substitution. J Microsc 139: 27–33

37. McDonald K, Morphew MK (1989) Preservation of embryo ultrastructure by high-pressure freezing and freeze-substitution. Proc 47th Electron Microsc Soc Am, pp 994–995
38. Mersey B, McCully ME (1978) Monitoring the course of fixation of plant cells. J Microsc 114: 49–76
39. Mims CW, Richardson EA (1989) Ultrastructure of appressorium development by basidiospore germlings of the rust fungus *Gymnosporangium juniperi-virginianae*. Protoplasma 148: 111–119
40. Mims CW, Roberson RW, Richardson EA (1988) Ultrastructure of freeze-substituted basidiospores of *Gymnosporangium juniperi-virginianae*. Mycologia 80: 356–364
41. Mineyuki Y, Gunning BES (1988) Streak time-lapse video microscopy: analysis of protoplasmic motility and cell division in *Tradescantia* stamen hair cells. J Microsc 150: 41–55
42. Moor H, Bellin G, Sandri C, Akert K (1980) The influence of high pressure freezing on mammalian nerve tissue. Cell Tissue Res 209: 201–216
43. Moor H, Kistler J, Muller M (1976) Freezing in a propane jet. Experientia 32: 805
44. Moor H, Riehle U (1968) Snap-freezing under high pressure: A new fixation technique for freeze-etching. In: Bocciarelli DS (ed) Proceedings of the Fourth European Regional Conference on Electron Microscopy, Rome, Vol 2, pp 33–34
45. Muller M, Meister N, Moor H (1980) Freezing in a propane jet and its application in freeze-fracturing. J Mikros 36: 129–140
46. Newhouse JR, Hoch HC, MacDonald WL (1990) Virus-like particles in hyphae and conidia of European hypovirulent (ds-RNA-containing) strains of *Endothia (Cryphonectria) parasitica*: Can J Bot 68: 90–101
47. Newhouse JR, Hoch HC, MacDonald WL (1983) The ultrastructure of *Endothia parasitica*. Comparison of virulent with a hypovirulent isolate. Can J Bot 61: 389–399
48. O'Brien TP, Kuo J, McCully ME, Zee S-Y (1973) Coagulant and non-coagulant fixation of plant cells. Aust J Biol Sci 26: 1231–1236
49. Pfaller W, Rovan E (1978) Preparation of resin embedded unicellular organisms without the use of fixatives and dehydration media. J Microsc 114: 339–351
50. Phillips TE, Boyne AF (1984) Liquid nitrogen-based quick freezing: Experiences with bounce-free delivery of cholinergic nerve terminals to a metal surface. J Electron Microsc Tech 1: 9–29
51. Pscheild P, Schudt C, Plattner H (1981) Cryofixation of monolayer cell cultures for freeze-fracturing without chemical pre-treatments. J Microsc 121: 149–167
52. Sewall TC, Roberson RW, Pommerville JC (1989) Identification and characterization of Golgi equivalents from *Allomyces macrogynus*. Exp Mycol 13: 239–252
53. Steinbrecht RA, Müller M (1987) Freeze-substitution and freeze-drying. In: Steinbrecht RA, Zierold K (eds) Cryotechniques in biological electron microscopy. Springer, Berlin Heidelberg New York Tokyo, pp 149–172
54. Studer D, Müller M (1989) Cryofixation of thick (500 μm) biological specimens by high pressure freezing. Proc 47th Electron Microsc Soc Am pp 732–733
55. Tucker BE, Hoch HC, Staples RC (1986) The involvement of F-actin in *Uromyces* cell differentiation: The effects of cytochalasin E and phalloidin. Protoplasma 135: 88–101
56. Van Harreveld A, Crowell J (1964) Electron microscopy after rapid freezing on a metal surface and substitution fixation. Anal Rec 149: 381–386
57. Mendgen K, Welter K, Scheffold I (1990) High pressure freezing of infected plant leaves. In: Mendgen K, Lesemann D-E (eds) Electron microscopy of plant pathogens. Springer, Berlin Heidelberg New York Tokyo (this Vol)
58. Welter K, Müller M, Mendgen K (1988) The hyphae of *Uromyces appendiculatus* within the leaf tissue after high pressure freezing and freeze substitution. Protoplasma 147: 91–99

Chapter 2 NICK D. READ

Low-Temperature Scanning Electron Microscopy of Fungi and Fungus-Plant Interactions

1 Introduction

Low-temperature scanning electron microscopy (LTSEM) involves direct examination of frozen-hydrated material at low temperature in the scanning electron microscope (SEM). Biological specimens prepared and examined in this way typically exhibit superior preservation over those prepared by procedures which necessitate examination of dried material at ambient temperature. This is because frozen-hydrated specimens: (a) are rapidly immobilized and stabilized by cryofixation; (b) retain most or all of their water; and (c) are not exposed to chemicals, fixatives or solvents. Other advantages of LTSEM include ease and speed of preparation and versatility with regard to the range of ways in which a single frozen specimen can be manipulated [6]. Although Echlin and coworkers published the first detailed study involving LTSEM of biological material in 1970 [10], LTSEM did not become routine in many laboratories around the world until the 1980's when various commercial cryopreparation systems became available [6, 20, 27, 29].

In this short review, the basic methodologies of LTSEM and its use in the study of fungi and fungus-plant interactions are discussed. Only a brief literature survey is provided because of space limitations.

2 Specimen Preparation

Specimen preparation for LTSEM involves three main operational phases: (a) cryofixation; (b) freeze-fracturing, dissection, partial freeze-drying and/or coating at low temperature; and (c) observation and analysis at low temperature in the SEM [6].

Institute of Cell and Molecular Biology, University of Edinburgh, King's Buildings, Mayfield Road, Edinburgh EH9 3JH, Scotland, UK

Samples to be cryofixed are mounted on a specialized stub using a low temperature adhesive like Tissue Tek II OCT compound (Lab-Tek Products, Division Miles Laboratories, Inc., Naperville, IL, USA). Stub design varies depending on the cryopreparation system employed and stubs are often modified by users for specific purposes. Cryofixation normally involves plunging the mounted specimen and stub into liquid or slushed nitrogen. Water condensation on the frozen specimen surface is minimized after removal from the coolant by maintaining the sample under vacuum and protecting it, where possible, beneath a cold trap or shroud. Unless purposefully heated after freezing, the temperature of the specimen is normally maintained below at least $-100\,^{\circ}C$ and when located on cold stages below $-150\,^{\circ}C$. Samples may be heated to temperatures between $-60\,^{\circ}C$ and $-90\,^{\circ}C$ using the SEM cold stage heater. This causes sublimation of superficial frozen water which might be masking the specimen surface or etches fracture faces. The specimen interior, however, remains frozen-hydrated. This process of partial freeze-drying can be directly visualized in the SEM. Samples treated in this way are thus referred to as being *partially freeze-dried* to distinguish them from those which are *fully frozen-hydrated*. Specimens are usually coated with a conductive material (e.g., gold, or lighter elements like C, Be, Al or Ni if X-ray microanalysis is to be performed) in the cryopreparation chamber before final examination in the SEM [6].

This abbreviated account of LTSEM specimen preparation is far from comprehensive and references 6, 20, 27, 29 should be consulted for more detailed descriptions. A very important aspect of the technique stems from the large number of options available with regard to manipulation of the specimen once frozen; numerous different types of procedures may be performed on one original sample (e.g., the sample may be uncoated, coated, unfractured, fractured, fully frozen-hydrated, partially freeze-dried etc.) [6].

A variety of detection systems may be used to image and analyse frozen-hydrated specimens. These include secondary electron detection (the most common method for morphological studies), backscattered electron detection (e.g., to detect, by means of atomic number contrast, heavy metals or colloidal gold labels) and X-ray microanalysis (for elemental analysis).

Important recent developments in LTSEM technology are: (a) *More rapid methods of cryofixation*. These include plunge freezing into liquid propane [21], high pressure freezing with liquid nitrogen [21] and slam freezing against a liquid nitrogen-cooled copper block [13]. (b) *Complementary fracturing*. This provides an improved perception of the three-dimensional nature of a specimen [4, 13]; (Figs. 4, 5). (c) *Long-term storage of frozen-hydrated specimens*. Up until recently it was not possible to store specimens in the frozen-hydrated stage for extended periods because the microscope and cryopreparation system are warmed up after a work session [6]. Devices which allow the storage of large numbers of frozen-hydrated specimens at low temperature for periods up to several months have now been built [30]; (R. Porter and A. Beckett unpubl.). This approach, coupled with the fact that the preparation time for LTSEM is extremely short (i.e., a few minutes to just over an hour), now makes it practicable to store large numbers of treatments from an experiment performed remote from the SEM (e.g., in another laboratory or in the field). It can also allow LTSEM to be used to monitor dynamic processes such as infection of a leaf by a fungus or fungal development in culture, with repeated sampling at short intervals from even the same material [6, 30]; (R. Porter and A. Beckett unpubl.). These types of studies are not easily performed using ambient-temperature scanning electron microscopy because the preparation time is usually long (e.g., 1–2 days) [6].

3 Rapid Immobilization and Stabilization

Cryofixation achieves much more rapid specimen immobilization and stabilization than does chemical fixation. For example, we estimated the times required to completely immobilize a *Sordaria* perithecium (300 μm wide at its base and 600 μm high) to be: (a) approximately 50 ms using cryofixation by plunging into slushed nitrogen; and (b) greater than 5 min by using chemical immersion fixation [6]. Ice crystal damage is inevitable in most samples examined by LTSEM because of their relatively large size (commonly $>10\,\mu$m and often <5 mm) (see next section).

The employment of cryofixation in LTSEM is important to: (a) suitably prepare a sample for subsequent processing and/or examination (e.g., for freeze-fracturing or examination in the frozen-hydrated state); (b) capture dynamic cellular events (e.g., that occur during spore discharge [17]; Figs. 13, 14); (c) stabilize material in which the structural integrity has been lost (e.g., necrotic leaf tissue, Figs. 11, 12); and (d) retain diffusible ions at sites close to those *in vivo* (i.e. for X-ray microanalysis, [11, 16] Zeyen, Chap 5, this Vol). Recently, however, we have provided evidence in leaf tissue that solutes may be extruded from cells during cryofixation for LTSEM [13].

4 Retention of Water

Water is the major structural component in most biological material; commonly 60–90% of a cell is water. This water has variable properties and there is a transition from the majority of water which is freezable to water which is unfreezable due to specific interactions with membranes, macromolecules and solutes. The latter is often loosely referred to as "bound" or hydration water [2, 14].

The immediate impression one usually gets when first looking at frozen-hydrated cells is that they mostly have a turgid, "life-like" appearance (Figs. 1, 2, 4–7, 9–11, 13–16). However, not all frozen-hydrated cells have this appearance. Cells

s
p
v
c
1
2
3

may collapse as a result of autolysis [24]; (Fig. 10) and cell collapse is a natural, reversible phenomenon in many spores [7]; (Fig. 8).

Drying causes shrinkage whilst freezing causes expansion of biological material. We have measured the volumes of spores after different methods of preparation. In one analysis we found that critical point-dried and fully freeze-dried urediospores of *Uromyces viciae-fabae* had mean volumes which were 40% and 59%, respectively, of the mean volume of fully frozen-hydrated spores. The latter will have inevitably expanded to some extent but this expansion (i.e., less than the 9% by which pure water expands in volume) will be significantly less than the shrinkage suffered by dried spores [7]. We have also found that the amount of shrinkage varies in different cell types [7]. Partial freeze-drying by using the SEM cold stage heater to sublimate water can also cause shrinkage, particularly if prolonged (N.D. Read unpubl.), although it is commonly less than in fully dried samples (cf. Figs. 2 and 3).

Drying can distort or cause wrinkling of cell surfaces. We have shown that shrinkage of urediospores of *U. viciae-fabae* during critical point-drying is significant in this. However, we have also provided evidence that hydration water, preserved in frozen-hydrated cells and important in maintaining the stability of a cell's macromolecular architecture, may have an important influence on at least urediospore surface texture [23]. LTSEM has been used to monitor contact-mediated changes in the surface texture of partially freeze-dried conidia of *Erysiphe graminis* although it was not clear how these changes occurred [15, 22]; (Kunoh et al. Chap. 17, this Vol.). However, in most cases the undistorted or non-wrinkled, turgid morphology of cells is retained after partial freeze-drying (Figs. 1, 2, 4–7, 9, 10, 13–16), suggesting that tightly held, hydration water plays an important role in giving a cell its turgid appearance.

Cell-cell spatial relationships are very well preserved in frozen-hydrated tissues and is especially evident after freeze-fracturing (Figs. 4–7, 12). This feature has proven very advantageous for studying fungus-plant interactions [1, 3, 5–7, 8, 9, 12, 19, 28, 29, 33].

Freeze-fractured, fully frozen-hydrated cells or extracellular materials tend to have a relatively featureless morphology (Figs. 4–6). When we etch them, however,

◄

Fig. 1. *Aspergillus niger*. Asexual reproductive apparatus. Note the conidiophore (*c*), vesicle (*v*), phialides (*p*) and conidia (*s*). Partially freeze-dried. Bar = 10 μm

Fig. 2. *Ascodesmis sphaerospora*. Ascogonial coils surrounded by extracellular material. Partially freeze-dried. Bar = 20 μm

Fig. 3. *Ascodesmis sphaerospora*. Ascogonial coils. Note that most of the extracellular material seen in Fig. 2 has been extracted in this preparation. Only insoluble components remain on the ascogonial coils. Also note that this specimen is considerably shrunken relative to that in Fig. 2. Glutaradehyde fixation, OsO_4 post-fixation, ethanol dehydration, critical point-dried. Bar = 20 μm

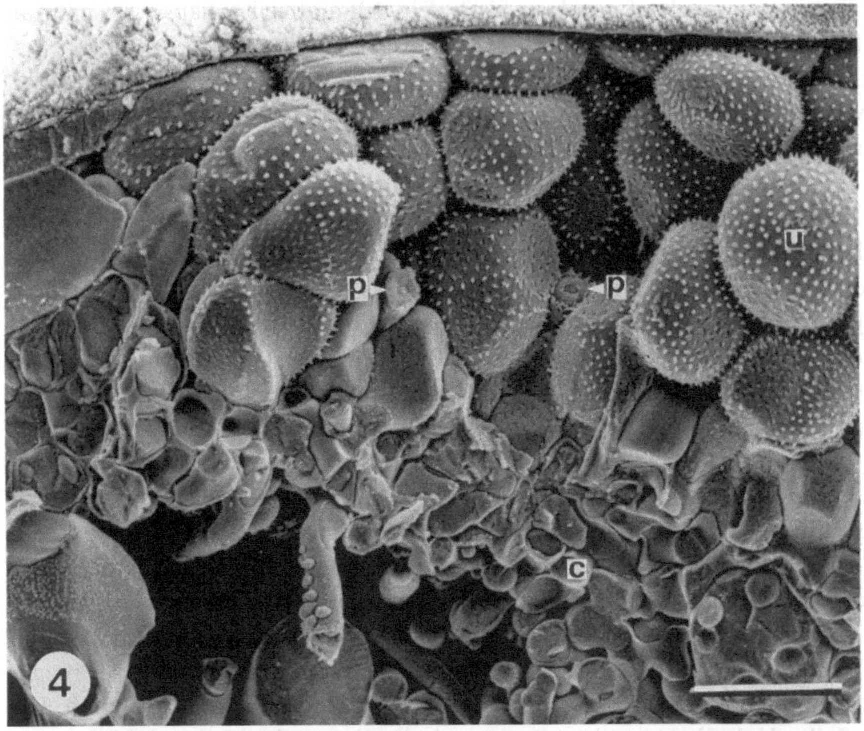
p
p
u
c
4

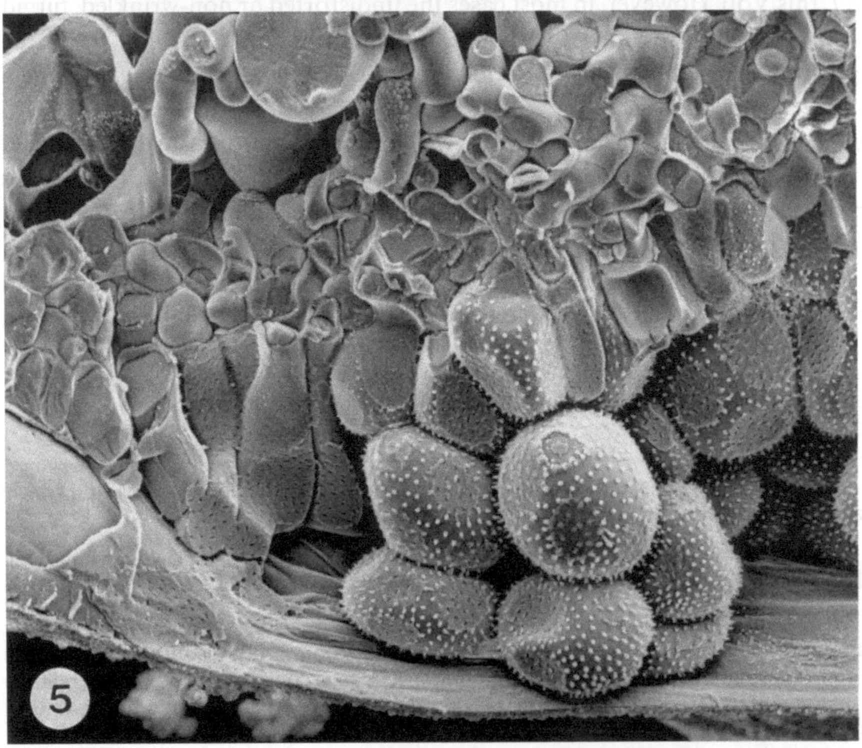
5

we find they contain artifactual segregation zones. These artifacts predominately arise from ice crystal formation during rapid freezing. The spaces between segregation zones represent ice crystals which have been subsequently removed during partial freeze-drying (Figs. 7, 9, 11, 12). Segregation zones contain the "biological matrix" (i.e., organelles, macromolecular structures, solutes and some hydration water) [6, 25, 26]. The extent of ice crystal damage determines the spatial resolution achievable within fractured cells using LTSEM. Nevertheless, useful structural information is often evident at the subcellular level. One useful feature of this artifact is that it can give one a qualitative idea of amounts of freezeable water in different parts of a cell; the more widely spaced the segregation zones, the higher the freezeable water content [6]. In Fig. 9 it is possible to differentiate between the vacuole, which has a high water content, and the surrounding cytoplasm containing organelles which has a lower water content. In necrotic leaf material the pattern of segregation zones becomes markedly changed from the patterns in healthy leaves (cf. Figs. 9 and 11). This presumably reflects altered water relations of necrotic tissues. It should be emphasized, however, that the precise pattern of segregation zones in a cell is impossible to predict because of the very complex kinetics of ice crystal growth [2].

5 No Exposure to Chemicals, Fixatives or Solvents

Fungal extracellular matrices (ECMs) tend to be labile components and often readily extractable with solvents (e.g., during chemical fixation and dehydration). Nevertheless these ECMs are often of great importance at the fungus-plant interface since they can be involved in fungal adhesion and provide a medium through which signals pass between the two partners.

An important advantage of LTSEM is that it avoids exposure of specimens to chemicals, fixatives or solvents. The dramatic effects of solvents on fungal material have been shown previously [5, 6, 24] and are illustrated by comparing Figs. 2 and 3. LTSEM can provide excellent preservation of extracellular matrices and secretions [4, 5, 8, 24, 25, 32]; (Figs. 4, 5, 9), ballistospore droplets [6, 17, 18] and leaf waxes at the host-pathogen interface [6, 24, 26]; (Fig. 8).

◄

Figs. 4 and **5.** *Puccinia striiformis/Hordeum vulgare.* Complementary fractures of a young uredium about to erupt through the leaf surface. Note the mass of coherent hyphae (*c*) which give rise to pedicels (*p*) upon each of which is borne a urediospore (*u*). Freeze-fractured, fully frozen-hydrated. Bar = 20 μm

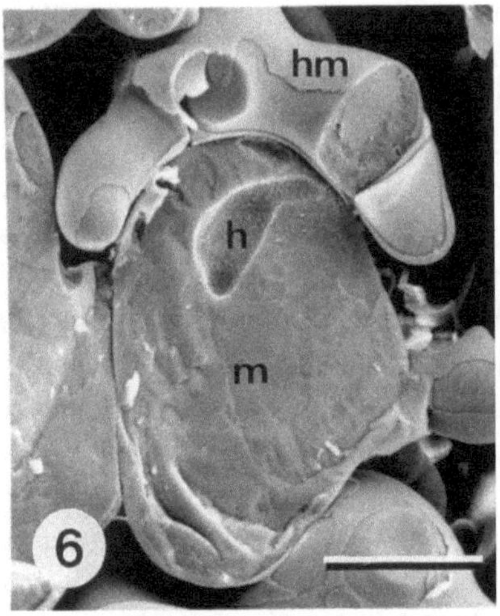

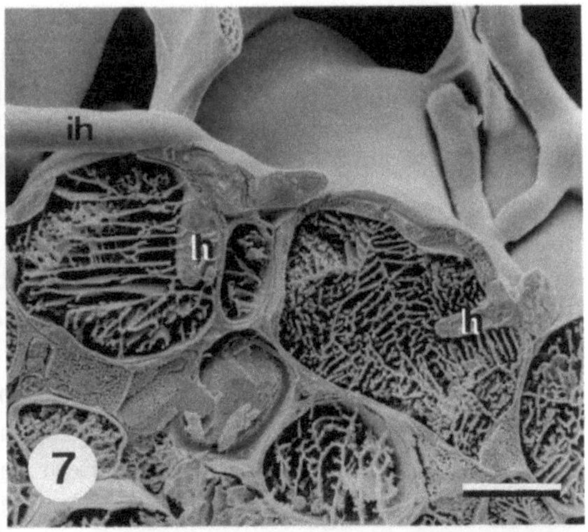

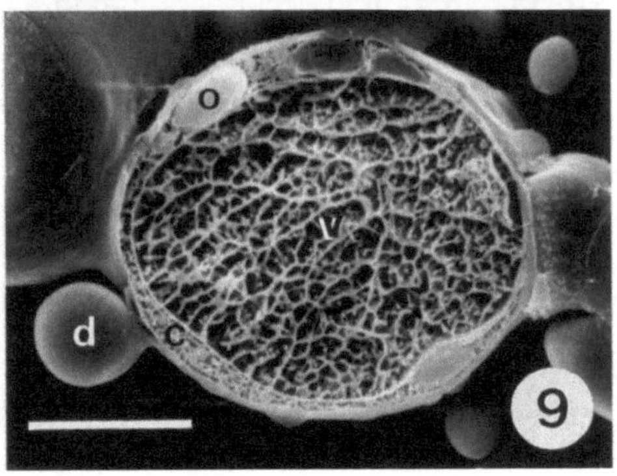

Fig. 6. *Puccinia striiformis/Hordeum vulgare* leaf. Partly fractured haustorial mother cell (*hm*) adhering to a mesophyll cell (*m*) within which the impression of a haustorium (*h*) can be seen. Freeze-fractured, fully frozen-hydrated. Bar = 10 μm

Fig. 7. *Phytophthora infestans/Solanum tuberosum.* Peg-like haustoria (*h*) penetrating cells of a leaf vein. Note the intercellular hyphae (*ih*) adhered to the outside of the vein in the leaf intercellular space. Also note segregation zones within fractured cells. Freeze-fractured, partially freeze-dried. Bar = 10 μm

Fig. 8. *Puccinia hordei/Hordeum vulgare.* Naturally collapsed urediospores on the surface of a leaf. This cell collapse is a reversible phenomenon which allows spores to reduce their water content. Note the erosion of the leaf wax around the urediospores. This dissolution probably resulted from the action of enzymes released from the ungerminated spores. Partially freeze-dried. Bar = 20 μm

Fig. 9. *Triticum aestivum.* Leaf mesophyll cell which have secreted unetchable, droplet-like exudates (*d*) in response to infection with *Puccinia striiformis.* Note the wider spacing of segregation zones in the large central vacuole (*v*) than in the surrounding cytoplasm (*c*) due to the former's higher freezable water content. Chloroplasts (*o*) are visible in the cytoplasm. Freeze-fractured, partially freeze-dried. Bar = 5 μm

Fig. 10. *Phytophthora infestans.* Oogonium (*o*) surrounded at it base by an antheridium (*a*). Note the considerable rupturing of the agar which is a common artifact in materials with a high water content (also see Figs. 13, 14). It arises from differential shrinkage of the specimen surface during partially freeze-drying. Also note the hypha (*h*) which is collapsed, probably naturally as a result of autolysis. Partially freeze-dried. Bar = 20 μm

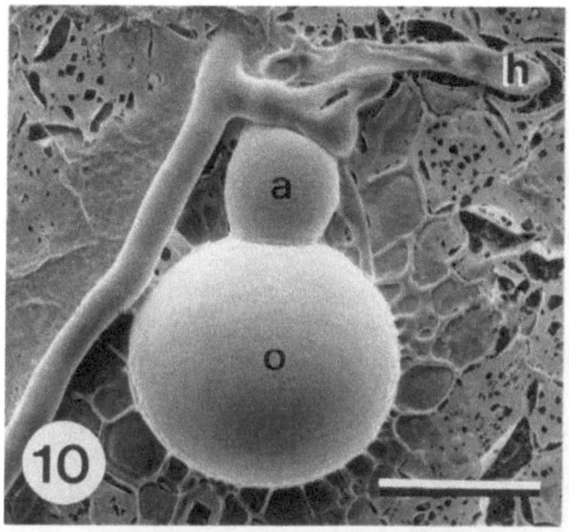

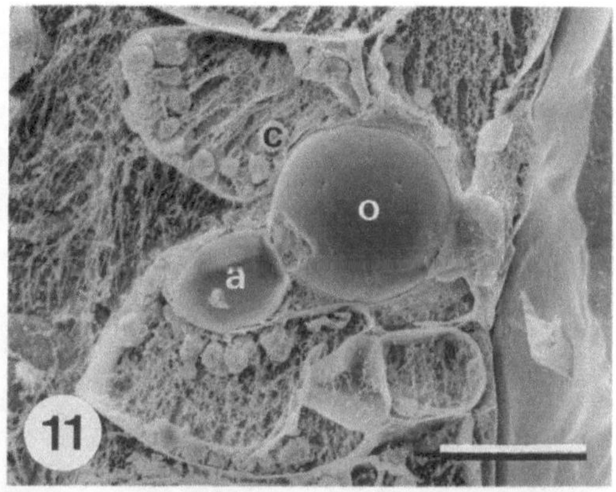

Fig. 11. *Phytophthora infestans/Solanum tuberosum.* Partly fractured oogonium (*o*) and antheridium (*a*) embedded in necrotic leaf tissue. Note that the pattern of segregation zones in the leaf cells is different from that in cells which are not necrotic (e.g., see Fig. 9). Organelles (*c*), probably chloroplasts, can be seen in the leaf mesophyll cells. This material (and also that in Fig. 12) was prepared by co-inoculating isolating potato leaves with A1 and A2 mating type strains isolated from Scotland (N.D. Read and J. Malcolmson unpubl.). Freeze-fractured, partially freeze-dried material. Bar = 20 μm

Fig. 12. *Phytophthora infestans/Solanum tuberosum.* Thirteen oospores (*o*), as revealed in this fracture plane, embedded in the vein region of a necrotic leaf. Freeze-fractured, partially freeze-dried material. Bar = 100 μm

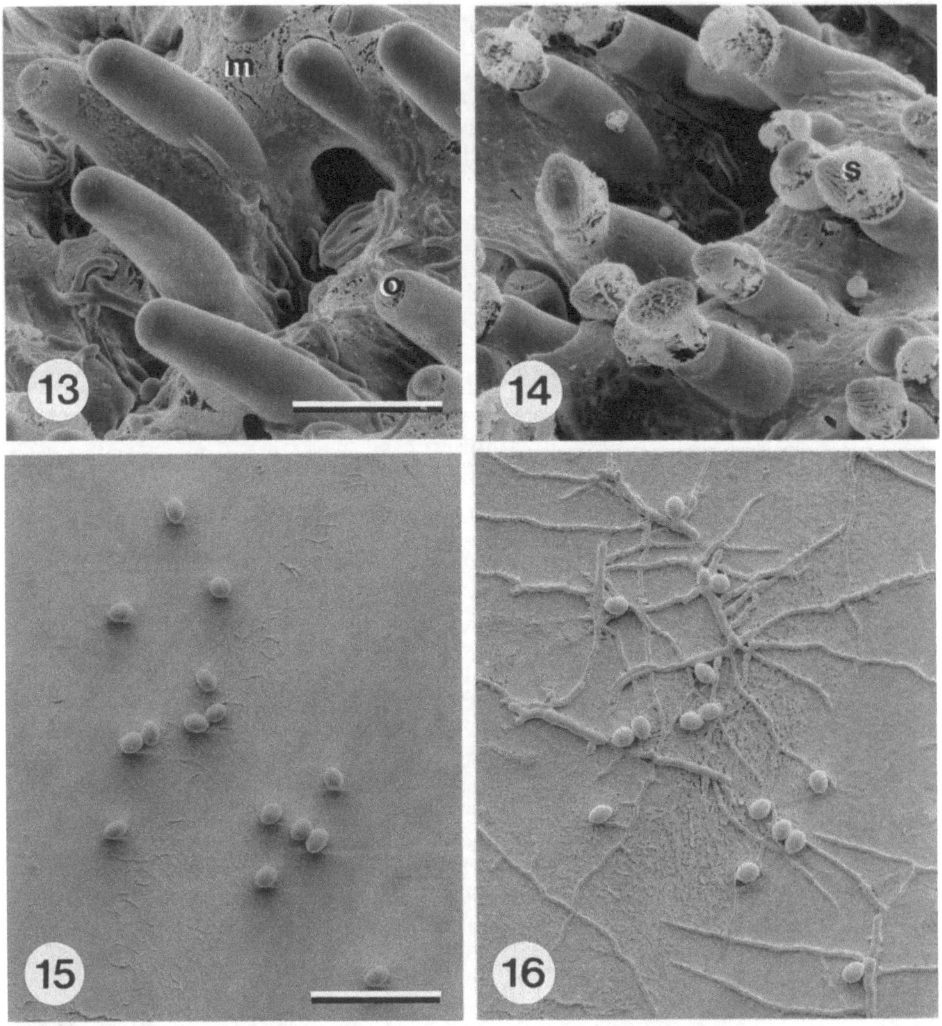

Figs. 13 and 14. *Ascobolus immersus.* Asci exhibiting different stages of ascospore discharge. Note ascus opercula (*o*), splitting around their rims and ascus sap (*s*) associated with the spore propagules. Also note the artifactual rupturing of mucilaginous extracellular material (*m*) which surrounds the asci and which mainly arose during partial freeze-drying. Partially freeze-dried material. Bar = 20 μm

Figs. 15 and 16. *Sordaria macrospora.* The same group of ascospores before and after germination on cellophane. After initial examination (**15**) the spores were incubated on agar medium for 40 h before they were examined after germination (**16**). Partially freeze-dried material. Bar = 100 μm

6 Artifacts Specific to Low-Temperature Scanning Electron Microscopy

Most of the common artifacts prevalent in ambient-temperature scanning electron microscopy are avoided by examining frozen-hydrated material; however, frozen-hydrated specimens are not artifact-free. LTSEM produces its own specific artifacts and most are related to the fact that frozen-hydrated specimens contain water. Common artifacts originate during cryofixation, partial freeze-drying, freeze-fracturing, specimen transfer and electron beam irradiation [6, 13, 24–26]. A virtually unavoidable artifact, usually only observed in freeze-fractured, partially freeze-dried material, is ice crystal damage (see above; Figs. 7, 9, 11, 12). Partially freeze-drying materials with a high water content (e.g., mucilages and agar) frequently causes surface rupturing (Figs. 10, 13, 14). Another artifact prevalent when working at high magnifications is beam damage [6, 24, 26].

7 Future Developments

LTSEM is beginning to reap the benefits of the latest developments in scanning electron microscopy and associated image processing techniques. Low electron dose, high resolution LTSEM is now possible with field emission electron guns operated at low accelerating voltages. This coupled with digital image storage and processing minimizes the problems of beam damage in frozen-hydrated specimens. Several laboratories are now successfully and usefully examining frozen-hydrated material at magnifications greater than $\times$ 100,000 (M. Müller pers. commun.; T. Müller, pers. commun.). Direct examination of frozen-hydrated specimens by high resolution LTSEM can bridge the gap between conventional LTSEM and the much higher resolution information obtainable from transmission electron microscope studies of freeze-etch replicas.

Another area ripe for future work is the use of LTSEM in conjunction with cytochemical procedures. One particularly useful approach will be to employ high resolution backscattered electron detection for the identification of colloidal gold-labelled antibodies and lectins on the specimen surface [31].

Recently we have discovered that some fungal spores can retain normal viabity after examination by electron microscopy (Figs. 15, 16). To our knowledge, this is the first time that any cells have been conclusively shown to be able to undergo growth and differentiation after specimen preparation and electron beam irradiation. The folklore of electron microscopy contains many tales of mites and similarly tough organisms walking shortly after examination in a SEM. However, we do not know of any published reports of these organisms undergoing growth and differentiation after the experience. Furthermore, in contrast to our results, only the outermost cells of these multicellular organisms will have been directly exposed to the electron beam and high vacuum. Our study opens up the possibility of LTSEM providing a means for monitoring the same spore before and after germination (Figs. 15, 16; N.D. Read and K.M. Lord, submitted).

In the future we can expect LTSEM to be employed much more as a powerful experimental tool rather than being used purely in descriptive studies. This will be facilitated by the short preparation time, the ability to freeze specimens remote from the SEM and by the availability of long-term storage devices.

Acknowledgements. I am very grateful to Lesley Campbell, Fiona Wilson, Dr. Oisín MacNamara and Dr. Jean Malcolmson for providing fungal and plant material. Thanks are also due to the Science and Engineering Research Council for a research grant (GR/D/33472).

References

1. Alexander C, Jones D, McHardy WJ (1987) Scanning electron macroscopy of cryofixed mycorrhizas of Sitka spruce *Picea sitchensis* (Bong.) Carr.: A comparison with critical point-dried material. New Phytol 105: 613–617
2. Bachmann L, Mayer E (1987) Physics of water and ice: implications for cryofixation. In: Steinbrecht RA, Zierold K (eds) Cryotechniques in biological electron microscopy. Springer, Berlin Heidelberg New York Tokyo, pp 3–34
3. Beckett A, Porter R (1982) *Uromyces viciae-fabae* on *Vicia faba*: scanning electron microscopy of frozen-hydrated material. Protoplasma 111: 28–37
4. Beckett A, Porter R (1988) The use of complementary fractures and low-temperature scanning electron microscopy to study hyphal-host cell surface adhesion between *Uromyces viciae-fabae*. Can J Bot 66: 645–652
5. Beckett A, Porter R, Read ND (1982) Low temperature scanning electron microscopy of fungal material. J Microsc 125: 193–199
6. Beckett A, Read ND (1986) Low-temperature scanning electron microscopy. In: Aldrich HC, Todd WJ (eds) Ultrastructural techniques for microorganisms. Plenum, New York, pp 45–86
7. Beckett A, Read ND, Porter (1984) Variations in fungal spore dimension in relation to preparatory techniques for light microscopy and scanning electron microscopy. J Microsc 136: 87–95
8. Beckett A, Woods AM (1987) The pattern of colony development and the formation of the uredinium of *Uromyces viciae-fabae* on *Vicia faba*. Can J Bot 65: 1998–2006
9. Brown DH, Rapsch S, Beckett A, Ascaso C (1987) The effect of desiccation on cell shape in the lichen *Parmelia sulcata* Taylor. New Phytol 105: 295–299
10. Echlin P, Paden R, Dronzek B, Wayte E (1970) Scanning electron microscopy of labile biological material maintained under controlled conditions. Scanning Electron Microsc/ 1970: 49–56
11. Galpin MF, Jennings DH, Oates K, Hobot JA (1978) Localization by X-ray micro-analysis of soluble ions, particularly potassium and sodium, in fungal hyphae. Exp Mycol 2: 258–269
12. Hammond KE, Lewis BG, Musa TM (1985) A systemic pathway in the infection of oilseed rape plants by *Leptosphaeria maculans*. Plant Pathol 34: 557–565
13. Jeffree CE, Read ND, Smith JAC, Dale JED (1987) Water droplets and ice deposits in leaf intercellular spaces: redistribution of water duing cryofixation for scanning electron microscopy. Planta 172: 20–37

14. Kellenberger E (1987) The response of biological macromolecules and supramolecular structures to the physics of specimen cryopreparation. In: Steinbrecht RA, Zierold K (eds) Cryotechniques in biological electron microscopy. Springer, Berlin Heidelberg New York Tokyo, pp 35–63
15. Kunoh H, Yamaoka N, Yoshioka H, Nicholson RL (1988) Preparation of the infection court by *Erysiphe graminis*. I. Contact-mediated changes in morphology of the conidium surface. Exp Mycol 12: 325–335
16. Marshall AT (1987) Scanning electron microscopy and X-ray microanalysis of frozen-hydrated bulk samples. In: Steinbrecht RA, Zierold K (eds) Cryotechniques in biological electron microscopy. Springer, Berlin Heidelberg New York Tokyo, pp 240–257
17. McLaughlin DJ, Beckett A, Yoon KS (1985) Ultrastructure and evolution of ballistosporic basidiospores. Bot J Linn Soc 91: 253–271
18. McLaughlin DJ, Beckett A (1987) Low temperature scanning electron microscopy of discharged basidiospores of *Coprinus cinereus*. Mycologia 79: 158–161
19. Mendgen K, Dressler E (1983) Culturing *Puccinia coronata* on a cell monolayer of the *Avena sativa* coleoptile. Phytopathol Z 108: 226–234
20. Müller T, Guggenheim R, Lueoend G, Dueggelin M (1986) On-line cryopreparation and cryomicroscopy in SEM with SCU 020. In: Imura T, Maruse S, Suzuki (eds) Proceedings of the XI international congress on electron microscopy, Kyoto, pp 2233–2234
21. Müller T, Guggenheim R, Düggelin M, Mestres P, Van Aelst AC, Heyser W, Kumpfer W (1988) Freeze etching and cryo scanning electron microscopy (CSEM) of plant material and animal tissues with SCU 020. In: Dickinson HG, Goodhew PJ (eds) Inst Phys Conf Ser No. 93, 3: 15–16
22. Nicholson RL, Yoshioka H, Yamaoka N, Kunoh H (1988) Preparation of the infection court of *Erysiphe graminis*. II. Release of esterase enzyme from conidia in response to a contact stimulus. Exp Mycol 12: 336–349
23. Read ND, Beckett A (1983) Effects of hydration on the surface morphology of urediospores. J Microsc 132: 179–184
24. Read ND, Porter R, Beckett A (1983) A comparison of preparative techniques for the examination of the external morphology of fungal material with the scanning electron microscope. Can J Bot 61: 2059–2078
25. Read ND, Beckett A (1985) The anatomy of the mature perithecium in *Sordaria humana* and its significance for fungal multicellular development. Can J Bot 63: 281–296
26. Read ND, Jeffree CE (1988) Common artifacts associated with biological material examined by low-temperature scanning electron microscopy. In: Dickinson HG, Goodhew PJ (eds) Inst Phys Conf Ser No 93, 3: 17–18
27. Robards AW, Sleytr UB (1985) Low temperature methods in electron microscopy. In: Glauert AM (ed) Practical methods in electron microscopy, Vol 10. Elsevier, Amsterdam
28. Sargent JA (1986) Cryo-preservation for scanning electron microscopy avoids artifacts induced by conventional methods of specimen preparation. Tissue Cell 18: 505–511
29. Sargent JA (1988) Low temperature scanning electron microscopy: advantages and application. Scanning Microsc 2: 835–849
30. van Gardingen PR, Jeffree CE, Grace J (1989) Variation in stomatal aperture in leaves of *Avena fatua* L. observed by low-temperature scanning electron microscopy. Plant Cell Environ 12: 887–897
31. Walther P, Kríz S, Müller M, Ariano BH, Brodbeck U, Ott P, Schweingruber ME (1984) Detection of protein A gold 15 nm marked surface antigens by backscattered electrons. Scanning Electron Microsc 3: 1257–1266
32. Williams MAJ, Beckett A, Read ND (1985) Ultrastructural aspects of fruit body differentiation in *Flammulina velutipes*. In: Moore DT, Casselton LA, Wood DA, Frankland JC (eds) Developmental biology of higher fungi. Cambridge University Press, Cambridge, pp 429–450
33. Williamson B, Duncan GH (1989) Use of cryo-techniques with scanning electron microscopy to study infection of mature red raspberry fruits. New Phytol 111: 81–88

Chapter 3 KURT MENDGEN, KLAUS WELTER, FRANK SCHEFFOLD and GERTRUDE KNAUF-BEITER

High Pressure Freezing of Rust Infected Plant Leaves

1 Introduction

Chemical fixation of plant leaves is a relatively slow process. Penetration of glutaraldehyde or osmic acid may extend over many minutes. In terms of cellular processes, this is a very long time, resulting in the introduction of numerous artifacts, such as undulation of membranes [23, 19]. This problem is even more pronounced in plants infected by pathogens. The hyphae or reproductive structures of fungal pathogens in plant tissue are variable in their structure and composition. The penetration of such cells by fixatives may be affected by the thickness or composition of their walls. For instance, the host-parasite interface of rust haustoria within plant cells appears quite distorted after chemical fixation [e.g. 18].

Rapid cryofixation is recommended as an alternative to chemical fixation [31, 28, 29, 6, 32, 37]. In plant pathology such methods have successfully been applied to infected epidermal cells [10, 5], fungal hyphae [12, 11, Hoch Chap. 1, this Vol) or fungal spores [24]. In samples of larger size, however, artificial segregation compartments destroy the tissue during freezing at ambient pressure. Slow heat transfer in biological tissue restricts the application of such methods to samples with a thickness of a few microns [33]. The depth of a well-preserved border zone in the tissue can be increased by applying pressure of at least 2000 bar before freezing is performed [27, 25]. With high pressure freezing (hpf) as introduced by Moor [25], complete root tips were frozen adequately [4, 14]. However, plant leaves have still proven difficult to freeze uniformly. The intercellular spaces are filled with air and in fungal infected tissues the intercellular mycelium may also trap small air spaces. Therefore, such leaves were previously infiltrated with water to avoid pressure-induced damage [15, 35]. However, only small areas were well preserved in most samples. Heat transfer appeared to be very uneven in such highly vacuolated tissue. To improve the situation, we studied rust-infected leaves of different host plants infiltrated with different concentrations of methanol, as described by Welter et al. [35], or hexadecene, as described by Studer et al. [34]. The experiments with samples

Lehrstuhl für Phytopathologie, Fakultät für Biologie, Universität Konstanz, Post Fach 5560, D-7750 Konstanz, FRG

Table 1. Variations of experiments performed to obtain optimal conditions for cryopreservation of rust infected plant tissues for ultrastructural study

No. of Experiment					
	Plunge Freezing				
1		*Uromyces vignae*			
	High pressure freezing		Width of specimen space	Composition of specimen chamber	Infiltration medium
2		*Uromyces vignae*	0.2 mm	Gold alloy	Tap water
3	*Phaseolus vulgaris*	*Uromyces appendiculatus*	0.3 mm	Aluminium	Tap water
4	"	"	0.3 mm	"	5% PVP
5	"	"	0.3 mm	"	5% Methanol
6	"	"	0.3 mm	"	8% Methanol
7	"	"	0.3 mm	"	10% Methanol
8	*Vicia faba*	*Uromyces fabae*	0.2 mm	Gold alloy	1-Hexadecene
9	"	"	0.5 mm	"	1-Hexadecene
10	"	"	0.2 mm	"	8% Methanol
11	"	"	0.3 mm	Aluminium	8% Methanol
12	*Triticum aestivum*	*Puccinia recondita*	0.2 mm	Gold alloy	1-Hexadecene
13	"	"	0.5 mm	"	1-Hexadecene
14	"	"	0.2 mm	"	8% Methanol
15	"	"	0.3 mm	Aluminium	8% Methanol
16	"	*Puccinia striiformis*	0.3 mm	"	1-Hexadecene
17	"	*Puccinia graminis*	0.2 mm	Gold alloy	0.05 M phosphate butter, ph 6.5 1% glucose

that were freeze substituted are listed in Table 1. Some results after hpf and freeze fracturing [15] are shown in addition. For comparison, infection hyphae of *Uromyces vignae* grown on membranes were plunge frozen and, after freeze substitution, processed similarly to the rust-infected leaves.

2 The Specimen Sandwich

A cross-section of the sandwiches used to enclose the leaf samples is shown in Fig. 1. They are simpler than those recommended by Craig et al. [3] and thinner than those suggested by Moor [25]. The platelets were made either of an alloy of 95% gold and 5% nickel or of aluminum. We found no differences in the preservation of fine structure using either gold platelets with 0.2 mm space (Table 1, Exp. 8, 10, 12, 14) or aluminum platelets with 0.3 mm space (Table 1, Exp. 11, 15). However, gold platelets with 0.5 mm space (Table 1, Exp. 9, 13) gave inferior results, with many segregation compartments occurring throughout the specimen.

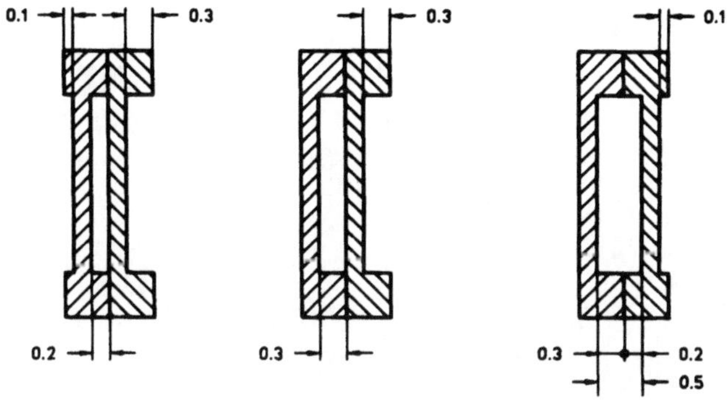

Fig. 1. Cross-sections through specimen chambers used for hpf of plant leaf tissue. The width of the specimen space is 0.2, 0.3 or 0.5 mm, the total thickness of the specimen chamber is 1 mm

3 Plunge Freezing vs. High Pressure Freezing

The reduced cooling rate compared to plunge freezing and the application of high pressure before freezing starts gave rise to concerns of possible damage to tissue caused by hpf [25, 32]. Similar to the results obtained by a prototype [25], our commercial instrument reaches a pressure of 2100 bar within 18 ± 0.5 ms. The pressure is maintained for 520 ± 15 ms. If the sample in the sandwich is replaced by a thermocouple, the drop in temperature, which begins at 2100 bar, takes 9.5 ± 0.4 ms to decrease from 273 K to 223 K. Since it can be assumed that the conditions in the thermocouple differ from those in a plant leaf [17], we do not know very much about the time course of the temperature in our samples. To obtain a rough estimate on the influence of hpf on the rust fine structure, we compared the results of the following experiments:
U. vignae (= *U. phaseoli* var. *vignae*) infection hyphae [21] were grown on scratched polyethylene sheets, plunge frozen as described by Howard and O'Donnell [13], freeze substituted and further processed as described by Welter et al. [35] (Table 1, Exp. 1). *U. vignae* infection hyphae raised as above were covered with water and, after hpf, (Table 1, Exp. 2) processed as above. *Puccinia graminis* f. sp. *tritici* sporogenous hyphae growing below the wheat leaf epidermis were, after infiltration, high pressure frozen and processed as above (Table 1, Exp. 17).

To compare results after freeze-substitution and freeze-fracturing, *U. appendiculatus* (= *U. phaseoli* var. *phaseoli*) sporogenous hyphae growing below the epidermis of a bean leaf were, after water infiltration, high pressure frozen and freeze fractured as described by Knauf and Mendgen [15].

The cytoplasm of the infection hyphae of *U. vignae* exhibited a comparable fine structure after plunge-freezing and hpf. The membranes of mitochondria and endoplasmatic reticulum were smooth and parallel (Fig. 2a, b). Also in the sporogenous hyphae of *P. graminis*, the mitochondria (Fig. 2c) looked very similar

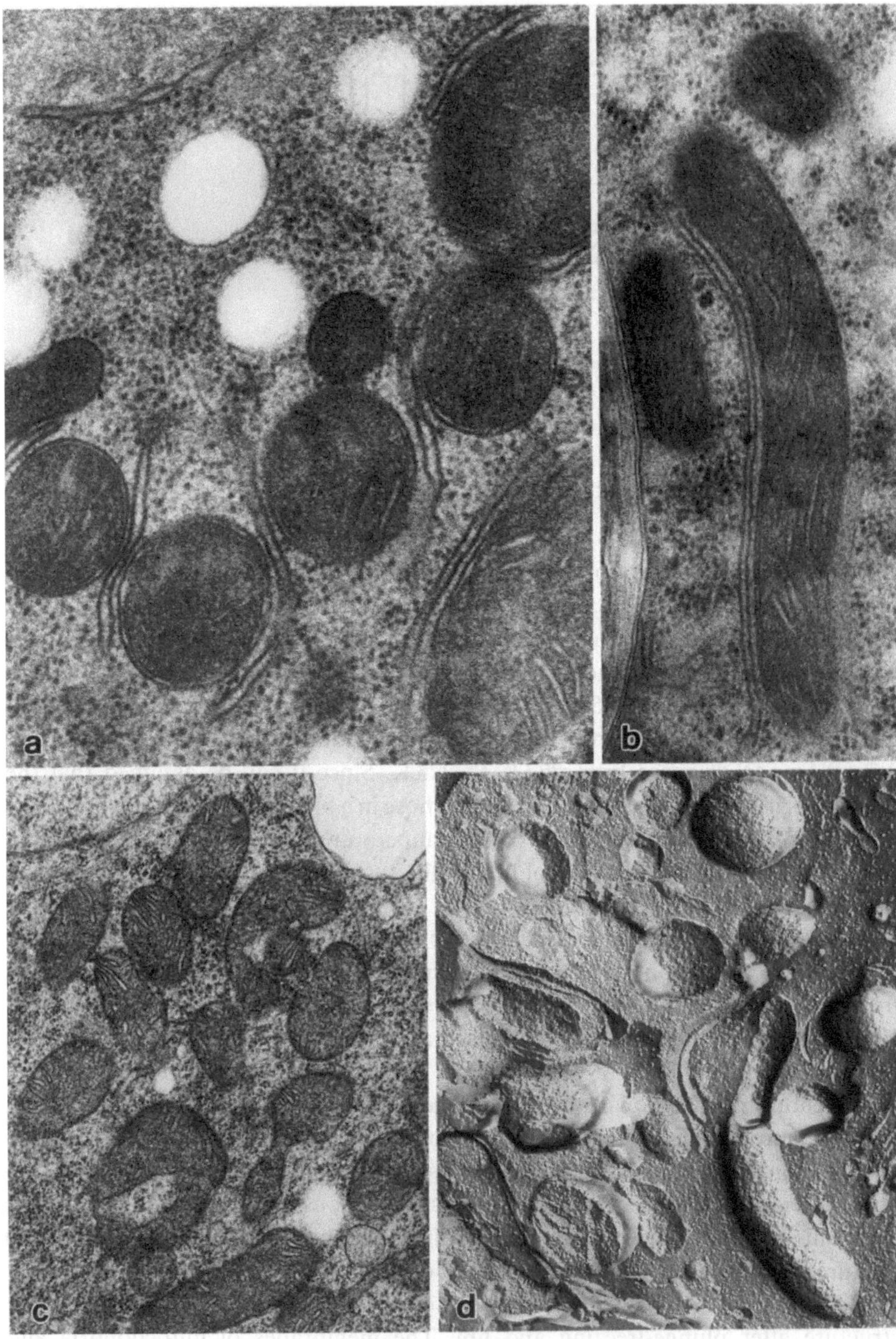
a
b
c
d

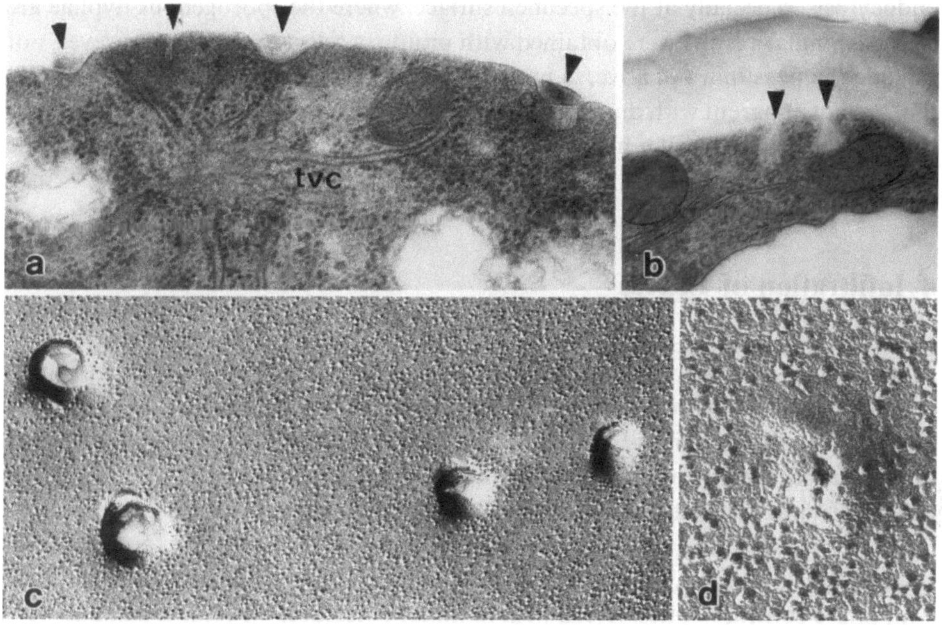

Fig. 3. Structures related to exo-endocytotic processes in hyphae of different rust fungi; **a** plunge freezing and freeze substitution of *U. vignae* infection hyphae; *arrow heads* secretion profile; *tvc* tubular vesicular complex (see 35), × 40000. **b** hpf and freeze substitution of *P. striiformis* sporogenous hyphae; *arrow heads* secretion profile, × 20000. **c** hpf and freeze fracturing of *U. appendiculatus* sporogenous cells, protoplasmic face, with various depressions, which may correspond to the depressions indicated in the cross-sections of **a**, × 52000. **d** hpf and freeze fracturing of *U. appendiculatus* sporogenous cells, protoplasmic face, with sectory opening, × 103000. **d** from [15]

to those of *U. vignae* in Fig. 2a and b. After hpf and freeze fracturing sporogenous tissue of *U. appendiculatus* (Fig. 2d) exhibited similar profils of ER and mitochondria. Obviously, the results of the different techniques are comparable, with the reservation that different, although related, rust fungi were used.

In addition, membrane profiles suggesting exo-endocytotic processes have been observed in infection hyphae after plunge freezing and freeze substitution (Fig. 3a), in sporogenous hyphae after high pressure freezing and freeze substitution (Fig. 3b) and in sporogenous hyphae after hpf and freeze fracturing (Fig. 3c, d). Possible secretory openings were observed after hpf (Fig. 3d).

From our results, it appears that the high pressure had no adverse affects on membrane structure, since their profiles were similar after plunge freezing or hpf. It seems that hpf is sufficiently rapid to immobilize cellular processes such as exo-

◄

Fig. 2. The cytoplasm in the infection hyphae of *Uromyces vignae* after **a** hpf, × 42000, and **b** plunge freezing, × 44000, followed by freeze substitution. The cytoplasm of *Puccinia graminis* sporogenous hyphae **c** after hpf and freeze substitution, × 25000, and the sporogenous hyphae of **d** *Uromyces appendiculatus* after hpf and freeze fracturing, × 20250

endocytosis, especially at the specimen surface, where the sporogenous hyphae are located. Similar results were obtained with pituitary cells (36). We do, however, not exclude the possibility of adverse pressure effects such as on membrane lipids (20) that are not apparent with our methods. Also, very rapid events such as may occur at the contact point of fusing membranes [17] may still not be immobilized by these methods.

4 Infiltration of Leaves

Infiltration of leaves was obtained under the mild vacuum of a water jet pump. During infiltration with the different media (Table 1), we checked with a microscope that no gas bubbles remained in the tissue. Leaf discs were cut out with a steel cork borer, placed between the platelets of the specimen chamber (Fig. 1, Table 1) and introduced into the instrument for hpf. The whole procedure was performed within 8 min. Processing followed as described elsewhere [35]. From the samples, 2 μm thick sections, about 1.5 × 0.2 mm large, were cut with glass knives and evaluated with phase contrast optics. Areas with minimal detectable structural damage were then selected for further study with the electron microscope.

After infiltration of bean leaves with water (Table 1, Exp. 3) or wheat leaves with buffer (Table 1, Exp. 17) serious damage, probably as a result of ice crystal formation, was ovious throughout the leaf parenchyma. Less than 10% of the mesophyll cells remained intact. In these few cells, haustoria exhibited a well-preserved fine structure. In the bundles, many phloem parenchyma cells remained intact (Fig. 4a). The cytoplasm in some intercellular hyphae was well preserved. Good results were obtained with all the hyphae of the sporogenous tissue. No ice crystal damage was evident in these hyphae except for very small areas in the nucleus (Fig. 2c). Infiltration of bean plants with 5% polyvinylpyrrolidone (PVP) did not improve the results significantly (Table 1, Exp. 4).

Infiltration with 5, 8 or 10% methanol in water (Exp. 5, 6, 7) as a cryoprotectant dramatically improved tissue preservation. After the application of 5% methanol some highly vacuolated cells were still damaged. Ten percent methanol induced shrinkage of some cells of the leaf parenchyma and also of some hyphae (Fig. 4b). With 8% methanol, good results were obtained throughout the leaf tissue including all fungal structures [35, 16]. These improved results using methanol were obtained with bean rust (*U. appendiculatus*) in beans (Exp. 6), broad bean rust

Fig. 4. Variations of pretreatment and consecutive processing of material after hpf. **a** Phloem parenchyma of bean (*Ph. vulgaris*), water infiltrated, × 14 700. **b** Bean leaf, 10% methanol infiltrated, × 2800. **c** Haustorium of *P. striiformis* in wheat (*T. aestivum*), 1-hexadecene infiltrated, × 17 000. **d** Wheat parenchyma, 1-hexadecene infiltrated, × 17 150. **e** *P. striiformis*, sporogenous cell, Lowicryl K 11 M embedded, × 11 150. **f** *U. phaseoli*, freeze etch replica of the haustorial host-parasite interface; *eme* extrahaustorial membrane; *fpm* fungal plasma membrane, × 88 000. **f** from [16]

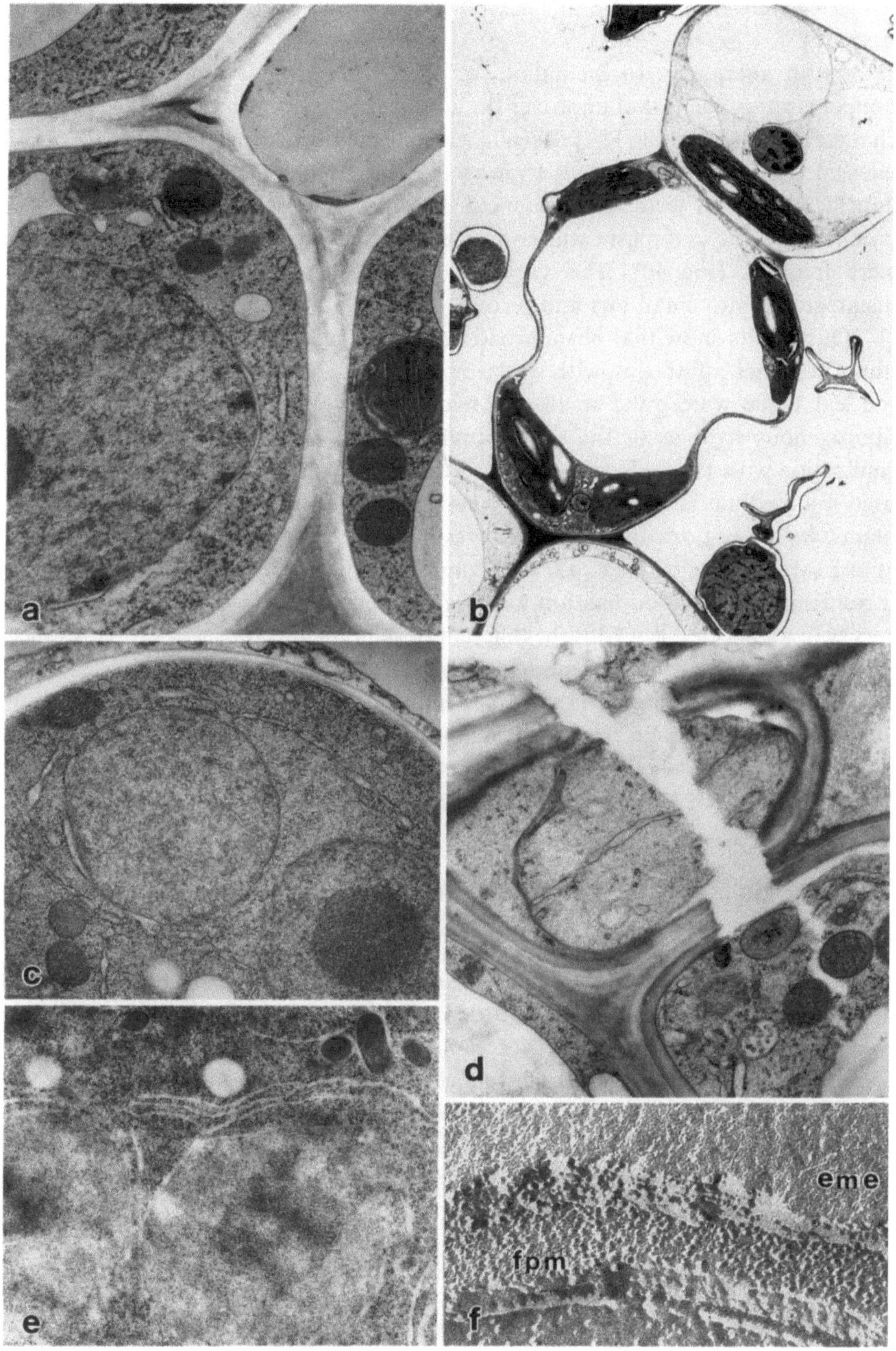
a
b
c
d
e
eme
fpm
f

(*U. vicicae-fabae*) in broad beans (Exp. 12), and brown rust (*P. recondita*) is wheat (Exp. 15).

As an alternative to methanol, we used 1-hexadecene for infiltration as a nonpenetrating agent that improves the contact between the metal of the specimen chamber and the sample [34]. Results with infected broad bean leaves (Exp. 8) or infected wheat leaves (Exp. 12) showed that nearly all intercellular hyphae and haustoria (Fig. 4c) were well preserved. In only 10–30% of the cells of the leaf parenchyma, the cytoplasm was conveniently fixed except that prominent cracks were frequent (Fig. 4d). The vacuolar content of the leaf parenchyma had disintegrated into small bits and pieces.

Our results show that bean, broad bean, or wheat leaves cannot be cryofixed uniformly after infiltration with water or buffer only. Well preserved areas within the leaf tissue were quite small and mainly restricted to the vascular bundles. Sporogenous hyphae of the corresponding rusts were always well preserved. Infiltration with 1-hexadecene improved the situation considerably and should be used if a medium is needed that does not penetrate the cells. However, with our plants, we did not obtain the excellent results realized with apple leaves by Studer et al. [34]. For practical reasons, we recommend infiltration with methanol because it results in overall good fixation with well-preserved fine structure. This makes a survey of an infected leaf with the different developmental stages of a fungus much easier. However, although methanol appears to cause no major structural alterations, we are well aware that it may influence cell physiology. This remains an important disadvantage. However, methanol exhibited excellent cryoprotective properties for some algae [26] and surprisingly even 10% of this additive has been "nontoxic" (≧75% survival after 15 min at 20 °C).

5 Freeze-Substitution and Embedding

The samples were either freeze-substituted in anhydrous acetone with 4% OsO_4 and embedded in Epon-Araldite [35] or freeze substituted in dried acetone alone for 3 days at 183 K and subsequently embedded in Lowicryl K 11 M at 223 K (− 50 °C) as recommended by the producer. Material embedded in Lowicryl K 11 M (Fig. 4e) was difficult to cut because sections tended to disintegrate in the water bath of the microtome knife and difficult to stain. Good contrast was obtained when the routine protocol for poststaining treatment with 3% uranylacetate for 30 min and lead citrate for 3 min was preceded by 2% OSO_4 in water for 10 min.

6 A Re-Evaluation of the Haustorial Host-Parasite Interface

An important element in biotrophic relationships are fungal haustoria (Harder and Chong, Chap. 18, this Vol). A number of uncertainties regarding the fine structure of haustoria have remained because of problems associated with chemical fixation. Osmotic shock during fixation tends to separate the zone between host and parasite

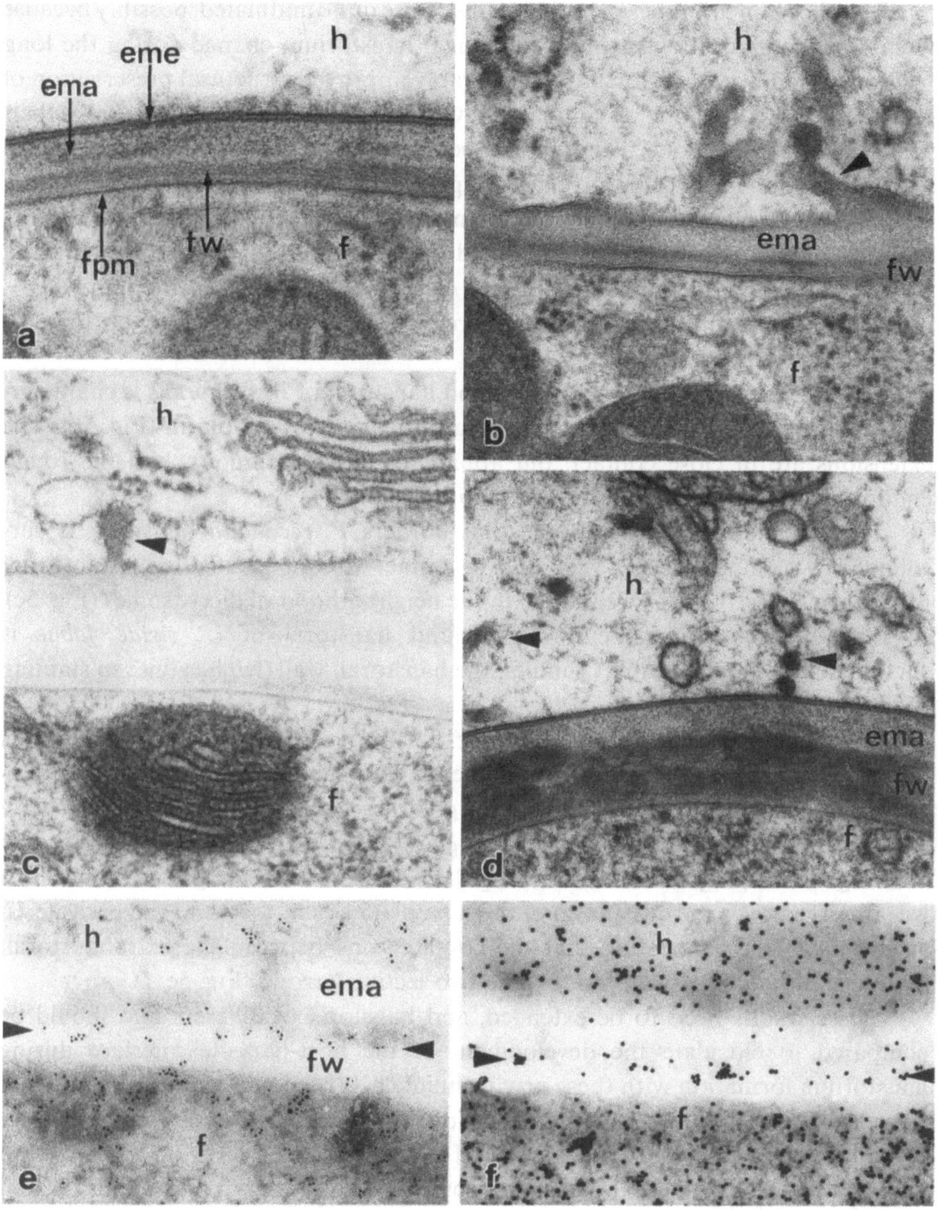

Fig. 5. The host-parasite interface of **a** *U. appendiculatus* and *Ph. vulgaris*, processed as indicated for experiment 6, × 41 600. **b** *P. striiformis* and *T. aestivum*; *arrow* extensions of the ema into the host cell protoplasts, exp. 16, × 64 300. **c** *P. recondita* and *T. aestivum*; *arrow head* "coated" pit, and fringe of material on the eme, exp. 15, × 66 400. **d** *U. fabae* and *V. faba*, exp. 11, K 11 M embedded, Con A-gold labeled; *arrow heads* indicate zone between ema and fw, × 64 000. **f** *U. fabae* and *V. faba*, exp. 11, K 11 M embedded, PNA-gold labeled; *arrow heads* indicate zone between ema and fw, × 54 000. *ema* extrahaustorial matrix; *eme* extrahaustorial membrane; *f* fungus; *fpm* fungal plasma membrane; *fw* fungal wall; *h* host plant

[1]. The plasmamembranes of both organisms are quite undulated, possibly because membrane configurations around the fungal haustorium change during the long fixation period. Cryofixation has considerably improved structural preservation of haustoria [16]. Therefore, we compared several rust fungi for special features at their intercellular contact zone: The interface between the bean rust (*U. appendiculatus*) haustoria and the host cell protoplasts (Figs. 4f and 5a) may be said to extend from the "coated" extrahaustorial membrane (eme) to the fungal plasmamembrane (fpm), with the extrahaustorial matrix (ema) and the haustorial cell wall (fw) intervening between these two membranes (Fig. 5a). A freeze etch replica of the fpm reveals numerous intramembrane particles, whereas no such particles occur on the eme (Fig. 4f).

The extrahaustorial membrane around haustoria of *P. striiformis* is characterized by large tubular extensions reaching into the host cytoplasm (Fig. 5b). The extensions are in close contact, but apparently not continuous with the rough endoplasmatic reticulum of the host plant (Fig. 5b).

The extrahaustorial membrane that borders *P. recondita* haustoria is surrounded by "coated" vesicles, very often in contact with the "coated" extrahaustorial membrane. These vesicles were often in the neighborhood of dictyosomes (Fig. 5c).

The extrahaustorial membrane around haustoria of *U. viciae fabae* is surrounded by small "coated" tubules; the haustorial wall (fw) has uneven staining properties (Fig. 5d).

A molecular characterization of the haustorial host-parasite interface was obtained with lectins. Sections from tissue without chemical fixation embedded in Lowicryl K 11 M were incubated with the gold-labeled lectin Concanavalin A (Con A-Gold) (Fig. 5e) and that of *Arachis hypogea* (PNA-Gold) (Fig. 5f) as described by Rohringer et al. [30] and Harder et al. [8]. A quite dense labeling with both lectins was observed over the cytoplasm of the host plant, *Vicia faba* and the parasite *U. viciae-fabae*. It is interesting that the contact zone between the extrahaustorial matrix and the fungal wall binds these two lectins (arrows, Fig. 5e, f).

These results need to be extended, and haustoria of different age should be compared to elucidate the development of the host-parasite interface during haustorium formation with these new techniques. It becomes evident already after these preliminary studies, that the extensions of the eme first described by Harder [7] and Chong and Harder [2] around haustoria of *P. coronata avenae*, with conventional techniques, exhibit typical conformations for each rust fungus [9] and may be used to differentiate rust fungi.

7 Conclusions

Our results have shown that the ultrastructure of fungal infections within leaf tissue can be studied using physical fixation by hpf. There is no evidence of major damage induced by the application of the high pressure if the tissue is properly pretreated. The samples may be studied after freeze fracturing or after freeze substitution, which may or may not include chemical fixation. Samples may be embedded in the conventional expoxy resins or in low temperature acrylic resin. The results are superior to those obtained with conventional techniques. Therefore, it is useful to verify previous results, particularly membrane contact zones or secretion events [22]. Furthermore, hpf combined with low temperature embedding is highly qualified for cytochemical studies.

However, some problems remain. The different components of the leaf and the fungus have different freezing qualities. Therefore, the samples should be as thin as possible. Infiltration of the intercellular space with water is not sufficient. Results were improved by infiltration with 1-hexadecene, but the best results for overall evaluation of leaf tissue were obtained by cryoprotection with methanol.

Acknowledgements. We thank Thomas Schmitz and Annette Klink for excellent technical assistance, D. E. Harder, Winnipeg, G. Knoll and H. Plattner, Konstanz, for reviewing the manuscript and Jutta Schlepper–Schäfer for a generous gift of PNA-Gold. The Deutsche Forschungsgemeinschaft helped with grants (Me 523/12, Me 523/13).

References

1. Allen FHE, Coffey MD, Health MC (1979) Plasmolysis of rusted flax: a fine-structural study of the host-pathogen interface. Can J Bot 57: 1528–1533
2. Chong J, Harder DE (1982) Ultrastructure of haustorium development in *Puccinia coronata avenae:* some host responses. Phytopathology 72: 1527–1533
3. Craig S, Gilkey JC, Staehelin LA (1987) Improved specimen support cups and auxiliary devices for the Balzers high pressure freezing apparatus. J Microsc 148: 103–106
4. Craig S, Staehelin LA (1988) High pressure freezing of intact plant tissues. Evaluation and characterization of novel features of the endoplasmic reticulum and associated membrane systems. Eur J Cell Biol 46: 80–93
5. Dahmen H, Hobot JA (1986) Ultrastructural analysis of *Erysiphe graminis* haustoria and subcuticular stroma of *Venturia inaequalis* using cryosubstitution. Protoplasma 131: 92–102
6. Gilkey JC, Staehelin LA (1986) Advances in ultrarapid freezing for the preservation of cellular ultrastructure. J Electron Microsc Tech 3: 177–210
7. Harder DE (1978) Comparative ultrastructure of the haustoria in uredial and pycnial infections of *Puccinia coronata avenae.* Can J Bot 56: 214–224
8. Harder DE, Chong J, Rohringer R, Mendgen K, Schneider A, Welter K, Knauf G (1989) Ultrastructure and cytochemistry of extramural substances associated with intercellular hyphae of several rust fungi. Can J Bot 67: 2043–2051
9. Heath MC, Bonde MR (1983) Ultrastructural observation of the rust fungus *Physopella zeae* in *Zea mays.* Can J Bot 61: 2231–2242
10. Hippe S (1985) Ultrastructure of haustoria of *Erysiphe graminis* F. sp. *hordei* preserved by freeze-substitution. Protoplasma 129: 52–61
11. Hoch HC, Staples RC (1983) Ultrastructural organization of the non-differentiated uredospore germling of *Uromyces phaseoli* variety *typica.* Mycologia 75: 795–824
12. Howard RJ, Aist JR (1979) Hyphal tip cell ultrastructure of the fungus *Fusarium:* improved preservation by freeze-substitution. J Ultrastruct Res 66: 224–234
13. Howard RJ, O'Donnell KL (1987) Methodological review. Freeze substitution of fungi for cytological analysis. Exp Mycol 11: 250–269
14. Kaeser W, Koyro H-W, Moor H (1989) Cryofixation of plant tissues without pretreatment. J Microsc 154: 279–288
15. Knauf G, Mendgen K (1988) Secretion systems and membrane-associated structures in rust fungi after high pressure freezing and freeze-fracturing. Biol Cell 64: 363–370

16. Knauf G, Welter K, Müller M, Mendgen K (1989) The haustorial host-parasite interface in rust-infected bean leaves after high-pressure freezing. Physiol Mol Plant Pathol 34: 519–530
17. Knoll G, Verkleij AJ, Plattner H (1987) Cryofixation of dynamic processes in cells and organelles. In: Steinbrecht RA, Zierold K (eds) Cryotechniques in biological electron microscopy. Springer, Berlin Heidelberg New York Tokyo, pp 258–271
18. Littlefield LJ, Heath MC (1979) Ultrastructure of rust fungi. Academic Press, New York
19. MC Cully ME, Canny MJ (1985) The stabilization of labile configurations of plant cytoplasm by freeze-substitution. J Microsc 139: 27–33
20. McDonald AG (1984) The effects of pressure on the molecular structure and physiological functions of cell membranes. Philos Trans R Soc, Lond, Ser B 304: 47–68
21. Mendgen K, Schneider A, Sterk M, Fink W (1988) The differentiation of infection structures as a result of recognition events between some biotrophic parasites and their hosts. J Phytopathol 123: 259–272
22. Mendgen K (1989) Secretion in the parasitic phase of rust fungi. In: Lugtenberg B (ed) Molecular signals in microbe-plant symbiotic and pathogenic systems. NATO Series, Springer, Berlin Heidelberg New York Tokyo, pp 281–288
23. Mersey B, McCully ME (1978) Monitoring the course of fixation of plant cells. J Microsc 114: 49–76
24. Mims CW, Richardson EA, Timberlake WE (1988) Ultrastructural analysis of conidiophore development in the fungus *Aspergillus nidulans* using freeze-substitution. Protoplasma 144: 132–141
25. Moor H (1987) Theory and practice of high pressure freezing. In: Steinbrecht RA, Zierold K (eds) Cryotechniques in biological electron microscopy. Springer, Berlin Heidlberg New York Tokyo, pp 175–191
26. Morris GJ, Canning CE (1978) The cryopreservation of *Euglena gracilis*. J Gen Microbiol 108: 27–31
27. Müller H, Moor H (1984) Cryofixation of thick specimens by high pressure freezing. In: Revel J-P, Barnard T, Hagis GH (eds) The science of biological specimen preparation. SEM, AMF O'Hare, IL 60666, pp 131–138
28. Plattner H, Bachmann L (1982) Cryofixation: a tool in biological ultrastructural research. Int Rev Cytol 79: 237–304
29. Robards AW, Sleytr UB (1985) Low temperature methods in biological electron microscopy. In: Glauert AM (ed) Practical methods in electron microscopy, Vol 10. Elsevier, Amsterdam
30. Rohringer R, Chong J, Gillespie R, Harder DE (1989) Gold-conjugated arabinogalactan-protein and other lectins as ultrastructural probes for the wheat/stem rust complex. Histochemistry 91: 383–393
31. Sitte H (1979) Cryofixation of biological material without pretreatment. A review. Mikroskopie (Wien) 35: 121–131
32. Sitte H, Edelmann L, Neumann K (1987) Cryofixation without pretreatment at ambient pressure. In: Steinbrecht RA, Zierold K (eds) Cryotechniques in biological electron microscopy. Springer, Berlin Heidelberg New York Tokyo, pp 87–113
33. Steinbrecht RA, Zierold K (1987) Cryotechniques in biological electron microscopy. Springer, Berlin Heidelberg New York Tokyo
34. Studer D, Michel M, Müller M (1989) High pressure freezing comes of age. Scanning Microscopy, Suppl 3: 253–269
35. Welter K, Müller M, Mendgen K (1988) The hyphae of *Uromyces appendiculatus* within the leaf tissue after high pressure freezing and freeze substitution. Protoplasma 147: 91–99
36. Draznin B, Dahl R, Sherman N, Sussman KE, Staehelin A (1988) Exocytosis in normal anterior pituitary cells. J Clin Invest 81: 1042–1050
37. Dahl R, Staehelin A (1989) High-pressure freezing for the preservation of biological structure: Theorie and practice. J Electr Micr Techn 13: 165–174

Chapter 4 JEAN-PAUL LATGÉ

Cytochemistry of Fungal Surfaces: Carbohydrate Containing Molecules

1 Introduction

Specific *attachment* of microorganisms to host tissues is mediated by receptor-ligand interaction at the outer layers of both fungal and host cells. However, unlike bacteria, adhesion of fungi is poorly documented and very few molecules of the fungal surface involved in adhesion processes have been identified [25, 41]. This is due in part to insufficient knowledge of the chemistry of the outer layer of fungal walls [51]. Amongst the numerous treatments used to extract the outer layer, water washings, ultrasonication and/or incubation with detergents, denaturing agents, solvents, enzymes, acids, and bases have typically been employed [10, 11, 13, 34, 36]. Successful removal of the outer layer is dependent on both the fungal species and the propagule. Analysis of fungal cell wall extracts have demonstrated that outer layers contain a complex array of (glyco) proteins, polysaccharides, lipids (lipoproteins), minerals, and pigments. Results on the composition of the outer layer vary depending on the extraction method used and the organism studied. Moreover, some of the external molecules are expressed differentially during fungal growth [12, 49].

Extraction of the outer layer can be monitored a posteriori if the treatment has removed a morphologically distinct feature present on the untreated fungal propagule. For example, removal of rodlet fascicles from fungal spores is an indication of the probable extraction of the spore surface layer [10, 11]. However, the disappearance of the rodlet structure can be due solely to the destruction of the rodlet configuration in the absence of any extraction of this layer, or to the coating of these structures with internal cell components released during extraction. At present, it seems more appropriate to construct specific cytochemical probes on the basis of the chemical or immunological properties of the compounds isolated. If these probes exclusively label the *surface layer*, they may be used to monitor the extraction efficiency of the outermost components. Alternatively, positive labeling of the inner wall would indicate either the existence of similar components in the

Unité de Mycologie, Institut Pasteur, 28, rue du Dr. Roux, F75724 Paris Cedex 15, France

external and internal layers or the extraction of internal components which are contaminating extracts of the outer components. Release of inner *wall compounds* can be due either to active excretion of these compounds through wall pores or the disruption of inner wall linkages so that internal components may be translocated to the outer wall domain. To address this problem, probes specific to inner components should be produced; obtaining components characteristic of the inner structural layer free of those from outer layers is always easier than the reverse. Another approach to the study of fungal surfaces is the use of commercially available probes with known chemical specificity. These probes can give an indication of the chemical nature of the components present on the surface, thus providing a basis for selecting suitable chemical methods for subsequent isolations.

Most molecules present on the surface of the fungal cell wall are carbohydrate-containing molecules. Although the mechanisms are still poorly understood, it is well recognized that carbohydrates and their receptors participate in cell surface interactions between fungal pathogens and their respective plant or animal host [28, 35, 41, 49]. This review presents some of the techniques that can be applied to the study of external fungal polysaccharides.

2 EM Probes for External Carbohydrates

With the exception of the periodic acid-thiocarbohydrazide-silver proteinate (PATAg) method based on the oxidation of sugars by periodic acid, all other methods for detection of carbohydrate-containing molecules are based on the affinity existing between two molecules: the target molecule and the ligand coupled to enzymatic or particulate dense markers [4, 18, 43, 44, 48]. Four types of ligand can be used: polycations, lectins, enzymes, and antibodies (Table 1). The most popular enzyme used for visualization of cell glycoconjugates is horse radish peroxidase in conjunction with diaminobenzidine (DAB)–H_2O_2–OsO_4 treatment. Amongst the particulate markers, mercury, iron carbohydrate, latex microspheres, ferritin, and colloidal gold have been employed (Figs. 1–4). In recent years, colloidal gold has become the most popular electron-dense marker. Monodisperse colloidal gold solutions can be easily obtained by a variety of methods in a size range between 5 and >50 nm and are well suited for TEM or SEM studies for single as well as multiple marking experiments [20, 23, 26, 29, 39, 40]. Under appropriate conditions, gold particles can be labeled with a variety of macromolecules presumably through a noncovalent binding process. Factors influencing the adsorption process, the stability of the probe, and the effect of particle size on labeling density have been reviewed [14, 17, 21, 22, 33, 42]. However, three major problems may be encountered with *colloidal gold.* (a) Because the macromolecules must be dissolved in water or very dilute buffer before coupling, some proteins aggregate, so that the newly formed probe fail to disperse. Formation of such aggregates can be minimized

Table 1. A review of the methods applied to localize fungal polysaccharides

Target	Method
1. Sugars with free vicinal glycols	
Non oxidized	PTA at pH1
Periodic acid oxidized	Thiocarbohydraside + silver proteinate Biotin hydrazide + streptavidin – Au
2. Anionic sites	Cationized Fer, chitosan-Au, polylysine – Au
3. Lectin hapten	
Mannose	ConA – Au, conA + glc BSA – Au, ConA + mannosyl – Fer, conA + HRP – Au, ConA + Mannan – Au
N-Acetyl glucosamine	WGA – BSA – Au, WGA + chitobiosyl – Fer, WGA + glcNAc – BSA – Au, WGA + ovomucoid – Au
Galactose	RCA – Au, RCA – BSA – Au, RCA + guaran – Au, RCA + DC – Au,
Fucose	UeA – Au
N – Acetyl galactosamine	HpA – Au, SbA – Au
Sialic acid	LfA + fetuin – Au
Galacturonic acid	Ad – Au
4. Enzyme substrate	
Chitin	Chitinase – Au
Cellulose	Cellulase – Au
$\beta(1 \rightarrow 3)$ glucan	$\beta(1 \rightarrow 3)$ glucanase – Au, $\beta(1 \rightarrow 3)$ glucanase + RA or MA enzyme – Au
Galacturonane	Pectinase – Au
5. Antigens	
Mannan	RA mannan + prot A – Au, MA mannan + GAMIg – Au, RaA mannan + GARaIg – Au
$\beta(1 \rightarrow 3)$ glucan	RA laminaribiose-edestin + prot A – Au
$\beta(1 \rightarrow 6)$ glucan	RA gentiobiose – edestin + prot A – Au
Galactan	RaAGM + GARaIg – Au
Fucose GP	MAGP + GAMIg – Au

Abbreviations Ad: *Aplysia depilans* lectin; Au: colloidal gold, conA: *Canavalia enisformis* lectin; DC: desialated ceruloplasmin; Fer: ferritin; HpA: *Helix pomatia* lectin; HRP: horse radish peroxidase; GM: galactomannan; GP: glycoprotein; Lf A: *Limax flavus* lectin; MA: mouse monoclonal antibody anti-; Prot A: protein A; PTA: phosphotungstic acid; RA: rabbit polyclonal antibody anti-; RaA: rat monoclonal antibody anti-; RCA: *Ricinus communis* lectin; SBA: *Glycine max* lectin; UeA: *Ulex europaeus* lectins; WGA: *Triticum vulgaris* lectin.

by diluting the ligand in a nonionic detergent such as Chaps (Sigma, 1 mg/ml), which does not interfere with coupling (Latgé unpubl.). (b) Obtaining stable complexes with the ligand may sometimes be difficult. (c) Coupling may result in the loss of biological activity of the ligand. The use of "sandwich" techniques can overcome all of these problems. Incubation of the cell with a nonconjugated ligand is followed by an incubation with a visualizer specific for the ligand (enzyme, lectin, glycoprotein, polysaccharides, or antibodies) which can be easily coupled to the electron-dense marker [4, 18, 23, 44]. In these two step methods, better stereochemical accessibility

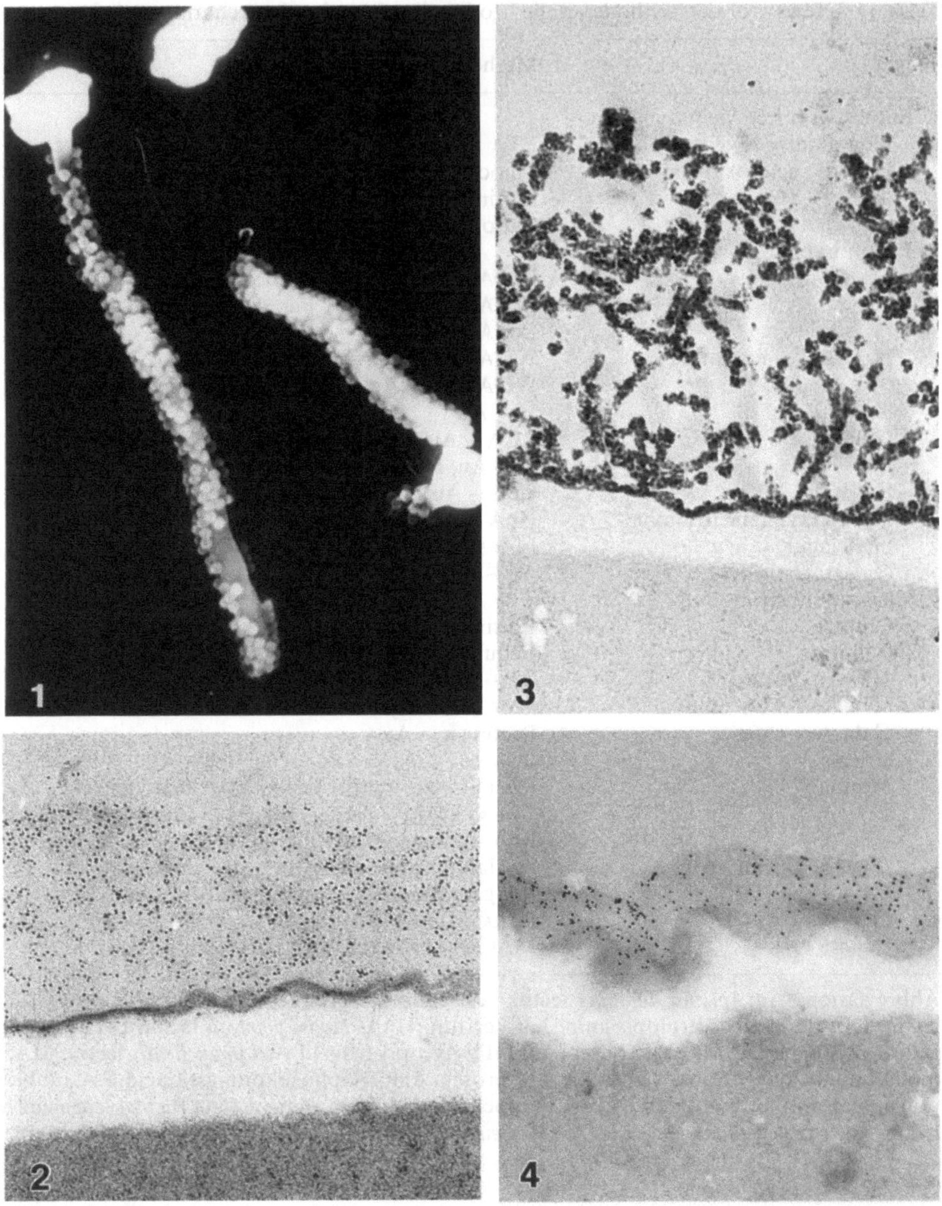

Figs. 1–4. **1** Localization of fibrinogen binding sites on *Candida albicans* germ tubes using fibrinogen conjugated latex particles (courtesy of G. Tronchin). **2** ConA-receptors in the mucus layer of *Conidiobolus obscurus* visualized by ConA-mannosyl ferritin. **3** Labeling of mucus fibrils of *Conidiobolus obscurus* with horse radish peroxidase. **4** Exocellular lectin of *Conidiobolus obscurus* demonstrated with Glc-BSA neoglycoprotein conjugated with colloidal gold

will be obtained because the intermediary probe will now protrude from the original binding site. Because of the usual multivalent nature of the intermediary probe, the number of gold-labeled particles and the density of the marking will thus be increased.

Periodic acid cleaves carbohydrate molecules bearing neighboring free hydroxyl groups. Aldehydes formed are usually blocked with *thiocarbohydrazide* to form carbohydrazones that give electron-opaque complexes with silver proteinate [38, 42, 44, 46] (Fig. 6). Streptavidin and biotinyl hydrazide reagents can also be used to visualize periodate-generated aldehydes (Latgé unpubl.). In this case, after periodate oxidation, the material is incubated in a solution of *biotin hydrazide*. Biotinyl hydrazones are visualized by streptavidin coupled to colloidal gold. Phosphotungstic acid at low pH can be used also for polysaccharidic staining without an oxidation step [44] (Fig. 5). This method is very quick, and good staining can be obtained after a 2-min incubation in 1N HC1 (Latgé unpubl.).

The net surface charge of a fungal cell is negative [26, 27], mainly due to phosphate groups of glycoproteins, glycuronic acid groups of polysaccharides, or carboxylic groups of amino acids. Fungal surface anionic groups can be localized with cationized ferritin, polylysine gold complexes, or chitosan gold complexes

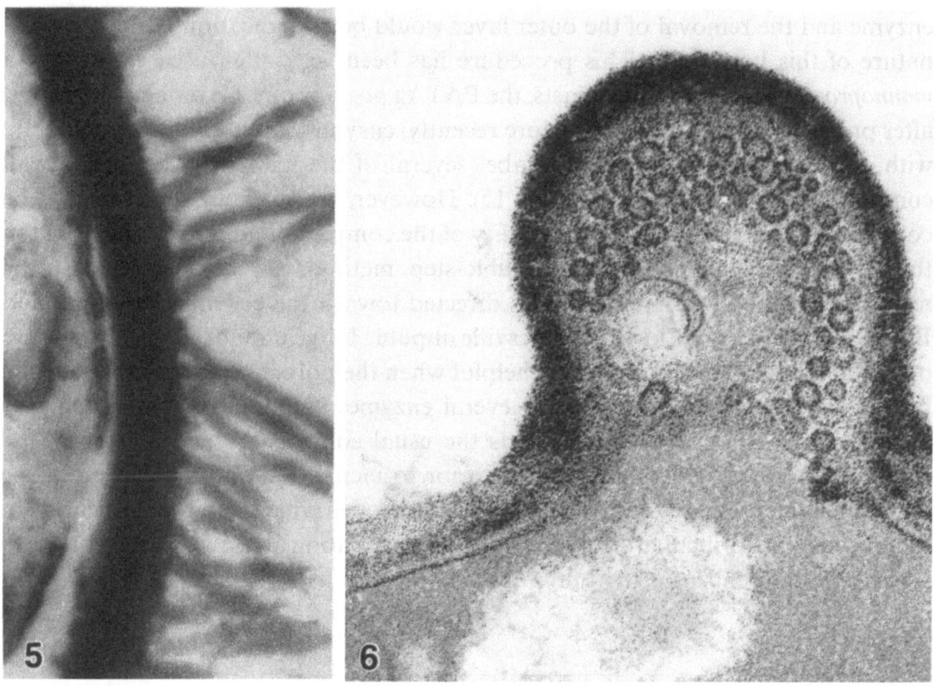

Figs. 5 and 6. 5 Phosphotungstic acid staining for capsular polysaccharides of *Cryptococcus neoformans*. **6** PATAg reaction on yeast bud of *Paracoccidioides brasiliensis*

[28, 29, 45, 50] (Figs. 7–9). Control cells are preincubated with homologous or heterologous unlabeled polycation. The disadvantage of the methods used to detect anionic sites or alcohol groups with or without selective oxidation is that of specificity. Unlike the three other types of ligand used (lectins, enzymes, antibodies), cationic or alcohol probes do not discriminate the nature of the molecules labeled. However, they may be very useful, for example, in following surface changes during aging or under different growth conditions [27, 28, 38].

Lectins are sugar-binding proteins or glycoproteins which are the most widely used ligands for labeling specific sugar residues on the fungal surface [4, 6, 8, 18, 29, 31, 32, 36] (Figs. 2, 10–13). Direct methods where the lectin is directly coupled to colloidal gold, or indirect methods where the marker is complexed to a secondary reagent having a high affinity for the lectin, have been used. Controls usually include (a) omission of the lectin, (b) preadsorption of the lectin with the inhibitory sugar and incubation of the sample with the mixture sugar-lectin, and (c) incubation with unlabeled lectin prior to labeled conjugates. These controls are especially necessary in the case of indirect labeling. They will allow one to monitor for the presence of endogenous lectins which can interfere with the labeled secondary reagent. In the case of positive controls, the presence of endogenous lectins can be assessed with neoglycoproteins conjugated with gold [35] (Fig. 4).

Enzymes have been used for a long time to assess the chemical identification of surface components. Initially, fungal propagules were incubated with a specific enzyme and the removal of the outer layer would be an indication of the chemical nature of this layer [29]. This procedure has been very often used to identify a *mannoprotein* coat on various yeasts; the PATAg positive outer layer being removed after proteolytic treatment [38]. More recently, enzymes have been labeled directly with gold and have been used to label several of the main polysaccharide wall components [2, 4, 5, 9, 16] (Figs. 14, 15). However, for enzymes which cannot be coupled to gold because of the instability of the complex or an interaction between the enzymatic site and gold, a double-step method can be performed using monoclonal or polyclonal antibodies directed toward the enzyme (Pendland and Boucias, University of Florida, Gainesville unpubl.; Latgé unpubl.) (Fig. 15). The use of an intermediary antibody is also helpful when the polysaccharidic substrate can be degraded only by a complex of several enzymes acting either sequentially or simultaneously. For one-step methods the usual controls are (a) addition of the substrate to the enzyme gold complex prior to incubation and incubation of the material in the presence of substrate, (b) addition of enzyme inhibitors or heating of the enzyme, and (c) incubation with BSA-gold. Incubation at a low temperature or at a pH very different from the optimum also may be used as control. However, in

▶

Figs. 7–9. Localization of anionic sites on *Cryptococcus neoformans* yeast using cationized ferritin **(7)** or polylysine gold **(8)** on *Candida albicans* yeast using chitosan gold (**9** courtesy of M. Horisberger)

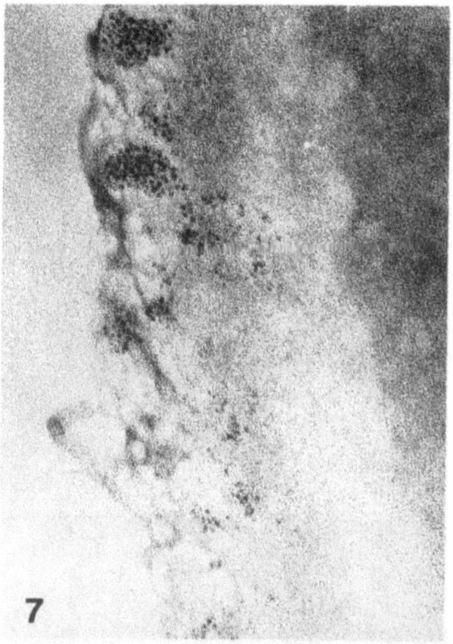

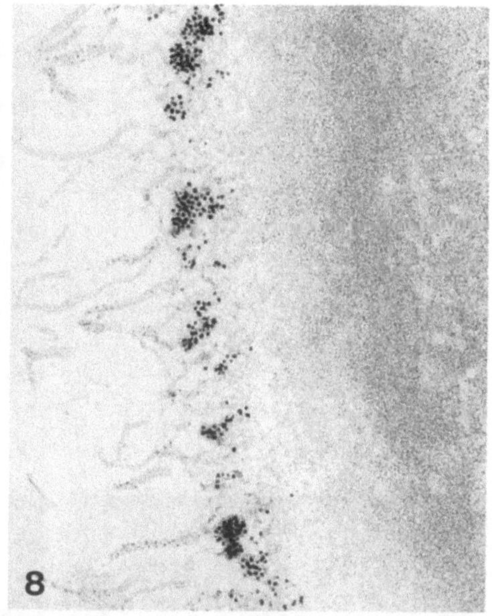

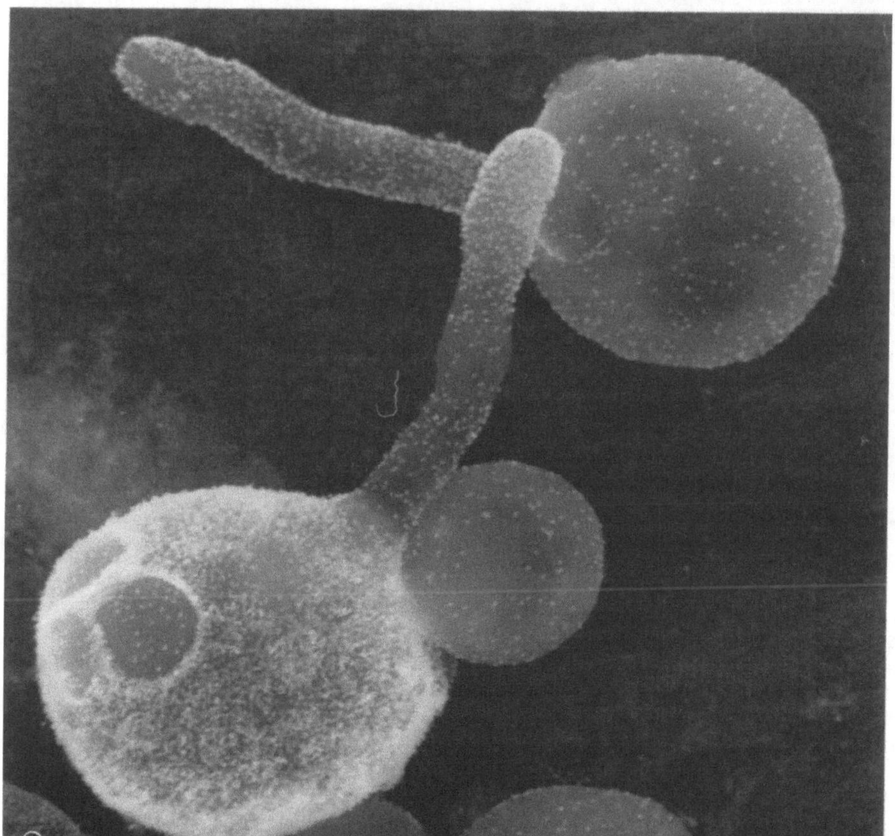

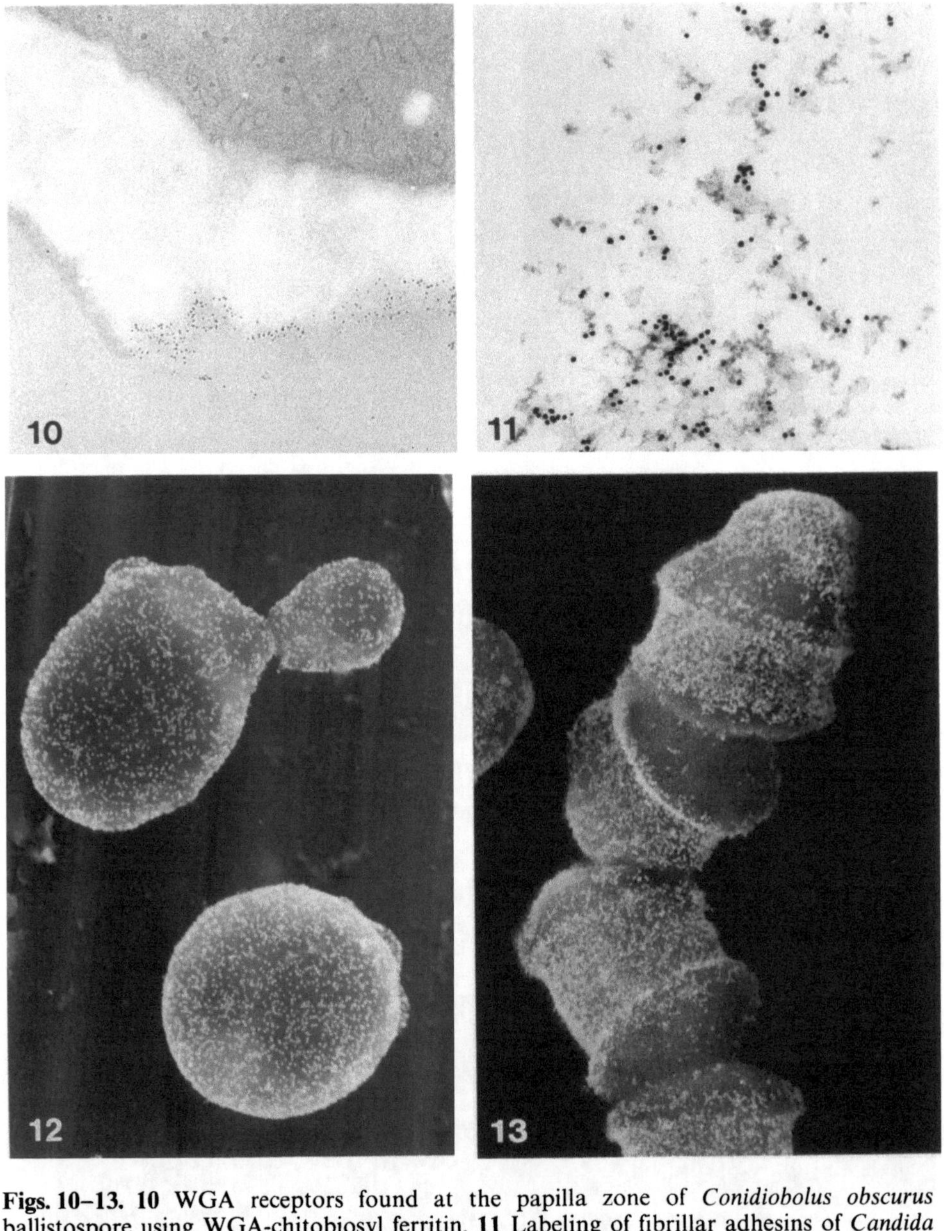

Figs. 10–13. 10 WGA receptors found at the papilla zone of *Conidiobolus obscurus* ballistospore using WGA-chitobiosyl ferritin. **11** Labeling of fibrillar adhesins of *Candida albicans* after adhesion of the yeast to plastic and labeling with conA-gold. **12** Localization of mannoproteins with conA-gold on *Candida albicans*. **13** Surface marking of galactomannan on *Schizosaccharomyces pombe* using RCA-gold (**11** courtesy of G. Tronchin; **12**, **13** courtesy of M. Horisberger)

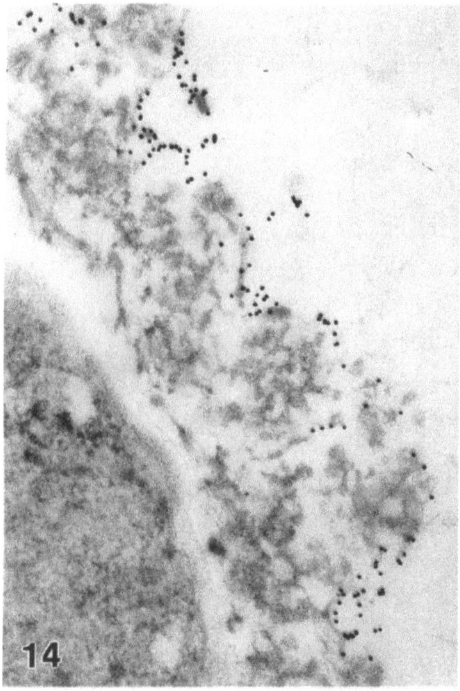

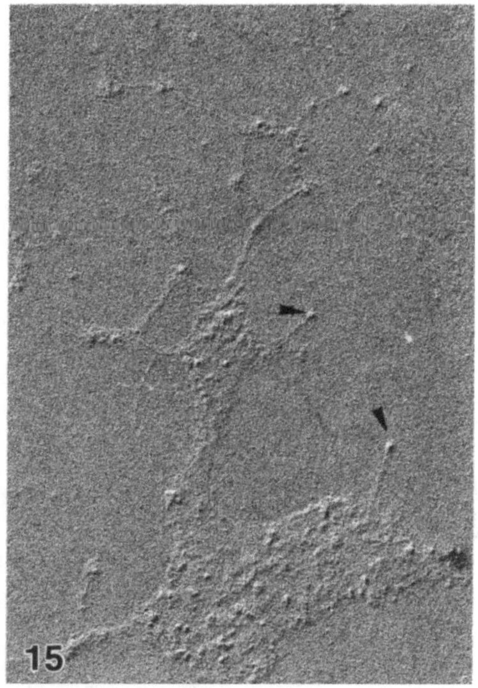

Figs. 14 and **15.** **14** Identification of $\beta(1 \rightarrow 3)$ glucan on the exocellular material produced by *Nomuraea rileyi* using $\beta(1 \rightarrow 3)$ glucanase gold. (Courtesy of J. Pendland). **15** Identification of $\beta(1 \rightarrow 3)$ glucanes synthesized by a mixed membrane fraction of *Aspergillus fumigatus* using sequential $\beta(1 \rightarrow 3)$ glucanase-rabbit anti-$\beta(1 \rightarrow 3)$ glucanase-Prot A gold and Pt shadowing (*arrowheads* indicate gold particles)

certain cases, these conditions may permit attachment of the enzyme to the substrate without substrate degradation and may favor labeling. Similarly, a short heating of the enzyme, destroying the activity of the enzyme without destroying the binding site, may be helpful (B. Vian, Ecole Normale Supérieure, Paris, pers. commun.). In the case of two step methods, pre-immune rabbit serum should be employed. Even though all controls have been carefully run, the use of glycosidases may lead to erroneous results for mainly two reasons: accessibility problems of the enzyme to the substrate, and presence of unexpected activities in the enzyme preparation used. These types of problems, which can also occur with the other method of labeling, should be specially checked whenever enzymes are used [4, 18].

Immunoelectromicroscopy has become a valuable tool in the study of microbial antigens and consequently in the localization of external fungal components, recognized as antigens by specific antibodies (Ab) [7, 19, 23, 30, 31]. One advantage of the Ab is their very narrow molecular specificity, especially in the case of monoclonal antibodies [12, 15] (Figs. 16–22). For example, one of the MAb we used in our study of *A. fumigatus* wall galactomannan is specific for four $\beta(1 \rightarrow 5)$-linked

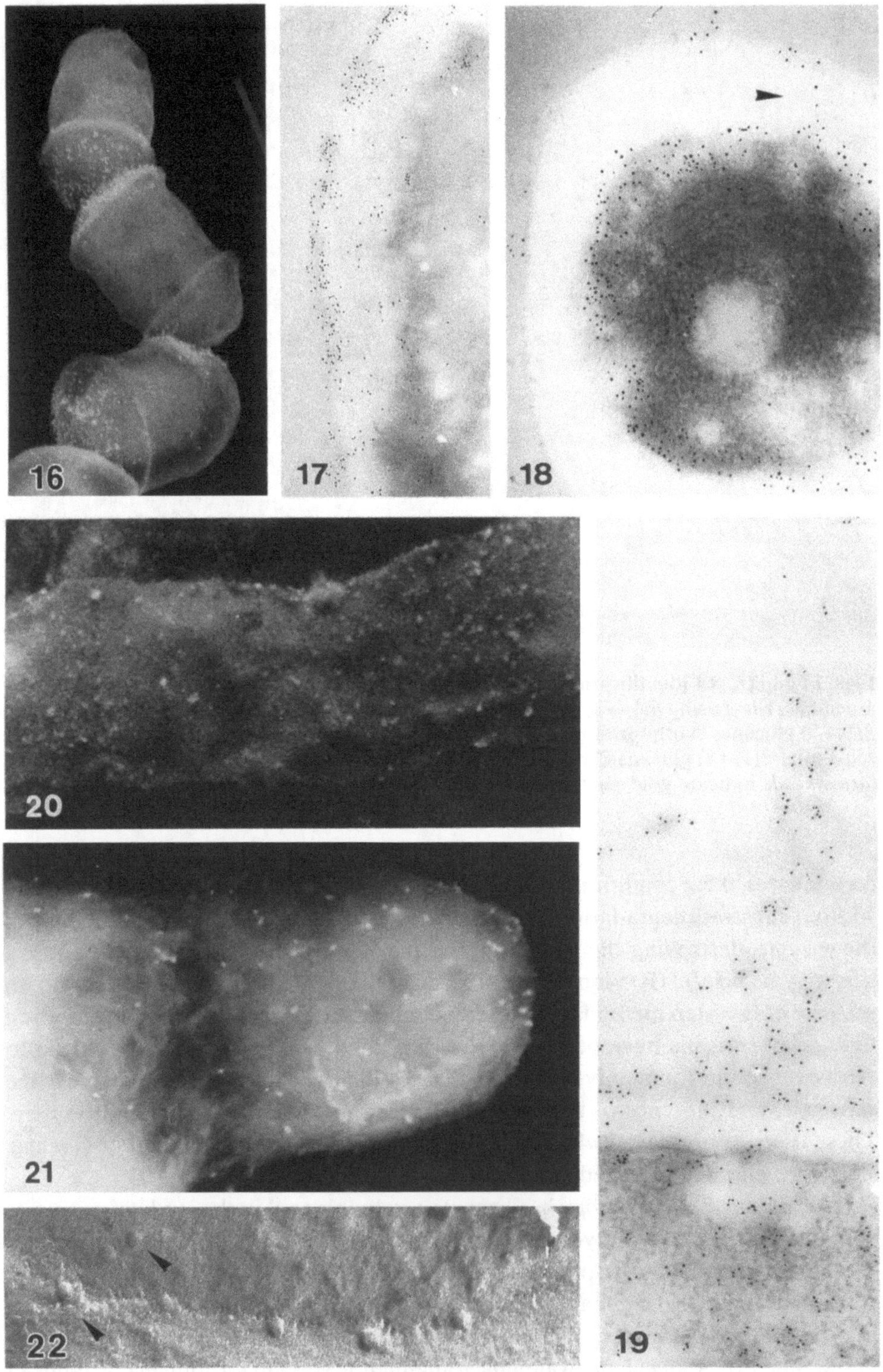
16
17
18
20
21
22
19

galactofuranoses (Figs. 19–22). Ab bound to external antigens can be easily detected with protein A, protein G, or streptavidin-gold complexes, all of which are easy to prepare and possess a high affinity for immunoglobulins [3]. When enzymes or lectins cannot be coupled to gold, antibodies remain the usual intermediate for sandwich methods. Controls are simple: (a) use of nonimmunized animal serum, (b) use of the specific antiserum adsorbed overnight at 37 °C with the antigen studied, (c) incubation of the fungal material with protein A gold or streptavidin-gold, (d) in the case of double labeling, use of a serum from a heterologous animal.

Although very rarely used, the gold labeled antigen method developed by Larson will be presented here because it may be very helpful to study fungal surface determinants [39, 40]. In this method, fungal material is incubated successively with a specific polyclonal antiserum and a solution of antigen-labeled gold particles. Because of their bivalent nature, specific IgGs will bind to both the antigen present on the fungus surface as well as the gold labeled antigen. This method has been rarely used, but offers certain advantages in cases where antigens are able to elicit antibodies when present in a complex crude mixture but not after purification. It is also very useful in assessing the identity between internal and external wall components.

3 EM Methods

Because of the absence of endocytotic phenomena in fungi, studies of surface wall components do not require a fixation step before probe application [47]. When fixation is used before applying the probe, as is the case in post-embedding techniques with hydrophilic resins, or in studies with wall-less propagules, fixation should be short and mild (30 min in 2.5% p-formaldehyde +0.1% glutaraldehyde) preserving the antigenic or chemical configuration of the target molecules [1, 36]. However, unlike the glycoproteins, polysaccharides are very "resistant" molecules and can be fixed with 2.5% glutaraldehyde without altering their structure.

Scanning electron microscope studies of cell surface-bound gold markers are currently being made using *back-scattered electron imaging* (BEI) (Figs. 9, 12, 13, 16, 20). BEI requires only a carbon coating of the specimen and good definition of 30 nm

◄

Figs. 16–22. 16 Immunogold marking of $\beta(1\rightarrow3)$ glucan at the surface of *Schizosaccharomyces pombe* using sequentially an anti-$\beta(1\rightarrow3)$ glucan rabbit antiserum and protein A gold (Courtesy of M. Horisberger). **17, 18** Localization of mannan at the surface of *Candida albicans* yeast using a mouse monoclonal antibody and an anti-mouse IgM-Au5 (**18** excretion of the mannoproteic antigen through the wall, *arrowhead*). **19–22** Localization of galactomannan at the surface of *Aspergillus fumigatus* germinating conidia using a rat monoclonal antibody visualized with anti rat-Au5 (TEM **19**) rabbit anti rat-Ig + anti rabbit – Au30 (SEM **20**; STEM **21**; replica TEM **22**)

gold particles can be easily attained with conventional SEM. However, recent progress in scanning electron microscopy engineering allows a resolution at the SEM level very close to TEM.

TEM studies will be applied to isolated surface components or to whole cells. The use of isolated molecules and Pt shadowing allows labeling at a supramolecular level and is particularly interesting in the case of disymetric molecules. When cells are surface-labeled in the whole-mount approach with pre-embedding methods, only a small portion of the labeled surface can be seen (Figs. 2, 7, 10, 14). This disadvantage can be overcome by the use of Scanning Transmission Electron Microscopy (STEM) where whole cells can be seen under the EM (Fig. 21). Replica techniques can be another alternative to study the complete distribution of labeling sites on a fungal surface [24] (Figs. 15, 22). Post-embedding methods are well adapted to study the overall distribution of external compounds in the inner and outer layer and their biosynthetic pathways within the cell [15, 33, 36, 37, 43] (Figs. 6, 17, 18, 19). These techniques may also be very helpful in the case of accessibility problems due either to the physical configuration of the fungal surface or to the size of the complex electron dense marker-ligand.

4 Conclusion

Although this review deals mainly with human pathogens and carbohydrate fungal structures, these strategies can be applied to any receptors present on the cell surface, irrespective of their chemical composition. It emphasizes that the identification of surface determinants will require the combined use of chemical, cytological, and immunological techniques.

Acknowledgments. I am particularly grateful to D. Boucias for critical reviewing of this manuscript, M. Horisberger, G. Tronchin and J. Pendland for providing illustrations and M. Cormier for typing the manuscript. TEM, STEM and MEB pictures have been obtained during collaborative studies with MC Prevost and M. Lesourd.

References

1. Beauvais A, Latgé JP, Vey A, Prevost MC (1989) The role of surface components of *Entomophaga aulicae* in the cellular immune response of *Galleria mellonella*. J Gen Microbiol 152: 229–236
2. Bendayan M (1984) Enzyme-gold electron microscopic cytochemistry: a new affinity approach for the ultrastructural localization of macromolecules. J Electron Microsc Tech 1: 349–372
3. Bendayan M, Garzon S (1988) Protein G-gold complex: Comparative evaluation with protein A-gold for high resolution immunocytochemistry. J Histochem Cytochem 36: 597–607

4. Benhamou N (1988) Ultrastructural localization of carbohydrates in the cell walls of two pathogenic fungi: a comparative study. Mycologia 80: 324–337
5. Benhamou N, Ouellette GB (1986) Use of pectinases complexed to colloidal gold for the ultrastructural localization of polygalacturonic acids in the cell walls of the fungus *Ascocalyx abietina*. Histochem J 18: 95–104
6. Benhamou N, Ouellette GB (1986) Ultrastructural localization of glycoconjugates in the fungus *Ascocalyx abietina*, the scleroderris canker agent of conifers using lectin-gold complexes. J Histochem Cytochem 34: 855–867
7. Benhamou N, Ouellette GB, Gardiner RB, Day AW (1986) Immunocytochemical localization of antigen-binding sites in the cell surface of two ascomycete fungi using antibodies produced against fimbriae from *Ustilago violacea* and *Rhodotorula rubra*. Can J Microbiol 32: 871–883
8. Benhamou N, Gilboa-Garber N, Trudel J, Asselin A (1988) A new lectin-gel complex for the ultrastructural localization of galacturonic acids. J Histochem Cytochem 36: 1403–1411
9. Berg RH, Erdos GW, Gritzali M, Brown RD (1988) Enzyme-gold affinity labelling of cellulose. J Electron Microsc Tech 8: 371–379
10. Boucias DG, Pendland JC, Latgé JP (1988) Non specific factors involved in attachment of entomopathogenic Deuteromycetes to host insect cuticle. Appl Environ Microbiol 54: 1795–1805
11. Bouziane H, Latgé JP, Mecheri S, Fitting C, Prévost MC (1989) Release of allergens from *Cladosporium* conidia. Int Arch Allergy Appl Immunol 88: 261–266
12. Brawner DL, Cutler JE (1986) Variability in expression of cell surface antigens of *Candida albicans* during morphogenesis. Infect Immun 51: 337–343
13. Briza P, Winkler G, Kalchhauser H, Breitenbach M (1986) Dityrosine is a prominent component of the yeast ascospore wall. J Biol Chem 261: 4288–4294
14. Bullock GR, Petrusz P (1983) Techniques in immunocytochemistry Vol II. Academic Press, London
15. Cailliez JC, Poulain D (1988) Analyse cytologique de l'expression d'un épitope porté par les glycoprotéines excrétées par *Candida albicans*. Ann Inst Pasteur Microbiol 139: 171–188
16. Chamberland H, Charest PM, Ouellette GB, Pauzé FJ (1985) Chitinase-gold complex used to localize chitin ultrastructurally in tomato root cells infected by *Fusarium oxysporum* f.sp. *radicis-lycoperoici* compared with a chitin specific gold-conjugated lectin. Histochem J 17: 313–321
17. De Roe C, Courtoy PJ, Baudhuin P (1987) A model of protein-colloidal gold interactions. J Histochem Cytochem 35: 1191–1198
18. Erdos GW (1986) Localization of carbohydrate-containing molecules. In: Aldrich HC, Todd WJ (eds) Ultrastructure techniques for microorganisms. Plenum, New York, pp 399–420
19. Erdos WG, West CM (1989) Formation and organization of the spore coat of *Dictyostelium discoideum*. Exp Mycol 13: 169–182
20. Frens G (1973) Controlled nucleation for the regulation of particle size in monodisperse gold suspensions. Nature Phys Sci 241: 20–22
21. Geoghegan WD (1988) The effect of three variables on adsorption of rabbit IgG to colloidal gold. J Histochem Cytochem 36: 401–407
22. Geoghegan WD, Ackerman GA (1977) Adsorption of horse radish peroxidase, ovomucoid and anti-immunoglobulin to colloidal gold for the indirect detection of concanavalin A, wheat germ agglutinin and goat anti-human immunoglobulin on cell surfaces at the electron microscopic level: a new method, theory and application. J Histochem Cytochem 11: 1187–1200
23. Hodges GM, Southgate J, Toulson EC (1987) Colloidal gold—a powerful tool in scanning electron microscope immunocytochechemistry: an overview of bioapplications In: Albrecht RM, Hodges GM (eds) Biotechnology and bioapplication of colloidal gold. Scanning Microscopy International, Chicago, 1: 1–17

24. Hohenberg H, Bohn W, Rutter G, Mannweiler K (1986) Plasma membrane antigens detected by replica techniques. In: Mueller M, Becker RP, Boyde A, Wolosewick JJ (eds) Science of Biological specimen preparation for microscopy and microanalysis. SEM Inc AMF O'Hara, Chicago, pp 235–244
25. Hohl HR, Balsiger S (1988) Surface glycosyl receptors of *Phytophthora megasperma* f. sp. *glycenea* and its soybean host. Bot Helv 98: 271–277
26. Horisberger M (1981) Colloidal gold: a cytochemical marker for light and fluorescent microscopy and for transmission and scanning electron microscopy. Scanning Electron Microsc 11: 9–31
27. Horisberger M, Clerc MF (1988) Chitosan-colloidal gold complexes as polycationic probes for the detection of anionic sites by transmission and scanning electron microscopy. Histochemistry 90: 165–175
28. Horisberger M, Clerc MF (1988) Ultrastructural localization of anionic sites on the surface of yeast hyphal and germ tube forming cells of *Candida albicans*. Eur J Cell Biol 46: 444–452
29. Horisberger M, Rosset J (1977) Colloidal gold, a useful marker for transmission and scanning electron microscopy. J Histochem Cytochem 25: 295–305
30. Horisberger M, Rouvet-Vauthey M (1985) Cell wall architecture of the fission yeast *Schizosaccharomyces pombe*. Experientia 41: 748–750
31. Horisberger M, Vonlanthen M (1977) Location of mannan and chitin on thin sections of budding yeasts with gold markers. Arch Microbiol 115: 1–7
32. Horisberger M, Vonlathen M, Rosset J (1978) Localization of galactomannan and wheat germ agglutinin receptors in *Schizosaccharomyces pombe*. Arch Microbiol 119: 107–111
33. Kellenberger E, Dürrenberger M, Villager N, Carlemalm E, Wurtz M (1987) The efficiency of immunolabel on Lowicryl sections compared to theoretical predictions. J Histochem Cytochem 35: 959–969
34. Latgé JP, Bouziane H, Diaquin M (1988) Ultrastructure and composition of the conidial wall of *Cladosporium cladosporioides*. Can J Microbiol 34: 1325–1329
35. Latgé JP, Monsigny M, Prevost MC (1988) Visualization of exocellular lectins in the entomopathogenic fungus *Conidiobolus obscurus*. J Histochem Cytochem 36: 1419–1424
36. Latgé JP, Cole GT, Horisberger M, Prevost MC (1986) Ultrastructure and chemical composition of the ballistospore wall of *Conidiobolus obscurus*. Exp Mycol 10: 99–113
37. Newman GR, Hobot JA (1987) Modern acrylics for post embedding immunostaining techniques. J Histochem Cytochem 35: 971–981
38. Paris S, Prevost MC, Latgé JP, Garrison RG (1986) Cytochemical study of the yeast and mycelial cell walls of *Paracoccidioides brasiliensis*. Exp Mycol 10: 228–242
39. Polak JM, Van Norden S (1983) Immunocytochemistry. Practical applications in pathology and biology. Wright-PSG, Bristol
40. Polak JM, Varndell IM (1984) Immunolabelling for electron microscopy, Elsevier, Amsterdam
41. Ralton JE, Smart MG, Clarke AE (1987) Recognition and infection processes in plant pathogen interactions In: Kosuge T, Nester EW (eds) Plant-Microbe interaction, 2, Macmillan, New York
42. Rambourg A (1971) Morphological and histochemical aspects of glycoproteins at the surface of animal cells. Int Rev Cytol 31: 57–114
43. Roth J (1987) Light and electron microscopic localization of glycoconjugates with gold-labeled reagents. In: Albrecht RM, Hodges GM (eds) Biotechnology and bioapplication of colloidal gold. Scanning Microscopy International, Chicago, 1, pp 53–62
44. Schrevel J, Gros D, Monsigny M (1981) Cytochemistry of cell glycoconjugates. Prog Histochem Cytochem 14: 2
45. Skutelsky E, Roth J (1986) Cationic colloidal gold-A new probe for the detection of anionic cell surface sites by electron microscopy. J Histochem Cytochem 34: 693–696
46. Thiéry JP (1967) Mise en évidence des polysaccharides sur coupes fines en microscopie électronique. J Microsc Paris 6: 987–1018
47. Thomopoulos GN, Schulte BA, Spicer SS (1987) Post embedment staining of complex

carbohydrates: influence of fixation and embedding procedures. J Electron Microsc 5: 17–44
48. Tronchin G, Robert R, Bouali A, Senet JM (1987) Immunocytochemical localization of in vitro binding of human fibrinogen to *Candida albicans* germ tube and mycelium. Ann Inst Pasteur Microbiol 138: 177–187
49. Tronchin G, Bouchara JP, Robert R, Senet JM (1988) Adherence of *Canida albicans* germ tubes to plastic: ultrastructural and molecular studies of fibrillar adhesins. Infect Immun 56: 1987–1993
50. Vorbrodt AW (1987) Demonstration of anionic sites on the luminal and albuminal fronts of endothelial cells with polylysine- gold complex. J Histochem Cytochem 35: 1261–1266
51. Wessels JGH (1986) Cell wall synthesis in apical hyphal growth. Int Rev Cytol 104: 37–79

Chapter 5

Richard J. Zeyen

Analytical Electron Microscopy in Plant Pathology: X-Ray Microanalysis and Energy Loss Spectroscopy

1 Introduction

Electron microscopes were developed to obtain high resolution images of very small subcellular objects, and have given plant pathologists enormous amounts of information concerning parasites and host-parasite interactions. Electron microscopes are of great value in plant pathology, as this book illustrates, and it is difficult to envision doing either applied or fundamental pathology research without them. Moreover, electron microscopes yield more than detailed morphological and cytological information, because interaction of beam electrons with specimens creates numerous signals useful in detecting and identifying *chemical elements* in specimens. When electron microscopes are fitted with instrumentation to detect these various signals they become versatile and powerful analytical instruments, adding greatly to the overall goal in plant pathology of relating in situ structure(s) and function(s), or dysfunction(s).

Two electron microscope-based technologies, useful for obtaining elemental chemical composition information from plant pathological specimens, are described in this chapter; *X-Ray Microanalysis* (XRMA), which is widely used in biological sciences; and *Electron Energy Loss Spectroscopy* (EELS), which is not yet widely used but is attracting attention. These two were chosen because they are complementary technologies [1]. Both identify chemical elements found in the periodic table; however, XRMA is most useful for elements whose atomic number (Z) is above 11 (Na), while EELS is more sensitive for elements with lower atomic numbers, Z = 10, (Ne). EELS, in addition to yielding elemental determinations, can give information on the binding states of some chemical elements. XRMA can be used in both stationary and scanning modes since the electron beam is used indirectly for X-ray excitation, while EELS measures direct energy loss from monoenergetic beam electrons passing through the specimen in a transmitted beam mode (Fig. 1). Since this chapter contains only brief introductory information on

Department of Plant Pathology, University of Minnesota, 495 Borlaug Hall, 1991 Upper Buford Circle, St. Paul, MN 55108, USA

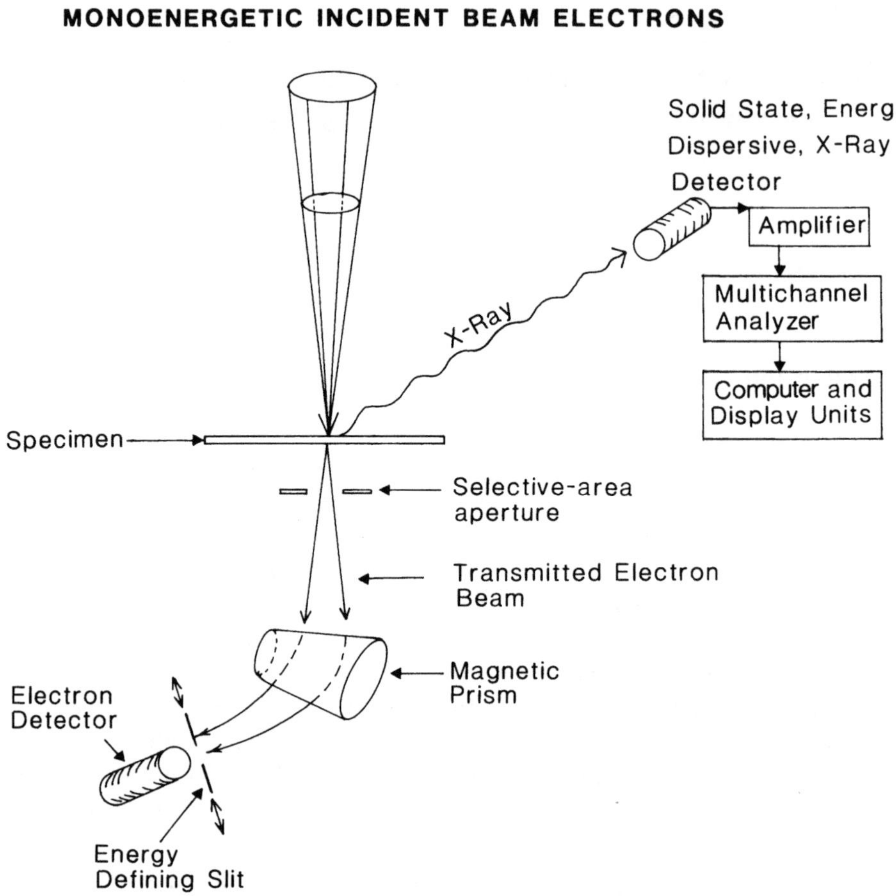

Fig. 1. Schematic diagram of a monoenergetic beam of electrons striking and passing through a specimen. The beam excites specimen atoms (ionization) producing elemental specific, *characteristic* X-rays which enter an energy dispersive, solid state detector creating electrical pulses linearly proportional to the energy of the X-ray detected. X-ray signal pulses are then amplified, passed to a multichannel analyzer and may either be displayed as spectra (histograms of X-ray energy) or dots, or by color coded images (mapping) (Fig. 2), and then saved for later computerized analysis. This technology is known as energy dispersive X-ray microanalysis (XRMA)
The portion of the beam passing through the specimen has lost energy proportional to the chemical elements ionized in the specimen. These transmitted electrons are selected by a deflection system (not shown), pass through a selective area aperture, and enter a uniform magnetic field prism which separates them according to their energy levels. High energy electrons curve through the magnetic field with greater radii than do low energy electrons, and thus are physically separated. The energy-separated electrons can be allowed or denied access to the electron detector by an energy-defining slit. Thus the range of characteristic energy loss electrons can be detected and these are directly related to the chemical elements they have ionized. This technology is known as electron energy loss spectrometry (EELS)

these technologies, readers are urged to consult the listed references [1, 8, 11, 12, 16, 18], and references contained therein, for more detailed explanations.

2 XRMA Principles

Among the most common signals generated by electron beam/specimen interactions are X-rays, which as part of the electromagnetic spectrum have both wavelength and energy properties. X-rays are produced by two interactions. First, high energy beam electrons can lose energy, as X-rays, by deceleration while passing near atomic nuclei of specimen atoms. These X-rays are termed *continuum* or background and are not useful for determination of chemical elements. The second interaction happens when high energy beam electrons collide with electrons residing in inner orbital shells of specimen atoms (K, L, M shells and their subshells, as designated by the Bohr model of atoms), causing native specimen electrons to be ejected and leaving the inner shells incompletely filled [13, 16]. Atomically this is an unstable, ionization, condition, and electrons from higher orbital shells move inward to fill the inner shell vacancies and relax the atom. When outer shell electrons fall to inner shell positions they must lose energy, and do so in the form of X-ray photons whose energy is equal to the energy difference between the inner and outer shells. Because each chemical element has a different nuclear charge, electron shell energies of elements differ. X-rays produced by shell transitions differ among chemical elements and are termed *characteristic X-rays* for each element. The *total* characteristic X-rays produced by each chemical element are unique, and when either their wavelength or energy properties are measured the element of origination can be identified. Thus, characteristic X-rays are, by analogy, the fingerprints of chemical elements. Just as it is considered inconclusive to attempt to identify an individual human by partial fingerprints, it can be equally inconclusive to attempt to identify a chemical element by a single wavelength or energy characteristic, since some X-rays produced from different elements can overlap [1, 8, 11–13]. Therefore, it is often important, especially when using higher beam accelerating voltages (over 12 kV) and specimens containing both high and low Z elements, to look at several characteristic X-rays of an element, or elements having overlapping X-ray lines in their atomic fingerprints, before making a final identification of the particular element in question.

The production of X-rays from a specimen is dependent upon several factors, among these are: (1) the initial energy of the incident beam electrons (accelerating voltage used to produce the monoenergetic beam); (2) the fluorescent yield or number of X-ray quanta emitted per electron shell ionization of specimen atoms, which decreases with low Z number elements; and (3) the interaction volume or area beneath the specimen surface where the beam electrons penetrate, spread, and interact with specimen atoms (greater in less dense or light elements). These and other factors, like positioning of the X-ray detector for the best angle of X-ray collection, determine the constraints and spatial resolution of X-ray microanalysis and become important when attempting advanced quantitative elemental determinations. The theory and practice of advanced quantitative elemental analysis is complex, requires the use of appropriate standards which mimic the texture and elemental composition of specimens, requires a number of advanced physical, mathematical and statistical assumptions, and should be approached with greater understanding than is conveyed here [7, 12].

3 XRMA Detectors

Because X-rays have both wavelength and energy properties, either property can be used to measure them. If wavelength is used, the technique is termed *wavelength-dispersive* (WD). In WD, a crystal is used which reflects or disperses only a particular X-ray wavelength, to the exclusion of all other wavelengths, and reflected X-rays of known wavelengths are then reflected towards and registered using a proportional counter. WD systems require several reflective, rotational crystals of differing physical properties to identify wavelengths from the range of chemical elements found in biological materials. The crystal spectrophotometer demands a restrictive geometric configuration effective only with very flat specimens [8, 11, 16]. While WD is the most accurate and sensitive method of measuring X-rays, it is somewhat cumbersome, ineffective with bulk samples, and therefore is rarely used for biological microanalysis.

Since the wavelength of an X-ray photon is inversely proportional to its energy level (measured in electron volts), a popular and versatile method of measuring X-ray energies was developed and is termed *energy-dispersive* X-ray microanalysis (ED). ED systems are almost exclusively used for analysis of biological specimens because a single energy absorptive crystal detector can, in theory, measure all elements of interest in the periodic table, and accomplish this simultaneously. ED detection systems consist of a solid-state semiconductor detector, an electronic amplifier, a multichannel analyzer, and a computer and display system (Fig. 1). The essential component of an ED system is the solid-state semiconductor detector, usually made of a silicon (Si) crystal diffused with a small number of electron-donating lithium (Li) ions which reduce defects in the silicon crystal. This crystal detector (a 10- or 30-mm diameter crystal is commonly used) is usually isolated from the microscope vacuum by a thin sheet of beryllium (Be), forming a window protecting the crystal from contamination. Unfortunately, this Be window absorbs weak or *soft* X-rays produced by light chemical elements with Z less than 11 (H, He, Li, Be, B, C, N, O, F, Ne). When an X-ray passes through the Be window and strikes the active region of the Si (Li) crystal, its energy is released as a series of collision events causing a voltage pulse of current directly proportional to the energy of the entering X-ray. These voltage pulses are amplified, sorted, digitized, entered, and stored in a multichannel analyzer so that the complete X-ray spectrum (actually a frequency histogram of pulses) is accumulated and displayed (Fig. 2). Thus, the ED system simultaneously takes all X-rays entering the detector and accumulates a spectrum showing the peaks of all measurable elements in a specimen [7, 10, 15]. In recent times, ED detectors have been used without the protection of the Be window, or by using an ultra-thin plastic or diamond window, to enable the entry of low energy X-rays produced by low Z number elements. This usage is called *windowless* or *light window* ED and competes with EELS technology, in certain respects, for the analysis of the low Z, light elements B, C, N, O, F, and Ne [1, 8, 11].

In addition to detection of X-rays, the display, long-term storage, and mathematical analysis of X-ray data is very important. Advances in microprocessors, small computers, and electronic data storage, when coupled with the availability of commercial software packages designed for XRMA, have brought the technology to *user-friendly* status. For most XRMA work commercial software and accompanying computers are adequate; however, they cannot completely substitute for user understanding of background theory.

4 Applications to Plant Pathology—XRMA

XRMA has been applied to numerous biological problems, and its application to plant pathology and botany has been the subject of recent reviews [4, 16]. Therefore, this discussion is brief, and for convenience divided into detection of insoluble elements left after conventional specimen preparation procedures (using fluids which may remove soluble elements), and detection of soluble or diffusible elements requiring either in situ precipitation or cryopreservation specimen procedures [7, 11, 18].

Insoluble Elements. Historically, plant pathologists combined XRMA technology with conventional specimen preparation procedures using liquid fixatives, liquid dehydrating agents, standard embedments, or critical point drying in the case of bulk SEM preparations. Under these conditions XRMA is useful only where the elements of interest are bound in a liquid-insoluble manner, in or on the structures of interest (i.e., host cell walls, fungal structures, etc.). Furthermore, it has been necessary to account for X-rays produced by the infusion of chemical elements contained in fixatives, buffers, embedments, and the various heavy metals used in evaporative coating of bulk specimens to provide needed electrical and thermal conductivity. Specimens used for XRMA should, when possible, be coated with evaporated carbon (Z = 12) or light metals up to Ni (Z = 28) for electrical and thermal conductivity; however, heavy metal coatings can produce X-ray peaks that may overlap with elements of interest in the specimen, and can reabsorb X-rays from the subtending specimen, thus prohibiting them from getting to the detector. In

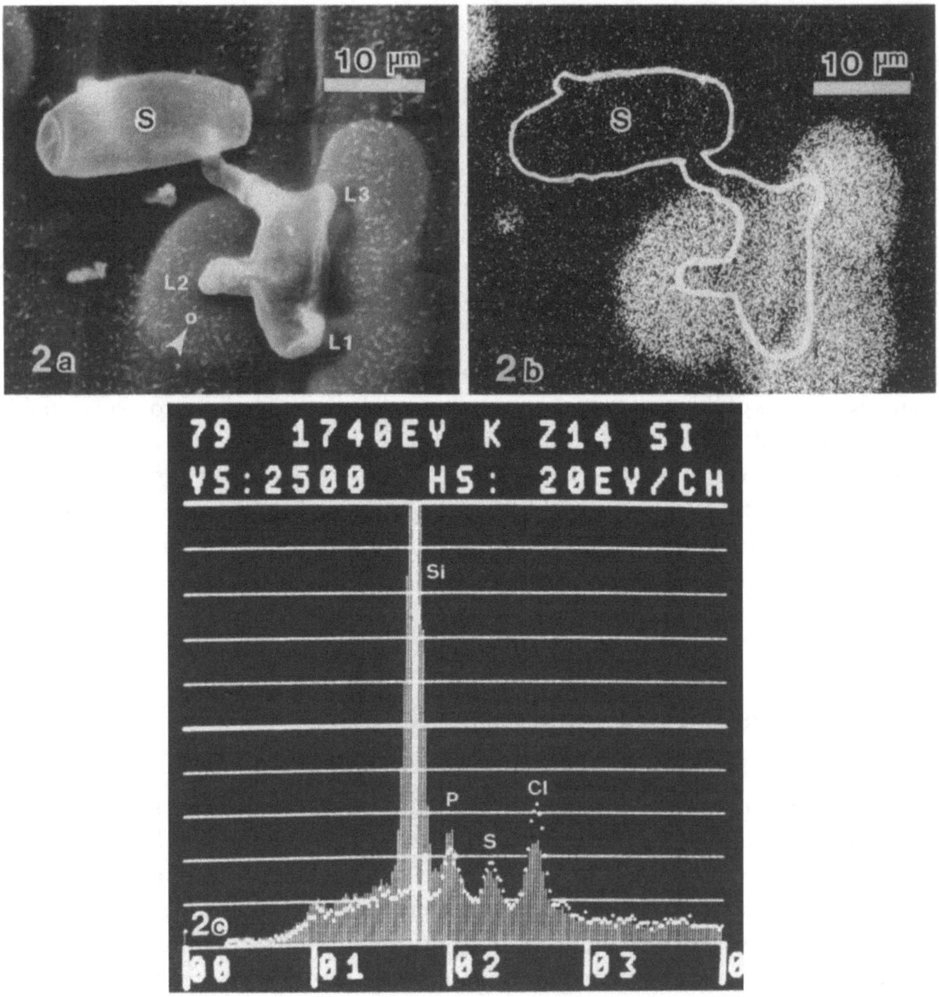

Fig. 2. This figure represents the two basic informational displays of X-ray data available using XRMA in conjunction with SEM. **a** is an SEM image of a germinated conidia of powdery mildew (*Erysiphe graminis*) attacking barley epidermal cells, prepared by glutaraldehyde fixation and critical point dried. The spore has produced an appressorial germ tube which has attempted three penetrations from separate lobes (*L1, L2, L3*). Each attempt at penetration has failed in conjunction with papillae deposited beneath the epidermal cell wall. Papillae are visible as lightened secondary eletron halo areas where emission is high due to greater density of papillae deposits. This density is due to localized, site specific deposition of silicon (Si) which accompanies the papilla response in leaves [15, 16]. The white arrow and open circle is where the scanning raster was stopped and a stationary electron beam placed for 100 s to produce the X-rays seen in the spectrum in **c**. **b** represents the K alpha X-ray signal of Si coming from the specimen shown in **a** and displayed as dots (40 000 dots used to make the photo). When the outline of the *E. graminis* conidia is superimposed on Si distribution it is evident that Si is directly related to the secondary electron halo areas associated with papillae. **c** is the X-ray frequency histogram (spectrum) taken from the halo area in **a** (arrow head and *open circle*). The *vertical lines* of the spectrum represent the halo site readings showing that Si is in a very high concentration relative to phosphorus (*P*), sulfur (*S*), and chlorine (*Cl*). Since

spite of the limitations of superimposing XRMA on conventional specimen preparations many discoveries have been made.

In host-parasite interaction studies, H. Kunoh and colleagues of Japan, as reviewed by Zeyen [16], were among the first to use XRMA with SEM to examine the elemental composition of epidermal cell walls under attack by several epidermal penetrating fungi, especially the powdery mildew fungus *Erysiphe graminis*. They studied elements in the halo response of barley leaves, where the epidermal walls directly beneath and to the sides of the fungal appressoria stain heavily with cotton blue and other light microscopic stains. They found insoluble deposits of Ca, Si, and Mn involved with the halo response, and later found these same elements in the papilla response subtending the halo area and located inside the epidermal cell wall. This research helped stimulate others to investigate the role of Si and Ca in two of the main cell defense mechanisms, the papilla response and the hypersensitive cell death response [5, 6, 17]. Subsequent research by Kunoh and others revealed that Ca may have a regulatory role in the cytoplasmic aggregate response leading to papilla deposition.

The role of Si in plants has been studied with XRMA, especially in grasses, where it apparently has several functions related to pathology. Si is taken up by roots of higher plants and translocated via transpiration stream transport as monosilicic acid and deposited in various cells and organs as insoluble amorphous silica, where it functions as a structural component [4, 6, 16, 17]. As a structural component, Si provides mechanical support for shoots, may facilitate passage of light through the epidermis to photosynthetic mesophyll, protects against water loss, and protects against insect feeding and parasite penetration [4]. The most studied effect of structural Si on protection against fungal parasites is in rice, where infection by the rice blast fungus, *Pyricularia oryzae*, and fungal brown spot, *Bipolaris oryzae*, is greatly diminished when plants are supplied with adequate amounts of Si. In rice, Si forms a barrier beneath the cuticle of epidermal cells, apparently making them highly resistant to fungal penetration, which may explain the need for high pressure fungal penetration structures as described in the chapter by R.J. Howard et al. A similar situation may exist in the barley-powdery mildew interaction where Si-supplied plants can become more resistant to penetration and colonization. Beyond its role in plant structure, Si is specifically deposited in certain rust fungal penetration structures [2] and during the papilla response of grasses [16, 17] (Fig. 2), and also appears when attacked cells undergo hypersensitive cell death responses to fungal parasite invasion [5, 6]. While it is unclear why Si is specifically

◀

liquid-based fixation and dehydration in ethanol were used in sample preparation, all elements seen are insoluble. The *white dot outline* imposed on the halo site spectrum represents a second spectrum taken from a nonattacked adjacent barley epidermal cell and is characteristic of normal cells, having very little Si content

deposited in parasitic fungi and in plant cells undergoing defense responses, a number of hypotheses have been advanced [5, 6, 16]. Si is known to form insoluble complexes with organic acids, di-and polyphenolics, and other organic compounds. Since phenolics are implicated in the papilla response and in hypersensitive cell death responses, an attractive hypothesis is that Si forms insoluble complexes with these compounds [6], so that the Si is easily detected in conventionally fixed specimens by using XRMA. In vitro synthesis of insoluble Si precipitates has been achieved in the presence of phenolics commonly implicated defense responses (D. Jiang and R.J. Zeyen unpubl.).

XRMA of conventionally fixed specimens has yielded information concerning plant diseases caused by abiotic agents, for certain metabolic disorders causing disease symptoms, and for determining distribution and effectiveness of wood preservatives and fungicides containing elements above $Z = 11$. Correlations between plants showing air pollution damage and accumulation of chemical elements in plants have been done using XRMA. Heavy metals and various salts can be considered abiotic agents of disease and their accumulation in plants showing toxic effects has been documented. Likewise, metal detoxification sites within plants and microorganisms have been identified [4, 16]. The distribution of copper-chrome-arsenate and pentachlorophenol preservative in treated wood have been correlated with success or failure of wood-rotting fungi. Likewise, distribution and redistribution of metallic-based fungicides on leaf surfaces and their effectiveness has been the subject of several investigations [16]. These and many other studies are possible because the chemical elements of interest become insoluble in or on plant specimens and can be located in conventionally fixed tissues [16].

There are several specialized applications where XRMA liquid-fixed and processed plant pathological specimens may be useful. Certain cytochemical methods of enzyme localization rely upon enzyme reactions leading to formation of insoluble lead phosphate or copper ferrocyanide deposits. These enzyme reaction deposits are electron-dense and visible with TEM; however, the deposits are sometimes confused with other electron-dense substances in the same sections. XRMA has been used to confirm the elemental composition of cytochemical, metallic-based enzyme reaction products in other biological systems and may be useful in plant pathology [13, 16]. XRMA can be used to identify and confirm antibodies or lectins containing metallic probes (Au and Fe) which are used to locate proteins, glycoproteins, saccharides, toxins, nucleic acids, and other molecules in fixed cells. Again, it is possible to confirm that electron-dense substances are indeed the labeled antibody or lectin probe of interest, and perhaps to map and quantify their distribution [13, 19].

Soluble or Diffusible Elements. Obviously with soluble or diffusible elements, conventional fixation procedures using liquid fixatives, buffer washes, and liquid dehydrating agents cannot be used because they remove liquid-soluble elements. Keeping soluble elements at their exact in situ locations is difficult because it requires specimen preparation procedures which insure that what XRMA is detecting is representative of the distribution in the living specimen. There are two approaches to keeping soluble, diffusible elements in place during specimen preparation: to add chemical substances that will cause soluble elements to form insoluble complexes, in situ precipitation, which can then be followed by conventional liquid-based specimen preparation protocols [9, 13, 16]. The second approach is to use cryofixation (low temperature) procedures to literally freeze soluble, diffusible elements in place within cells and tissues [7, 18].

In situ precipitation of soluble elements and XRMA applications have been recently reviewed [13, 16]. While these techniques yield valuable data on the location of soluble elements there are certain limitations which must be understood. The first is that not all soluble elements of interest in a specimen can be precipitated by known reagents, and some reagents simultaneously precipitate several soluble elements. The second limitation is that there may be translocation of elements within specimens during the precipitation process itself. Park et al. [9] studied the location of elemental Na in association with permeability changes in pear leaf cells attacked by the toxin-producing black spot fungus, *Alternaria alternata*, by using a mixture of osmium and potassium antimonate to attempt to precipitate Na in the form of Na-antimonates. While they experienced some difficulty with the limitations previously mentioned, they were able to use the technique in a satisfactory manner with appropriate control tissues and washings.

Freezing soluble and diffusible elements in situ by cryopreservation is a popular and perhaps the most definitive approach to XRMA of these elements. There are several methods of quickly freezing specimens [3, 7, 18, and N.D. Read Chap. 2, this Vol] to "fix" soluble elements, followed by freeze-drying or various freeze-substitution procedures, or by using frozen-hydrated specimens directly. Freeze-drying and freeze-substitution techniques are popular because specimens are returned to room temperature and the electron microscope in which the XRMA is performed does not require a specialized low temperature freezing stage. Freeze-dried specimens have been used extensively to study physiological disorders like "bitter pit of apple" involving Ca and Mg balances, "blossom-end rot of tomato", "brown spotting of Adzuki bean" and many others [16]. It has also been applied in general ecological problems involving microbial decay of leaves in acidified lakes [10]. The problem with freeze-drying is that it often causes specimen collapse and distortion, perhaps causing physical relocation of areas within the specimen. When specimens are small and cytoplasmic detail essential, freeze-substitution can be used. However, in the substitution process small losses of elements like Na, K, Ca, and Cl are known to occur. Nevertheless, freeze-substitution is widely believed to be a valid technique, but has not been extensively used for XRMA of plant pathological specimens. The use of frozen-hydrated specimens is theoretically the best procedure for keeping soluble elements in place, since specimens are examined in the frozen state [3]. However, few electron microscopes are equipped with both XRMA units and adequate freezing stages. Thus, there is little information concerning XRMA of frozen-hydrated plant pathological specimens.

5 Electron Energy Loss Spectrometry (EELS) Principles

Electron beams produced by electron microscopes are monoenergetic because they are generated using specific and constant accelerating voltages. The monoenergetic nature of beam electrons is the basis for EELS technology. For almost 50 years it has been known that monoenergetic beam electrons lose measurable and constant amounts of energy as they collide with specimen atoms and cause ionization of inner shell specimen electrons. As noted earlier, the amount of beam energy lost is specific for chemical elements ionized, so that the amount of loss yields information about the elemental composition of the specimen through which the beam passed. In addition, the loss of beam energy can also provide some information about the binding states of elements in a specimen, a factor now used in material sciences. When an EEL spectrometer is placed beneath a very thin specimen in an electron microscope, used in TEM mode, it is possible to measure the energy loss of beam electrons passing through the specimen (Fig. 1). Since the great majority of electrons passing through a very thin specimen can be collected, almost all the useful signal information is maintained. Thus, EELS can detect smaller amounts of an element because it measures a strong signal, and has better analytical resolution than X-ray microanalysis [1, 14, 15]. Additionally, the sensitivity for light elements (Z less than 11) is better than for windowless ED, making EELS the theoretical microanalysis technique of choice for light element analysis.

6 EEL Spectrometer Systems

Electron spectrometers have a long history, and several types have been used for electron energy loss work [1, 14, 15]. Electromagnetic spectrometers are the most commonly available from commercial sources, are well understood, and are often used in EELS systems. In these systems, beam electrons transmitted through specimen are selected on the basis of their scattering angle by a series of deflection systems, and pass through a selected area aperture located beneath the specimen. Once through the aperture a magnetic prism with a uniform magnetic field separates electrons according to their energy levels. Less energetic electrons curve to smaller radii than do more energetic ones when passing through the magnetic field. So by the differing radial paths of these electrons and through manipulation of the energy defining slit, electrons of specific energies are directed to enter the electron detector (Fig. 1). Thus, electrons of differing energies can be sorted and recorded, and are used for determining elemental composition of the specimen because their energy losses are specifically related to the chemical elements they ionized while passing through the specimen.

7 Applications to Plant Pathology—EELS

The same specimen preparation procedures and limitations for insoluble and soluble elements mentioned for XRMA apply to EELS. EELS is very effective with chemical elements lighter than Na, $Z = 11$, but can also be used effectively with elements as high as Ca, $Z = 20$. The use of EELS is limited to very thin specimens (less than 40 mm in thickness, in the gray interference color range) because the electrons must pass directly through the specimen without multiple scattering which occurs in thicker specimens. EELS cannot be used with bulk specimens and the requirement for ultrathin samples means that if cells and tissues are to be examined, plastic embedded specimens are necessary since only they can be cut thin enough for effective use.

Since EELS, when used for elemental analysis, is restricted to very thin specimens, its most appropriate use may be to study elemental composition of isolated subcellular fractions like filamentous proteins, membranes, wall fragments, nucleic acids, microdroplets of cell fluids, etc. Since EELS also gives information concerning the binding states of chemical elements, it may be possible to distinguish between bound and free ions within such specimens. To date, EELS has not been extensively used in plant pathology or other biological sciences primarily because most electron optical facilities used by biologists and plant pathologists do not have EELS equipment. However, EELS represents a technology that can detect low levels of light elements and give physical-chemical information simultaneously. With the current interest in EELS technology it is only a matter of time before it becomes more widely applied.

8 Summary

XRMA, when used with electron microscopy, is a powerful tool offering plant pathologists many unique opportunities to study the cellular chemistry and internal and external chemical microenvironments surrounding pathological processes. The analytical procedures surrounding XRMA are still developing and remain challenging for users, although the hardware and software now available greatly facilitate use. The most challenging aspect of XRMA for plant pathologists remains the maintenance of specimen integrity through pre-preparation, preparation and analysis processes. Nevertheless, XRMA has been applied to plant pathology for many years and has resulted in important new information unavailable by use of other techniques. Given the wide range of problems in host-parasite relationships, plant nutritional dysfunctions and many other disease and deterioration situations that invite in situ exploration of the chemical elements involved, the use of XRMA will undoubtedly increase. The use of XRMA to complement many existing cytochemical procedures is an area of pathological research that will continue to grow. The use of XRMA with cryopreservation of specimens will need to be expanded into the area of soluble or diffusible elements important to the understanding of plant disease situations.

EELS is a complementary technology to XRMA, most useful for light element analysis. Its application to biological specimens has been somewhat limited by the requirement of ultrathin specimens and by the fact that very few electron optical facilities have EELS equipment. Nevertheless, there is growing interest in using EELS, and plant pathology is one area of biology rich in problems suitable for its application.

XRMA and EELS are technologies originally developed to answer questions in physics and material sciences. Their application to the biological sciences has opened a new era of analytical electron microscopy of biological specimens. It is now possible to move beyond the descriptive use of electron microscopy and into the realm of investigating both structure and in situ function simultaneously. Analytical electron microscopy offers exciting promise and is being rapidly assimilated in all areas of biological research including plant pathology. The first 50 years of electron microscopy application was dominated by descriptive studies, and that is now changing so the second 50 years may well be dominated by analytical approaches.

References

1. Budd PM, Goodhew PJ (1988) Light-element analysis in the transmission electron microscope: WEDX and EELS. (Microscopy handbooks: 16), Royal Microscopical Soc, Oxford University Press, New York, pp 73
2. Chong J, Harder DE (1982) Ultrastructure of haustorium development in *Puccinia coronata* f. sp. *avenae*: cytochemistry and energy dispersive X-ray analysis of the haustorial mother cells. Can J Bot 72: 1518–1526
3. Echlin P (1986) The preparation and X-ray microanalysis of bulk frozen hydrated vacuolate plant tissue. J Microsc 141: 329–348
4. Harvey DMR (1986) Applications of X-ray microanalysis in botanical research. Scanning Electron Microsc III: 953–979
5. Heath M, Stumpf MA (1986) Ultrastructural observations of penetration sites of the cowpea rust fungus in untreated and silicon-depleted French bean plants. Physiol Mol Plant Pathol 29: 27–39
6. Koga H, Zeyen RJ, Bushnell WR, Ahlstran G (1988) Hypersensitive cell death, autofluorescence, and insoluble silicon accumulation in barley leaf epidermal cells under attack by *Erysiphe graminis* f. sp. *hordei*. Physiol Mol Plant Pathol 32: 395–409
7. Marshall AT (1988) Progress in quantitative X-ray microanalysis of frozen-hydrated bulk biological samples. J Electron Microsc Tech 9: 57–64
8. Morgan AJ (1985) X-ray microanalysis in electron microscopy for biologists. (Microscopy handbooks; 5), Royal Microscopical Soc, Oxford University Press, New York, pp 79
9. Park P, Ohno T, Nishimura S, Kohmoto K, Otani H (1987) Leakage of sodium ions from plasma membrane modification, associated with permeability change, in host cells treated with a host-specific toxin from a Japanese pear pathotypes of *Alternaria alternata*. Can J Bot 65: 330–339

10. Perry JA, Zeyen RJ, Newson M, Ahlstrand G (1989) X-ray microanalysis of leaf-litter decomposition in lakes. BioScience 39(4): 260–263
11. Roomans GM (1988) Introduction to X-ray microanalysis in biology. J Electron Microsc Tech 9: 3–17
12. Roomans GM (1988) Quantitative X-ray microanalysis of biological specimens. J Electron Microsc Tech 9: 19–43
13. Sumner AT (1988) Cytochemical applications of X-ray microanalysis. J Electron Microsc Tech 9: 99–112
14. Zaluzec NJ (1985) A beginners guide to energy loss spectroscopy in an analytical electron microscope: part I—basic principles. EMSA Bull 15: 72–80
15. Zaluzec NJ (1986) A beginners guide to electron energy loss spectroscopy: part II —electron spectrometers. EMSA Bull 16: 58–63
16. Zeyen RJ (1982) Applications of in situ microanalysis in understanding disease: X-ray microanalysis. Annu Rev Phytopathol 20: 119–142
17. Zeyen RJ, Carver TLW, Ahlstrand G (1983) Relating cytoplasmic detail of powdery mildew infection to presence of insoluble silicon by sequential use of light microscopy, SEM, and X-ray microanalysis. Physiol Plant Pathol 22: 101–108
18. Zierold K (1988) X-ray microanalysis of freeze-dried and frozen-hydrated cryosections. J Electron Microsc Tech 9: 65–82

Chapter 6 Michel Wurtz

The Fine Structure of Virus Particles

1 Introduction

Due to their limited number of genes, at least the simpler plant viruses (like many other viruses) are made of two components: (a) nucleic acid molecule(s) contained in a protein capsid made of one or perhaps two molecular species regularly arranged as structural elements or subunits. The possible designs for such structures, made by polymerization of a large number of identical subunits [2000 for tobacco mosaic virus (TMV)], are limited [1]: helical for rod or filamentous-shaped forms, and icosahedral for the isometric or spherical ones. Electron microscopy and X-ray or neutron diffraction are the preferred tools for establishing the fine structure of these assemblies.

2 Spherical Plant Viruses

Also called isometric viruses, their capsids are arranged with icosahedral symmetry following the quasi-equivalence theory of Caspar and Klug [1]. The role of electron microscopy here is to place the studied structure in one of the different classes of icosahedrons (determination of the triangulation number T), to propose a model compatible with the biochemical data, and to determine the internal organization of the nucleic acid.

This category of plant viruses has been intensively studied by electron microscopy, mainly by analyzing negatively stained suspensions of purified particles [2]. Direct interpretation of such micrographs may be difficult because of: (a) superposition of the upper and lower parts of the particle; (b) curvature of the surface; (c) irregularities in distribution of the stain; (d) variable orientation of the individual particles on the grid; (e) artifacts and distortions caused by preparation techniques and the damage caused by the electron beam, and (f) the holographic

Microbiology Department, Biozentrum, University of Basel, Klingelbergstrasse 70, CH-4056 Basel, Switzerland

noise in bright-field imaging mode in the electron microscope. In order to improve the quality of the micrographs, several methods were developed and one of them introduced by Horne and Pasquale-Ronchetti [3] allows the application of image processing technique to paracrystalline arrays of spherical viruses.

2.1 Cowpea Chlorotic Mottle Virus (CCMV)

Cowpea chlorotic mottle bromovirus will serve as a good example to describe viruses of icosahedral form. With a diameter of 26 nm, it consists of a protein capsid (180 molecules) containing single-stranded RNA segmented into four pieces. It was previously postulated [4] that the virion is made of 12 pentamers and 20 hexamers located on the surface of a T = 3 icosahedron.

To test this model, regular arrays (2-D crystals) of CCMV [5] were made from highly purified and concentrated virus suspensions (4 mg/ml) in ammonium molybdate, spread over a freshly cleft mica surface. After drying, the mica was coated with evaporated carbon, floated on aqueous 1% uranyl acetate, and picked up on a copper holey carbon filmed grid. Two kinds of array were observed in the electron microscope:

a) hexagonal with very close packing of the particle but in random orientation (Fig. 1);
b) tetragonal with well oriented particles (Figs. 2 and 3).

Computer filtration of two different micrographs, in which the particles have been penetrated by the stain, shows that the capsomer organization becomes clearly visible from tetragonal arrays (Figs. 2, 3). The stain-excluded image, with predominantly one-sided contrast, shows the pentagonal or hexagonal shape of the capsomers (Fig. 3c) and is compatible with a T = 3 icosahedron (Fig. 4a), but at the resolution attained (2.5 nm; Fig. 3b) the structural units (monomers) of the capsomer are not visible. The stain-penetrated tetragonal specimen (Fig. 2) shows more clearly the part of the capsid neighborhood involved in crystal formation (Fig. 2c). Bonding takes place between pairs of hexamers related by a twofold axis in both particles, relatively rotated by 90°.

Using neutron diffraction data and by analogy with the related brome mosaic virus, the authors have proposed a three-dimensional model (shown in Fig. 4b, c) made of ellipsoidal capsomers. They have assumed that preparative artifacts are not sufficiently extreme to alter the structure at this moderate resolution.

Fig. 1. Large hexagonal array of cowpea chlorotic mottle bromovirus (CCMV) prepared by the Horne and Pasquali-Ronchetti method [3]. **a** Intermediate stain penetration and random orientation of particles. **b** Optical diffraction. **c** Computer filtration. No substructure, but only the averaged capsid mass distribution is visible

Fig. 2. Large tetragonal array of CCMV. **a** The particles are penetrated by the stain and are well oriented. **b** Optical diffraction. **c** Computer filtration

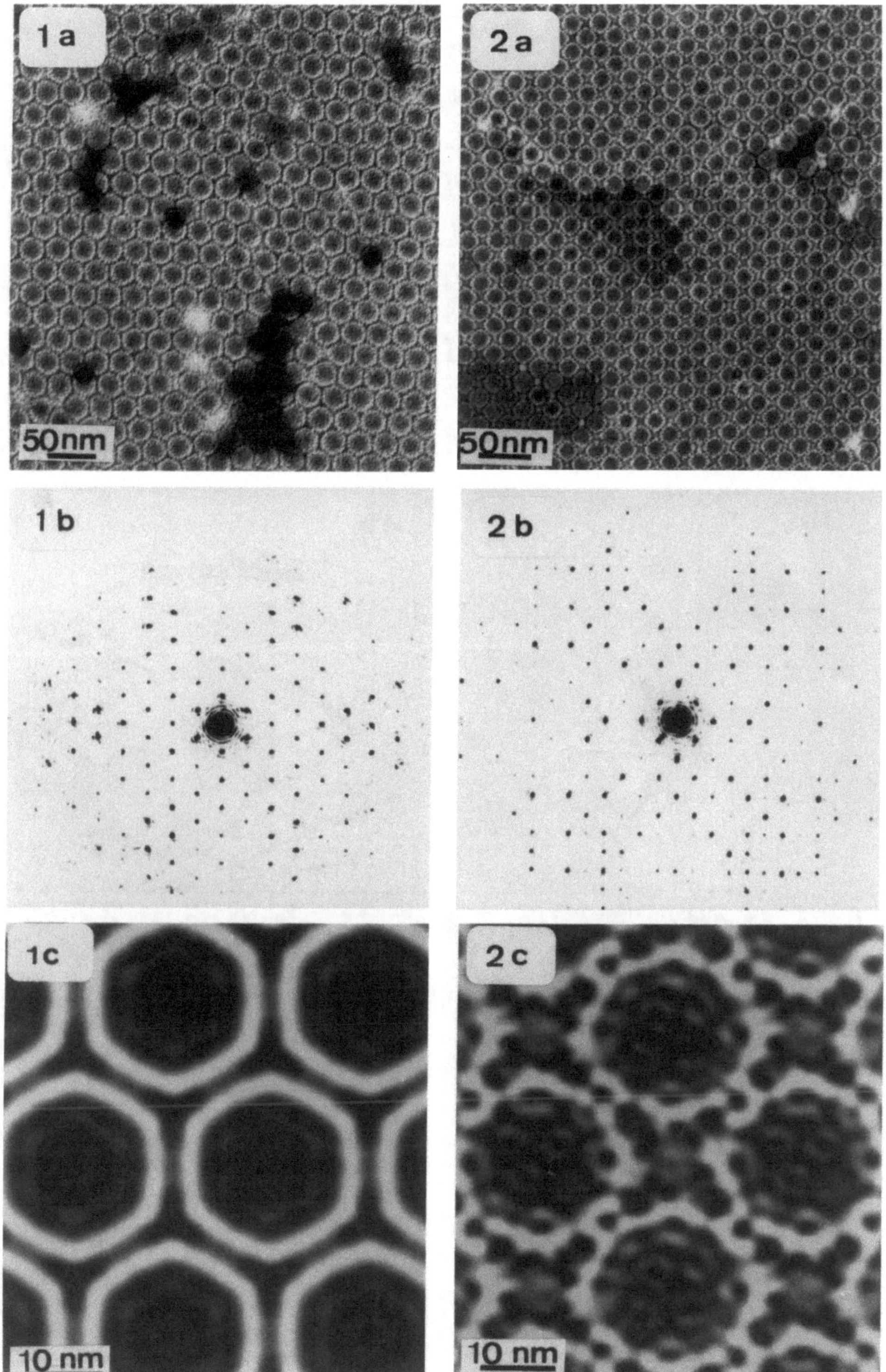
1a
2a
50nm
50nm
1b
2b
1c
2c
10 nm
10 nm

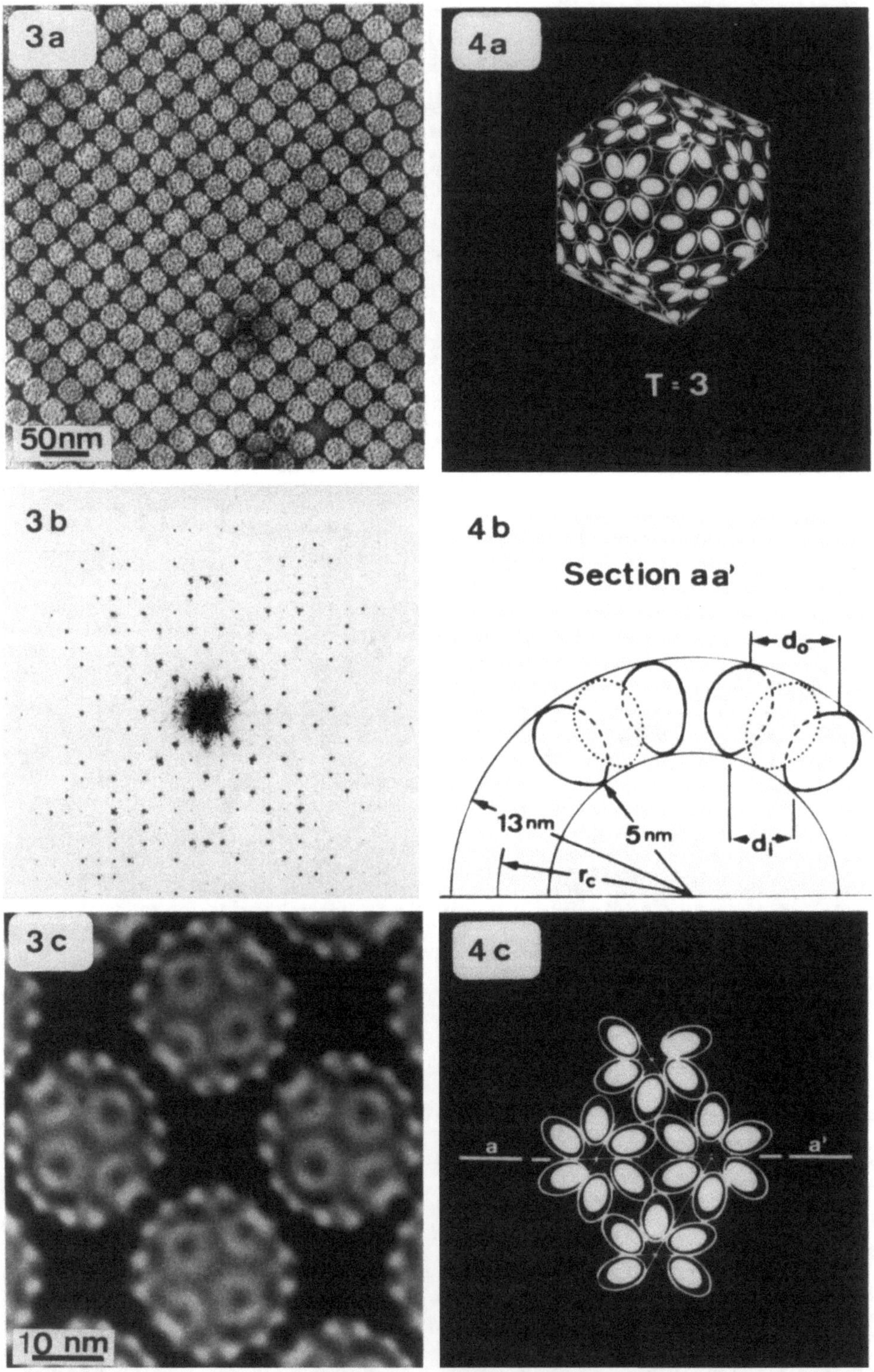
3a
50nm
4a
T=3
3b
4b
Section aa'
d_o
13nm
5nm
d_i
r_c
3c
10 nm
4c
a
a'

2.2 Beet Mild Yellowing Virus (BMYV)

Due to the difficulty of obtaining large amounts of purified beet mild yellowing virus (also an icosahedral virus) [6], it was not possible to apply the classical image processing technique used in the previous example. The physical properties of BMYV were unknown [7]. This virus has been studied with the scanning transmission electron microscope (STEM Vacuum Generator HB-5 [8] in our Institute), interfaced with a digital acquisition system as described by Engel et al. [9]. The digital image is made of picture elements (pixels); each pixel of the dark-field image gives information on the mass and chemical composition of the small volume irradiated with the very narrow electron beam.

Mass-mapping of unstained nucleoprotein structures is a classical technique, but only the features of the STEM made this method sufficiently precise and easy to use [10]. The validity of the method has been checked using tobacco mosaic virus as reference.

BMYV contains one single-stranded RNA molecule of 2×10^6 Da encapsidated in an icosahedral shell of 26 nm diameter (Fig. 5) made of one single type of polypeptide of 24×10^3 Da. Mass measurements from unstained darkfield STEM images (Fig. 6) yield a total mass of 6.5×10^6 Da (Fig. 7a). If we consider that BMYV capsids have a molecular weight of about 4.5×10^6 Da (STEM determination of empty particles: $4.0 \pm 0.4 \times 10^6$; Fig. 7b), we find that they are made of about 188 protein units. This is in good agreement with the same triangulation number as CCMV, $T = 3$, and 180 units to make the shell.

3 Rod-Shaped Plant Viruses

Isolates of beet necrotic yellow vein fusovirus (BNYVV) [11] which, like TMV, is a right-handed helix, contain four length classes of particles (85, 100, 265, and 390 nm) as shown in Fig. 8, related to the four RNA species of its genome (0.6, 0.7, 1.8, and 2.3×10^6 Da) [12]. This virus looks like TMV (Fig. 9a, b). By optical diffraction (Fig. 10), it was demonstrated that all the four classes of particles have the same helical symmetry. Their fine structure has been established by comparing them to TMV, as a standard, at every step of the study. Fourier analysis [13, 14] gives information about the molecular organization of the structure (Fig. 9b). It is a single-start helix (pitch 2.6 nm) which repeats after four turns. To determine the hand

◄

Fig. 3. Large tetragonal array of CCMV. **a** Micrograph of particles little penetrated by stain, and well oriented. **b** Optical diffraction of a selected area from **a**. **c** Computer filtration of **b**

Fig. 4. Models proposed for the particle structure. **a** Model icosahedron, class $T = 3$, to be compared with Fig. 3c. **b** Section of this model between two adjacent hexamers as shown in **c**: d_i is the capsomer diameter projected at the inner surface; d_o is the capsomer diameter projected at the outer surface; r_c is the radius where the maximal mass density of the capsid is located

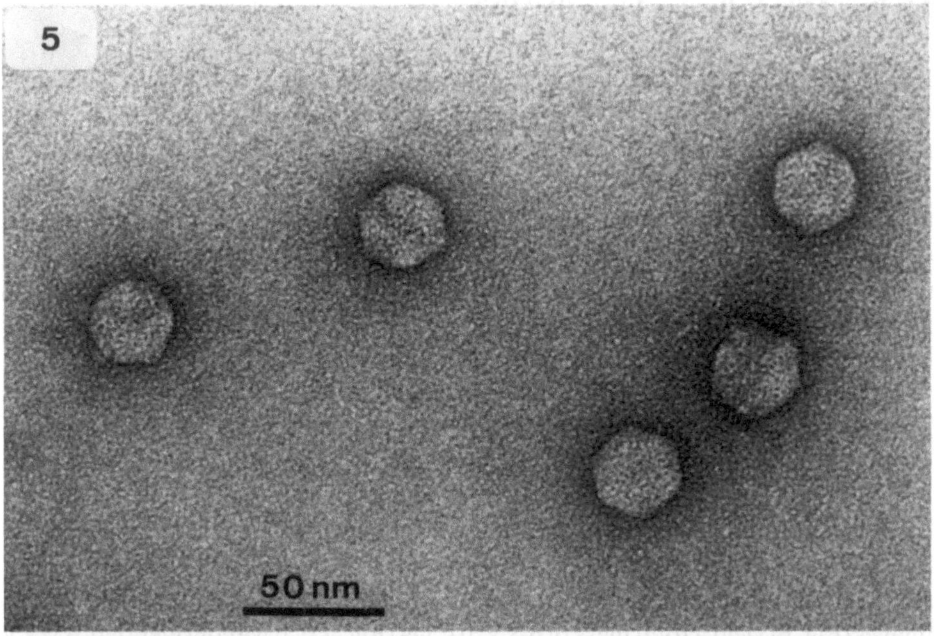
5
50 nm

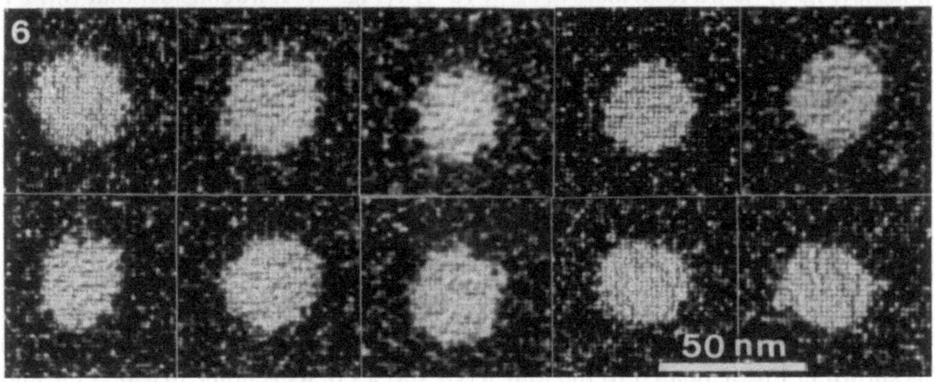
6
50 nm

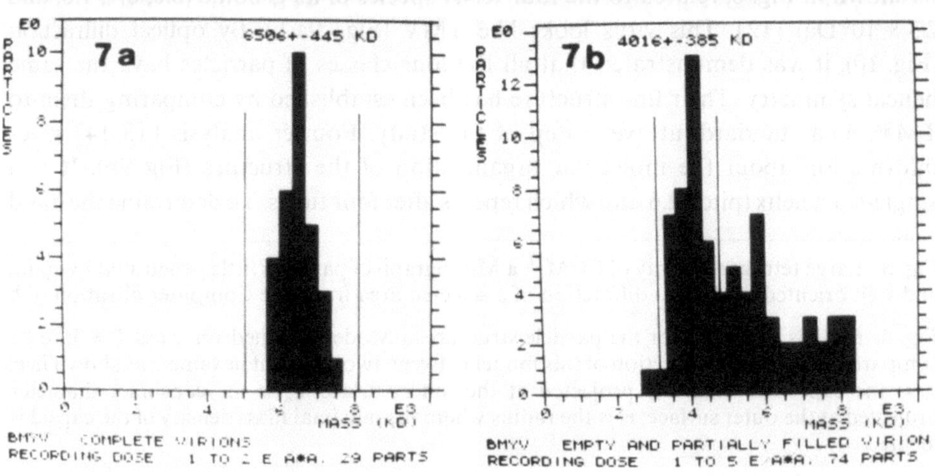
7a
6506+-445 KD
PARTICLES
MASS (KD)
BMYV COMPLETE VIRIONS
RECORDING DOSE 1 TO 2 E A*A. 29 PARTS
7b
4016+-385 KD
PARTICLES
MASS (KD)
BMYV EMPTY AND PARTIALLY FILLED VIRION
RECORDING DOSE 1 TO 5 E A*A. 74 PARTS

of this helix, we diffracted the image of a particle (Fig. 11) that had been contrasted on one side only, using the freeze-drying shadowing technique [15]. Only tungsten evaporated with an electron gun was able to give enough resolution to reveal the 2.6 nm periodicity of the particle. This conclusion was confirmed by using the method of Finch [16], based on analysis of diffraction patterns from two halves of a selected particle photographed before and after tilting by 6° in the electron microscope, as shown in Fig. 12.

The structures of those negatively stained particles which gave the best diffraction pattern were analyzed by computer filtration. The results shown in Fig. 13 allow an independent estimate of the number of subunits per helical turn. Although this figure is likely to be an overestimate since the outermost width was used, the result, 13 ± 1, is consistent with the solution calculated from the analysis of the first layer line.

The layer line rule [14] which was applied to the diffraction pattern is given by $l = 4n + 49m$. It shows that the four particles of different lengths are built from a single-start right-handed helix which repeats after four complete turns, involving 49 structure units. If the assumption of a one-to-one correspondence between these structure units and the coat protein molecules (MW 21 000) is correct, there are $12\frac{1}{4}$ subunits per turn. Their packing is illustrated in Fig. 13d in terms of a computer-generated model, which correlates remarkably well with the filtered electron micrographs (Fig. 13c). This structure is reminiscent in certain respects of TMV and the respective structural properties are most constructively compared as variations on a theme (Table 1).

Table 1. Some properties of TMV and BNYVV compared

	TMV	BNYVV
Length(s) in nm	300	390, 265, 100, 85
Pitch in nm	2.3	2.6
Nucleotides/protein	3	4
Coat protein M_r	17 400	21 000
Width in nm	18	20
Nucleotides/turn	49	49
Hand of the helix	Right	Right

Fig. 5. Conventional bright-field micrographs of purified negatively stained beet mild yellowing luteovirus (BMYV) particles. The hexagonal appearance indicates an icosahedral structure

Fig. 6. Gallery of unstained dark-field digital STEM pictures of BMYV selected for mass measurements. The absence of supporting stain explains the irregular shape of the particles

Fig. 7. Mass histogram obtained after application of the mass determination program on selected particles as shown in Fig. 6. The mass distribution presents a maximum at 6.5×10^6 and 4.0×10^6 Da full (**a**) and empty (**b**) particles, respectively

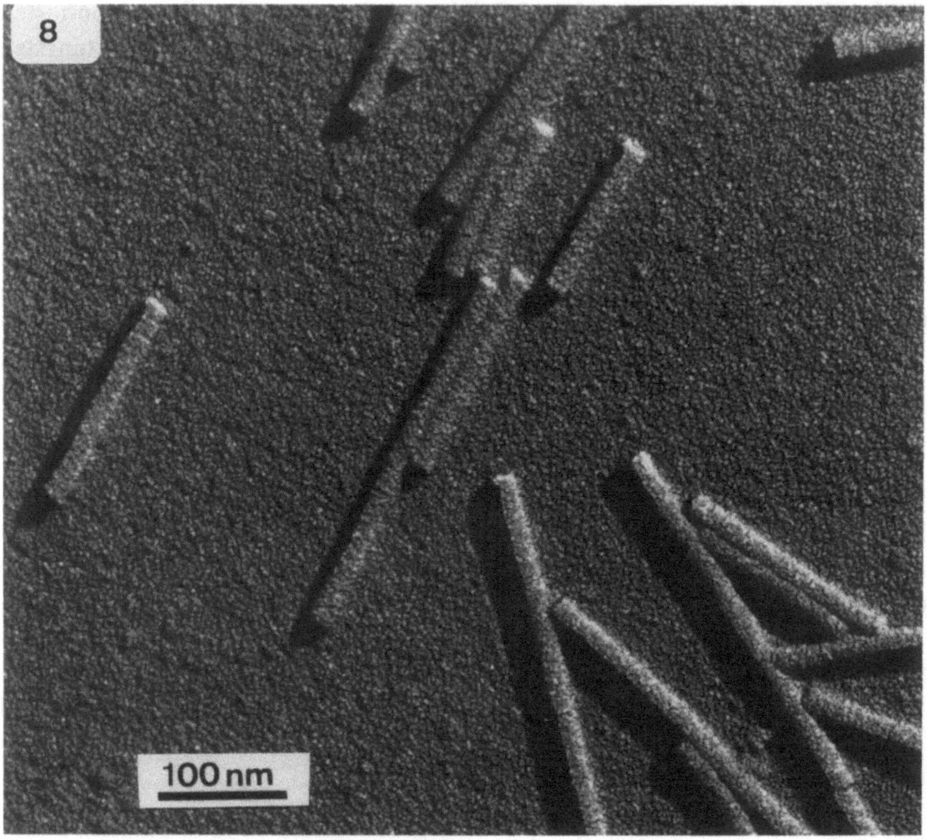
8
100 nm

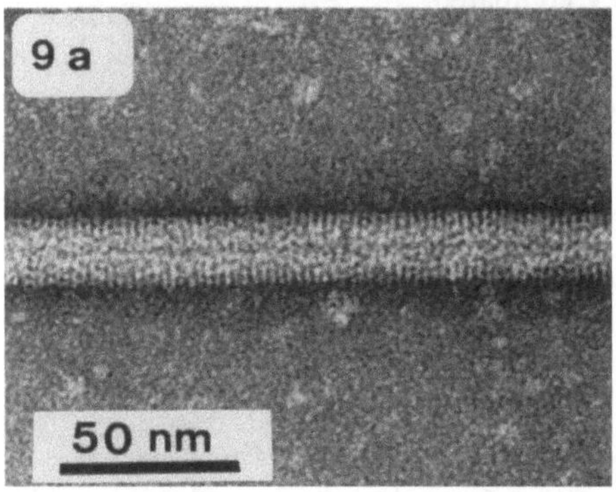
9 a
50 nm

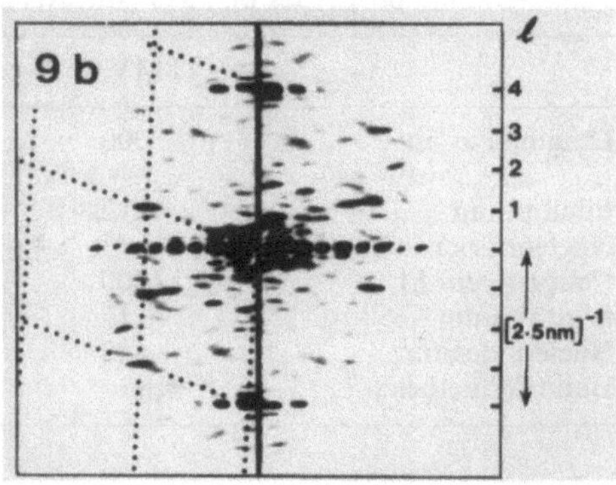
9 b
l
4
3
2
1
[2·5nm]⁻¹

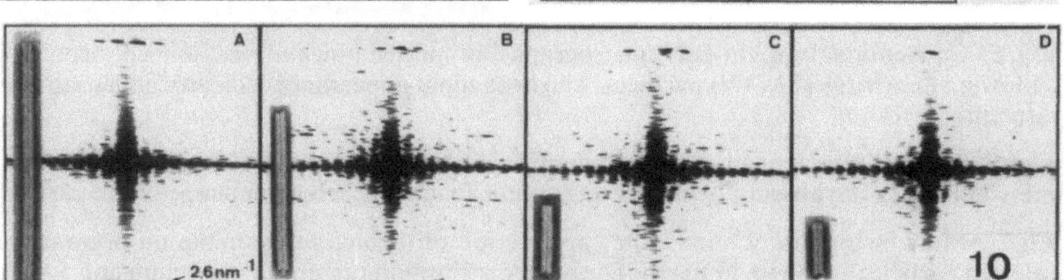
A
B
C
D
2.6nm⁻¹
10

Despite some differences in subunit packing and the number of nucleotides per coat protein molecule, it is noteworthy that the helical arrangement of single-stranded RNA (49 nucleotides per turn) is quite well conserved between these two viruses. This model proposed for the structure of BNYVV is consistent with all other data available on genetics and biochemistry.

4 Filamentous Plant Virus

Beet yellow closterovirus (BYV) has flexible filamentous particles 1250 nm in modal length (Fig. 14) [17], with helical cross-banding. We have reinvestigated this particle using the same methods as described for BMYV. The biochemical data give 25 000 Da for the mass of the protein subunit and 4.15×10^6 for that of the single-stranded RNA molecule. The total mass per unit length obtained with the STEM (Fig. 15) is $61\,200 \pm 5500$ Da/nm; the calculated total mass for the full length particle is thus 76.5×10^6. The optical diffraction pattern (Fig. 14b) shows a strong 3.7 nm^{-1} meridian spacing and a weak layer at 7.4 nm^{-1}. All these values are in agreement with those obtained by Bar-Joseph and Hull [18]. The ratio between the position of these two layer lines is about 2 (7.4/3.7), indicating that the helix is in register every two turns. By combining all available data, we find 8.57 (theoretically 8.5) subunits per turn; this means that 17 subunits repeat every two turns. These data are in good agreement with those of Brandes and Wetter [17].

5 Discussion

The fine structures established or confirmed for three structural types of plant viruses (spherical, rod-shaped, and filamentous) using new methods and instruments developed in our Institute, are good examples of the potential of high resolution electron microscopy and related techniques, such as image processing and mass determination of STEM digitalized images. Even when only small quantities of purified virus are available, precise data can still be obtained. Combining new preparation methods and imaging modes has progressively improved our know-

Fig. 8. Low magnification of platinum shadowed particles of BNYVV. Particles representative of the different length classes are visible

Fig. 9. High magnification of BNYVV negatively stained with uranyl formate. **a** The 2.6 nm striation is easily resolved. **b** Computer-generated diffractogram and its indexing

Fig. 10. Gallery of the four classes (*A–D*) of BNYVV particle and their similar optical diffractograms

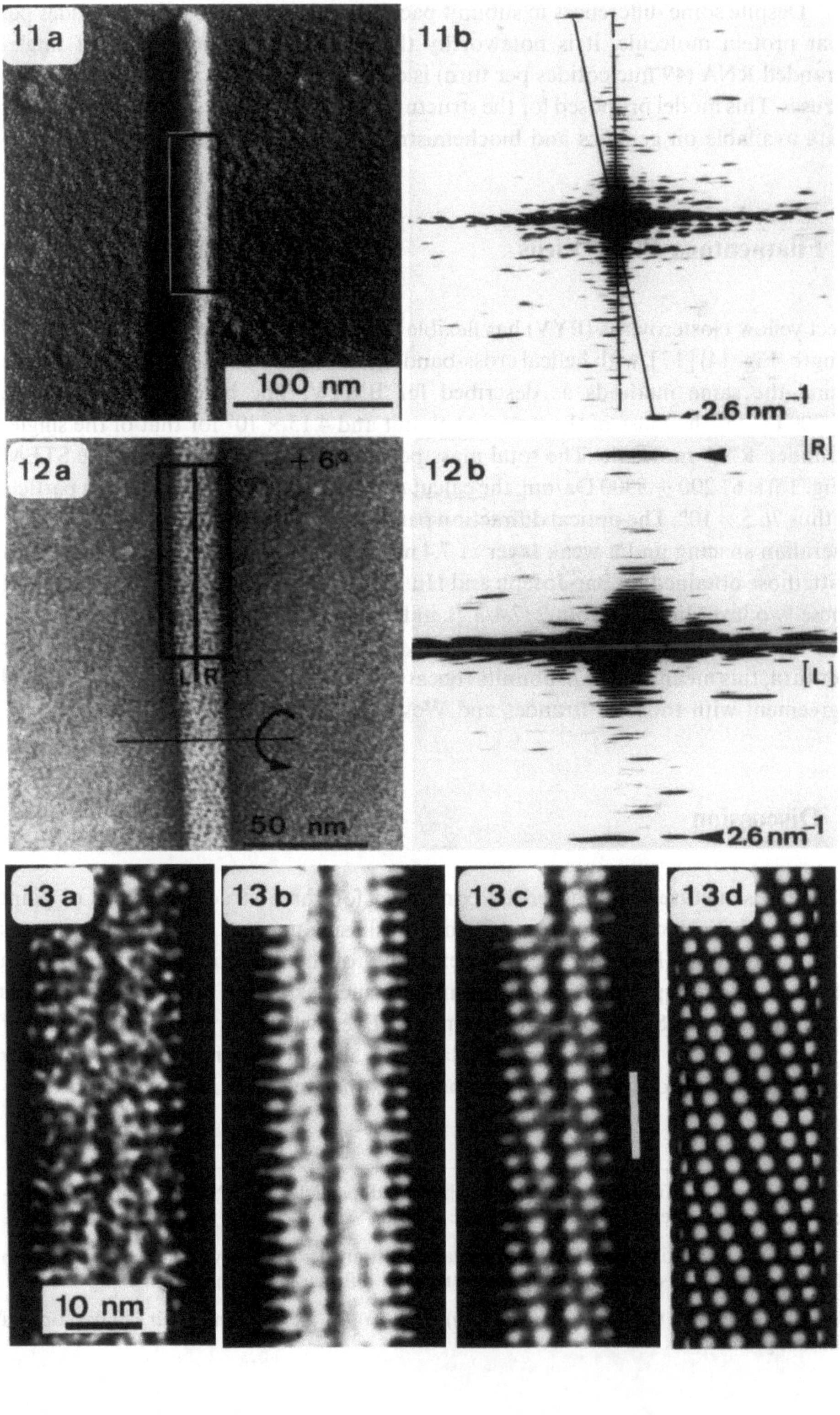
11a
100 nm
11b
2.6 nm⁻¹
12a
+6°
L R
50 nm
12b
[R]
[L]
2.6 nm⁻¹
13a
10 nm
13b
13c
13d

ledge of the structures of plant viruses, despite all the limitations involved, mainly the removal of water from the specimen prior to preparation for the EM.

Cryoelectron microscopy is now a routine method for maintaining intact the native aqueous environment of supramolecular structures [19]. It allows them to be studied in conditions very close to those existing in vivo, as shown in Fig. 17, a high resolution picture of TMV in the frozen hydrated state [20]. The main limitation of this technique is the electron beam sensitivity of the object.

In our laboratory, a group is developing new instruments and techniques [21]. They are working to apply the scanning tunneling microscope (STM) to biology. Again, TMV was the test object [22] to try this new technology (Fig. 18). The hope is to reduce the radiation damage and the artifacts associated with specimen dehydration. This will be possible in the near future if STM technology becomes applicable to aqueous specimens.

6 Conclusion

We have attempted to illustrate the contribution to plant virus studies of high resolution electron microscopy and associated new analytical methods. The results are interesting in three different ways. First, resolution is now at or close to the biomolecular level, and viruses offer us a challenge to interpret their organization and assembly in molecular terms. Second, a knowledge of the fine structure of viruses, both as passive objects and as active participants in pathological processes, will undoubtedly help us to combat them as disease agents. Third, classification and diagnostics are always necessary, and electron microscopy has always played and will continue to play a major part in these activities.

Acknowledgments. I would like to thank Drs. A.C. Steven, P.R. Smith, A. Engel and B.L. Trus for providing me with excellent processed images, Dr. J. Lepault for the frozen-hydrated and A. Stemmer for the STM pictures of TMV, Prof. T.A. Bickle for proofreading the manuscript, M. Zoller and H. Frefel for the photographic work, and E. Amstutz for preparing and typing the manuscript.

◄

Fig. 11. Freeze-dried BNYVV particles shadowed with tungsten. **b** Diffraction pattern of the window in **a**. The position of the 2.6 nm reflexion relative to the vertical axis is the same as for TMV, which is right-handed

Fig. 12. a Micrograph of negatively stained BNYVV, tilted by 6° around an axis perpendicular to the particle. **b** The handedness of the helix can be determined [16] by comparing the diffraction patterns of the two halves of the particle as shown on the picture

Fig. 13. Fine structure of BNYVV after computer filtration and modeling. **a** Original negatively stained micrograph. **b** Computer-filtered image of **a**, showing clearly the 2.6 nm repeat; only aperiodical noise has been removed. **c** Computer-filtered single-sided image of **a**. **d** Computer-generated model. The helical packing of subunits is in good agreement with the filtered image **c**. The 10.4 nm repeat is also indicated

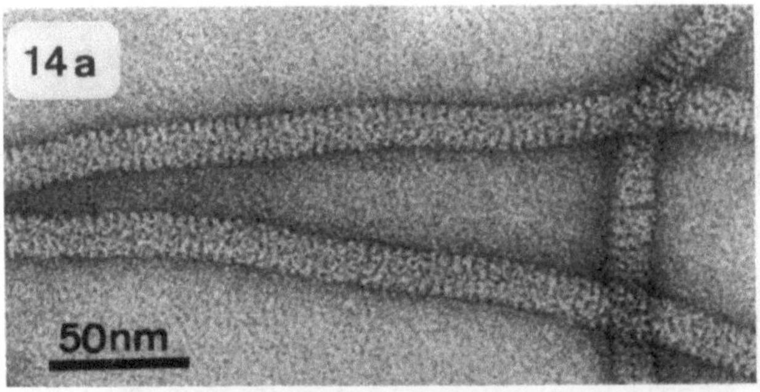
14a
50nm
14b

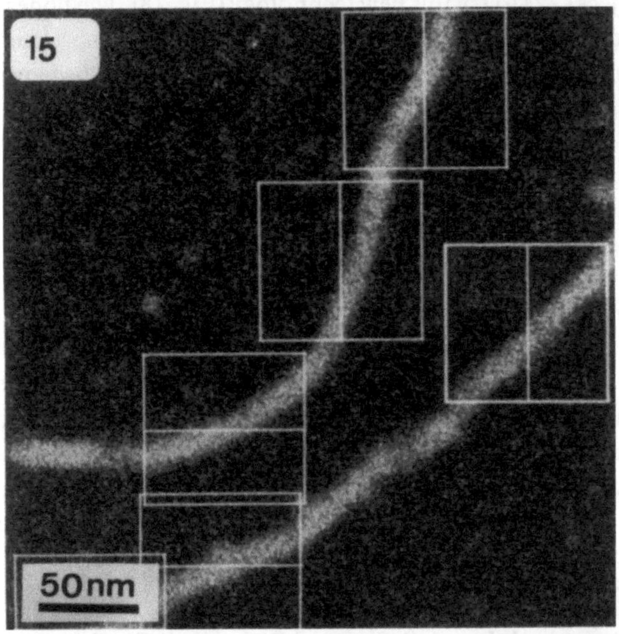
15
50nm

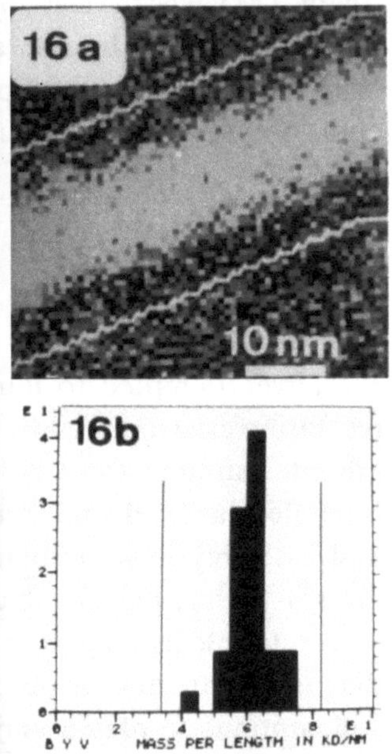
16a
10 nm
16b
MASS PER LENGTH IN KD/NM

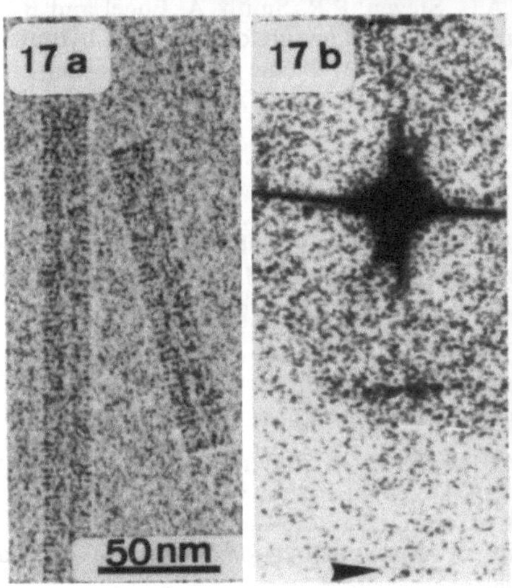
17a
50nm
17b

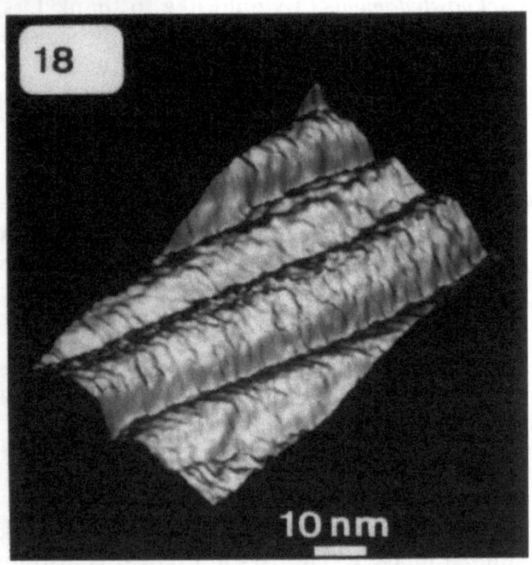
18
10 nm

References

1. Caspar DLD, Klug A (1962) Physical principles in the construction of regular viruses. Cold Spring Harbor Symp Quant Biol 27: 1–24
2. Finch JT, Klug A (1966) Arrangement of protein subunits and the distribution of nucleic acid in turnip yellow mosaic virus. II. Electron microscopic studies. J Mol Biol 15: 344–364
3. Horne RW, Pasquali-Ronchetti I (1974) A negative staining-carbon film technique for studying viruses in the electron microscope. I. Preparation procedure for examining icosahedral and filamentous viruses. J Ultrastruct Res 47: 361–383
4. Bancroft JB, Hiebert E, Rees MW, Markham R (1968) Properties of cowpea chlorotic mottle virus. Its protein and nucleic acid. Virology 34: 224–239
5. Steven AC, Smith PR, Horne RW (1978) Capsid fine structure of cowpea chlorotic mottle virus from a computer analysis of negatively stained arrays. J Ultrastruct Res 64: 63–73
6. Chevallier D, Engel A, Wurtz M, Putz C (1983) The structure and the characterization of a closterovirus, beet yellow virus and a luteovirus, beet mild yellowing virus by scanning transmission electron microscopy: optical diffraction of electron images and acrylamide get electrophoresis. J Gen Virol 64: 2289–2293
7. Chevallier D, Putz (1982) Detection of sugarbeet yellowing viruses in leaf extracts by enzyme-linked immunobent assay (ELISA). Ann Virol 133E: 473–481
8. Engel A, Dubochet J, Kellenberger E (1976) Some progress in the use of a STEM for the observation of biomacromolecules. J Ultrastruct Res 57: 322–330
9. Engel A, Christen F, Michel B (1981) Digital acquisition and processing of electron micrographs using a scanning transmission electron microscope. Ultramicroscopy 7: 45–54
10. Engel A (1979) Molecular weight determination by scanning transmission electron microscopy. Ultramicroscopy 3: 273–281
11. Steven AC, Trus BL, Putz C, Wurtz M (1981) The molecular organization of beet necrotic yellow vein virus. Virology 113: 428–438
12. Putz C (1977) Composition and structure of beet necrotic yellow vein virus. J Gen Virol 35: 397–401
13. Steven AC, Navia MA (1980) Fidelity of structure representation in electron micrographs of negatively stained protein molecules. Proc Natl Acad Sci USA 77: 4721–4725
14. Trus BL, Steven AC (1981) Digital image processing of electron micrographs—the PIC system. Ultramicroscopy 6: 383–386
15. Kistler J, Aebi U, Kellenberger E (1977) Freeze-drying and shadowing of a two-dimensional periodic specimen. J Ultrastruct Res 59: 76–86

Fig. 14. a Purified beet yellow closterovirus (BYV) negatively stained in uranyl formate, imaged in conventional bright-field. **b** Optical diffraction of a straight part of a well-preserved particle. The main reflexion at 3.7^{-1} nm and the layer line at 7.4^{-1} nm are clearly visible

Fig. 15. Unstained STEM dark-field image of BYV. No periodic information is now visible. The *boxes* are the areas selected for mass determination

Fig. 16. a Computer-transmitted track of a selected filament. **b** Histogram of calculated mass (in kDa/nm) with an average value of $6.12 \pm 0.55 \times 10^{-6}$ Da/nm

Fig. 17. a Micrograph of frozen hydrated TMV. **b** Optical diffraction. In this picture the resolution attains 1.15 nm due to the good preservation of the particle. (Courtesy of J. Lepault [20])

Fig. 18. Scanning tunnel microscopy (STM) image of parts of four TMV particles, after carbon-platinum coating. The 2.3 nm striation is visible. (Courtesy of A. Stemmer [22])

16. Finch JT (1972) The hand of the helix of tobacco mosaic virus. J Mol Biol 66: 291–294
17. Brandes J, Wetter C (1959) Classification of elongated plant viruses on the basis of particle morphology. Virology 8: 99–115
18. Bar-Joseph M, Hull R (1974) Purification and partial characterization of sugar beet yellow virus. Virology 62: 552–562
19. Dubochet J, Lepault J, Freeman R, Berriman JA, Homo JC (1982) Electron microscopy of frozen water and aqueous solutions. J Microsc 128: 219–237
20. Lepault J (1985) Cryoelectron microscopy of helical particles of TMV and T4 polyheads. J Microsc 140: 73–80
21. Stemmer A, Engel A, Häring R, Reichelt R, Aebi U (1988) Miniature-size scanning tunneling microscope with integrated 2-axes heterodyne interferometer and light microscope. Ultramicroscopy 25: 171–182
22. Stemmer A, Hefti A, Aebi U, Engel A (1989) Scanning tunneling and transmission electron microscopy on identical areas of biological specimens. Ultramicroscopy 30: 263–280

Chapter 7 ROBERT G. MILNE

Immunoelectron Microscopy for Virus Identification

1 Introduction

I am going to discuss immunonegative staining—the immunoelectron microscopy (IEM) of virus (or other pathogen-related) particles in suspension—with only short excursions into the topic of immune reactions on or in thin sections, as these are considered elsewhere in this book. As the field has been thoroughly reviewed, what I shall say will be more an informal commentary than an exhaustive survey, and I shall not necessarily follow each statement by a precise citation, or attempt to mention all the interesting papers. In places I shall use the term "grid" to mean support film.

As all those know, who have worked with viruses both in negative stain and in thin sections, resolution is much higher in negative stain, and intensity and fidelity of immunolabeling is also greater; you would never work with sections again if they did not furnish positional information that is lost when you make an extract. This loss means that in negative stain you can only work with structures that can be recognized out of context—virus particles, subviral components, or certain virus-induced inclusions; perhaps also mycoplasmas or their fragments. However, sometimes the same structures can be recognized both in vitro and in situ; and with the advantage of immune labeling we can, with luck, identify an antigen in both contexts. In that case we can obtain both positional or contextual information as well as high-resolution details of structure or antigen location. An interesting technique that I will not discuss, but which may prove a valuable compromise, is immunonegative staining of thin cryosections.

1.1 The early days

As our symposium was held to celebrate 50 years of electron microscopy in plant pathology, and also in order to place today's methods in context, I would like to

Istituto di Fitovirologia Applicata CNR, Strada delle Cacce 73, I-10135 Torino, Italy

remind you of some historical IEM landmarks. These were often the occasion for interesting reviews such as those of Williams (1954) [77], Valentine (1961) [71], Kleczkowski (1961) [31], Horne and Wildy (1963) [27], Lafferty and Oertelis (1963) [35], Horne (1967) [26], Almeida and Waterson (1969) [1], Ball (1971) [3], Doane and Anderson (1977) [17], and Milne and Luisoni (1977) [53]. There are also more recent commentaries [5, 29, 32, 34, 36, 47, 48, 51, 61, 62, 74].

It all began when two independent teams [2, 64] examined mixtures of tobacco mosaic virus and its antiserum in the electron microscope. These preparations were without benefit of added contrast, and the microscopes were not very advanced, but the already known rod-shaped particles were seen to be specifically clumped and to be approximately doubled in thickness.

Eighteen years passed before reports appeared of imaging individual antibodies attached to virus particles. The preparations in question were contrasted by metal shadowing, and clearly this technique did not offer sufficiently rewarding results. Valentine in 1961 [71] discusses negative staining of viruses and mentions antibodies, but not until 1962 were negatively stained virus-antibody complexes reported [e.g., 28]. This kind of work culminated in the review of Almeida and Waterson [1].

2 Leaf-Dip Serology

Leaf-dip serology [3, 4] became popular with plant virologists because it was simple and reliably demonstrated serological reactions in the electron microscope by negative staining. However, the method was inherently flawed because it consisted in mixing together a sap extract from an infected leaf, the antiserum diluted in buffer, and the negative stain. Conditions were made worse because of two factors. First, the sap-antiserum mixture was dried on the grid before being negatively stained, and as we know now, negative stain helps to support structures as they dry, minimizing distortion and, especially, flattening. This is not to mention the sticky mess resulting from drying even diluted plant sap and serum down on the support film.

Secondly, the serological reaction was reported to work much better in phosphate buffer (titer four two-fold steps higher) than in ammonium acetate buffer, but the latter was used since it sublimes on drying, whereas phosphate leaves crystalline deposits. A third factor working against the system was the use of PTA (neutralized phosphotungstic acid), although this was mixed with vanadomolybdate. As was then already known in part [20], PTA disrupts the particles of some viruses, especially those held together by protein-nucleic acid interactions (examples: alfalfa mosaic, cucumber mosaic, geminiviruses) or those containing lipid (rhabdoviruses, spotted wilt). PTA may also give poor stain distribution and lower resolution than some other stains such as uranyl acetate [47, 61, 62].

However, Ball and Brakke were able to demonstrate relatively clear results; moreover they used the method to titrate antisera—the first quantitative use of electron microscopy in this way.

3 Immune Complexes from Ouchterlony Plates

Meanwhile, in another part of the forest, a small but interesting development took place. Agar gel diffusion has long played a major part in virus diagnostics and titration of antisera, and the technique of cutting out the precipitin band, crushing the agar in a little buffer, and negative staining the immune complexes released has been reinvented several times. The first to describe the method were Watson et al. in 1966 [76]. Examples from our laboratory were the finding that whole closterovirus particles (grapevine virus A, 800 nm modal length) could migrate through agar to form a clear band [12], and the separation of the spherical particles of white clover cryptic viruses 1 and 2 to form two distinct bands, using a mixed virus preparation and a mixed antiserum (Figs. 1–3); [E. Luisoni and R.G. Milne unpubl.; 8, 9].

3.1 Clumping

As we have seen, the classical approach to immunoelectron microscopy in vitro was to mix antigen and antibody, and detect the resulting clumps, which resemble those in Figs. 2 and 3 [see e.g., 50]. This method is little used now by plant virologists, as the clumping effect is sometimes nonspecific, especially if the preparation has been pelleted and resuspended. Moreover, it requires relatively large concentrations of virus to be effective.

4 ISEM

In 1973 K.S. Derrick published his "serologically specific electron microscopy" method [14; see also 6, 16, 24, 59], which was later simplified, improved, and renamed more appropriately as immunosorbent electron microscopy (ISEM) [38, 39, 46, 53, 57, 63].

In this method, the grid is coated with a dilution of the antiserum, and rinsed; there follows a trapping phase in which the virus preparation is incubated with the grid, and virus particles are bound (Figs. 4 and 5). Even perhaps more important, [39], the layer of antibody and other serum constituents inhibits non-specific binding of structures, other than the viral antigen, that may be in the preparation. The result is selective immuno-purification of the virus on the grid.

It was also made clear [39] that when crude virus preparations are absorbed to grids without the benefit of ISEM, host constituents generally compete effectively with virus particles for an anchored site on the support film, so that many virus particles in the preparation are not retained on the grid. For similar reasons, coating grids with undiluted antiserum or indeed serum diluted less than about 1/1000 can inhibit efficient trapping, probably because with less diluted serum, proteins other

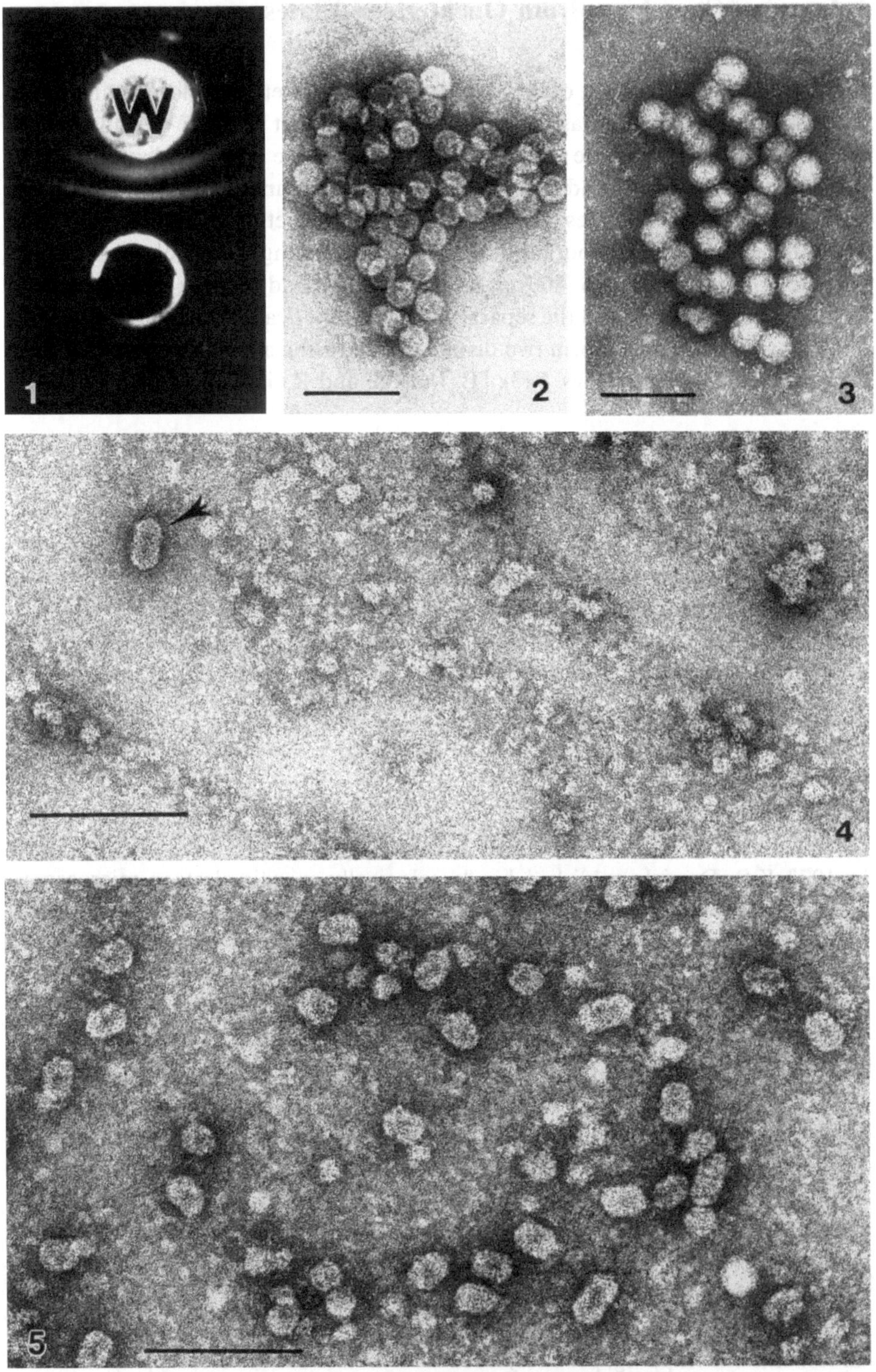
1
2
3
4
5

than the trapping antibodies preferentially occupy available sites on the support film [50].

In ELISA, antiserum coating times are of the order of 60 min or more, but with the strictly equivalent grid coating of the ISEM procedure, coating times of 5 min at room temperature have been shown to bind nearly all the effective antibody that is going to become attached [see 49]. This curiosity has not adequately been explained. However, with virus adsorption, the kinetics are different, and significantly more virus may continue to be trapped for several hours. Trouble may come if the virus particles detach or become degraded during long incubations (say, over 4 h). We routinely trap virus for only 15 min, while realizing that more could be had with longer incubation, if necessary.

Standard conditions will often give a satisfactory response with ISEM, but to optimize the system, a number of buffers, ionic strengths, and pH values should be tested, for both the antibody coating and virus trapping steps [11]. Additives such as EDTA, reducing agents, polyethylene glycol, or polyvinylpyrrolidone should also be considered.

The sensitivity of ISEM is generally comparable to that of ELISA, and it may detect 0.1–10 ng/ml of virus in volumes of a few μl [see 61]. Where the particle is especially stable or conspicuous, or where ELISA backgrounds are troublesome, ISEM can be appreciably more sensitive than ELISA [see e.g., 45]; in other cases ISEM has been considerably less sensitive, probably because free viral coat protein in the preparation competes in binding to the antibody-coated grid [46, 74].

An undeniable problem with ISEM for mass screening is its labor-intensive nature compared with ELISA. An advantage that hardly needs promoting to this readership is: seeing is believing. Just a very few virus particles observed on the grid give a clear positive result; this avoids the ELISA dilemma of trying to decide whether a given low level of absorbance above background is positive or not, and also avoids the requirements for a number of controls necessary to the ELISA test.

◀

Figs. 1–3. A gel double-diffusion plate (**1**, × 4) showing two bands resulting from interaction of a mixture (*upper well W*) of white clover cryptic viruses 1 and 2 (WCCV1, WCCV2) and an antiserum to both viruses (*lower well*). The bands from this plate were each cut out and crushed in buffer; material from the extracts was then adsorbed to grids and negatively stained in uranyl acetate. The lower band contained exclusively clumps of the more rapidly diffusing WCCV1 (ca, 30 nm in diameter **2**) and the upper band contained only clumps of WCCV2 (ca, 38 nm in diameter, **3**). Note that the relative positions of the bands also depends on the antigen/antibody ratios used in the test. Bars = 100 nm. **1** courtesy of Dr. E. Luisoni

Figs. 4 and **5.** Crude sap of a plant infected with Ourmia melon virus (OuMV) was adsorbed for 15 min to an untreated grid (**4**) or a grid precoated with OuMV antiserum diluted 1/1600, for 5 min (**5**). With limiting amounts of virus, the difference in particle counts was of the order of 10 000 times [41]. The one particle found on a 400-mesh grid square (**41**) is *arrowed*. Bar = 100 nm

4.1 Protein A-ISEM

It had been known for some time that protein A had the specific capacity to bind to the Fc portion or "tail" of certain IgG's, but Shukla and Gough in 1979 [23, 65] were the first to use this property in the context of ISEM, and show that trapping efficiencies could be thereby improved. When a grid coated with protein A is incubated with antiserum, the IgG's are trapped with their active Fab portions exposed. Thus we have both selection and orientation. Experiments in several laboratories (see reviews cited above) have shown that in practice the increased sensitivity obtainable with protein A (increase in numbers of particles trapped from a given preparation by a given antiserum) is really significant only when the concentration of virus is not limiting.

However, using protein A we can largely overcome the inhibitory effect of using, for coating, antisera that have not been highly diluted. The ability to use less diluted sera means that those of low titer can be used more effectively. (A similar result is obtained if the IgG fraction is isolated and used for coating). A second benefit is that a wider spectrum of heterologous antigens can be detected by ISEM using protein A [38, 75].

4.2 ISEM with dsRNA

Derrick [15] extended the ISEM technique to the detection of double-stranded RNAs. The method has been used to detect circular dsRNAs in viroid-infected plants [21], and to trap a dsRNA-containing mycovirus from *Agaricus bisporus* [13]. In our laboratory, some preliminary attempts to trap dsRNA's were not encouraging, but further work should be done, as there is reason to believe the method can work.

Fig. 6. Mixture of purified tobacco necrosis virus (TNV) and tomato bushy stunt virus (TBSV) absorbed to the grid, then treated with a saturating level of antibodies to TBSV. The TNV particles have remained clean, with sharp outlines, although antibody molecules are seen in the background. Bar = 100 nm

Fig. 7. A natural mixture of potyviruses from the wild European perennial cucurbit *Bryonia cretica*. The decorated particles have reacted with an antiserum to "white bryony mosaic virus"—supposed at that time to have been a carlavirus [see 55]. Note that one particle (near the center) is longer than normal and is free of antibody in part; such particles were not infrequent and probably arose by end-to-end aggregation of two virus particles from a mixed infection (or fragments of them) bearing different coat proteins. The phenomenon, detected only by EM decoration tests, also occurs with closteroviruses and may be responsible for anomalous transmission of one virus by vectors normally specific for another [see 56]. Bar = 100 nm

Fig. 8. a–f A panel of maize rough dwarf virus (Reoviridae) B-spiked subviral particles. All but the particle in **e**, which was untreated, were decorated with antiserum specific for the B spikes. Bar = 100 nm

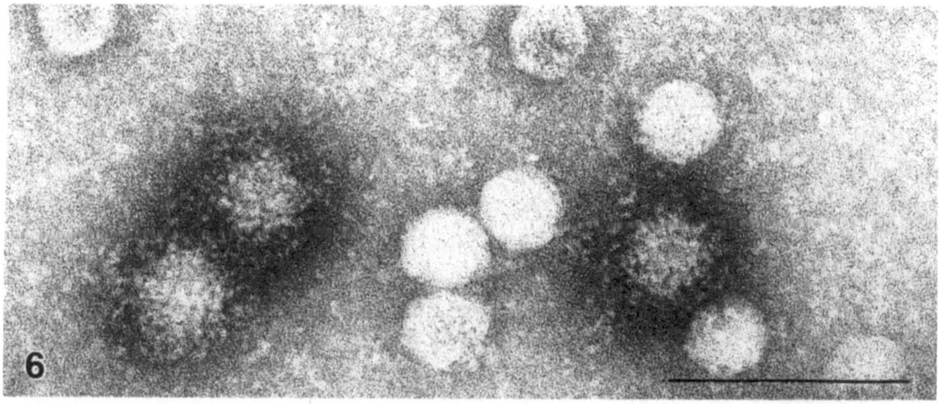
6

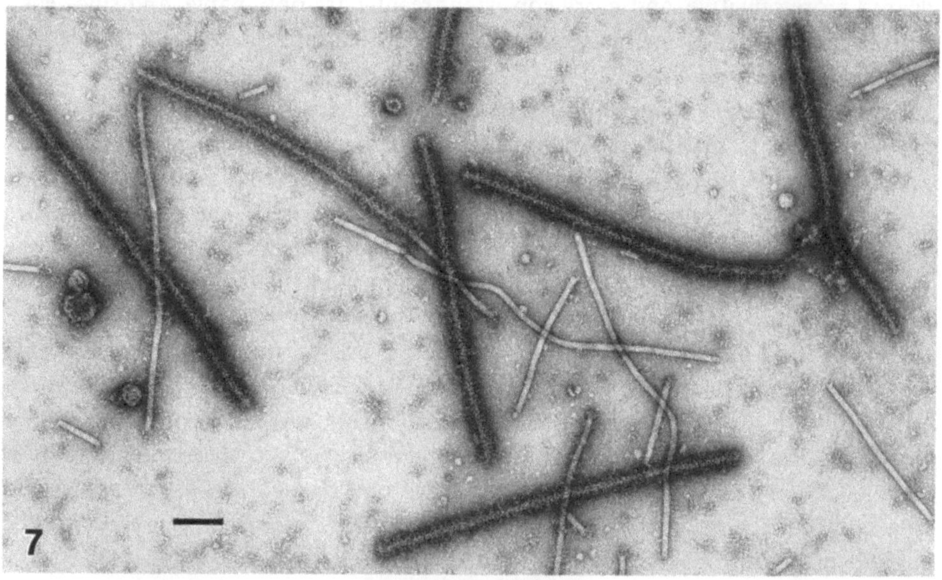
7

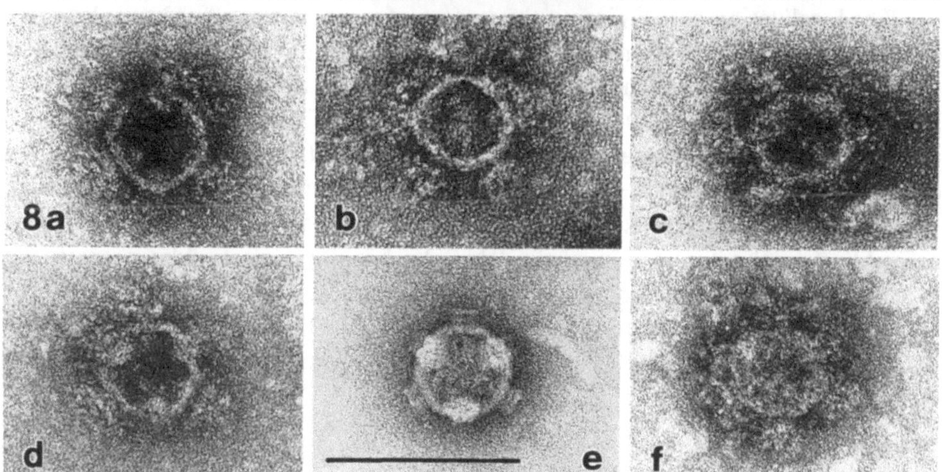
8a
b
c
d
e
f

5 Decoration

The clear advantages but also the evident defects of leaf-dip serology led Milne and Luisoni [52, 53] to develop the decoration method, only to discover later that they had been elegantly preempted by two groups working on bacteriophage structural proteins [70, 78]. In this method, the virus particles (or other structures) are first absorbed to the grid under whatever conditions are optimal for that process, and the grid is rinsed (a) to remove unwanted salts, proteins, sugars, impurities or buffers, and (b) to set up conditions which are optimal for the next phase: attachment of antibody to the virus. After antibody attachment, the grid is again rinsed, and then stained, usually with uranyl acetate. By separating and optimizing each step, much clearer, more sensitive, and more consistent results are obtainable, and since each virus particle is, ideally, well separated from the others, interpretation of the image is simplified (Figs. 6–8).

Uranyl acetate has worked satisfactorily in our hands, although the pH (about 4.2) seems close to that which would cause antigen-antibody separation. However, fixing the reacted particles with glutaraldehyde before applying uranyl acetate does not make any difference. It is interesting that a final rinse of the particles, before staining, with very slightly acidified distilled water, will immediately remove the decorating antibody—which however remains in place if the decorated particles have been fixed.

Decoration has become a popular method of identifying plant viruses, mainly because the result is (or should be) unequivocal and direct (Figs. 6, 7). Of all serological methods it is one of the most easily and safely interpreted because you see the virus particle and you see the antibody attached to it—and where the antibody is attached. False positives and false negatives are rare, though with some viruses and especially with poor or preliminary antisera, clear interpretation may not be possible in the murky preparations that result.

The localization of proteins on the surface of virus particles has been elegantly demonstrated by decoration in a number of cases. Yanagida and Ahmad-Zadeh [78] determined the position of certain gene products on the capsid of phage T4, and similar beautiful work was done by Tosi and Anderson [70] with phage 29. Luisoni et al. [44], as part of the serological analysis of the capsids of Fijiviruses (Reoviridae), showed that the B spikes and the inner capsids could be decorated with specific antisera (Fig. 8). Fukuda et al. [22] demonstrated the initiation of TMV rod assembly near the 3′ terminus of the RNA, by beginning encapsidation with the protein of one strain and completing it with the serologically different protein of a second strain. A similar approach earlier allowed Otsuki and Takebe [58] to demonstrate mixedly coated virus particles in protoplasts doubly infected by strains of TMV.

5.1 ISEM Plus Decoration

A natural step, once ISEM and decoration were established techniques, was to combine them [53], since ISEM cleans and concentrates the virus, presenting an optimized field for the decoration step (Figs. 10–12). Even where the numbers of virus particles are not limiting, ISEM is a very useful preliminary to decoration [see for example 54].

A further advantage of combining ISEM with decoration is that different antibodies can if desired be used for each phase. For example, a mixed antiserum to white clover cryptic viruses 1 and 2 was employed to trap both viruses from the crude sap of carrier plants, and antiserum to virus 2 was then used to decorate that virus but not the other, and show them to be distinct [8]. Similarly, a mixture of three cryptic viruses from hop trefoil (*Medicago lupulina*) was differentiated, although not all the viruses could be separated, and not all the relevant monovalent antisera were available [43]. Fukuda et al., referred to above [22], also trapped their mixedly coated TMV particles by ISEM before decorating them with the differentiating antisera.

Despite the above, the decoration technique is often (even usually) done badly, and then results are both less beautiful and less informative. The main problems (apart from human factors) may be among the following: poor antisera, insufficiently thorough washing steps, poor support films (e.g., use of uncarboned plastic films), use of PTA, drying of the grid before negative staining, or even persistence in the use of the classical but outmoded leaf-dip serology. Roberts [61, 62] has good advice on how to avoid these and other pitfalls.

One problem common to all serological tests, but sometimes easy to overcome using glutaraldehyde fixation, is that of antigenic change in the virus particle. It has become increasingly clear that viruses, especially perhaps filamentous ones, may be subject to partial hydrolysis in vitro or even in vivo, with consequences for the preparation of antisera and for their subsequent use. Shukla et al. [66, 67] have described the all too easy removal of the antigenically specific N-terminal end of potyvirus coat proteins, whereas Koeing et al. [33] described similar hydrolysis of the coat protein of a potexvirus. In vivo coat protein hydrolysis can be responsible for failures and anomalous results in the detection of bean yellow mosaic potyvirus in gladiolus [69], and antigen breakdown has been described in a mite-transmitted potyvirus [37]. Such problems are more likely with long incubation times involving unrinsed sap preparations, and may be responsible for the patchy or inconsistent appearance of decoration sometimes reported.

5.2 Quantitative Decoration

As we have seen, Ball and Brakke [3, 4] used leaf-dip serology to titrate antisera. The decoration method, being an improvement on leaf-dip serology, facilitates such

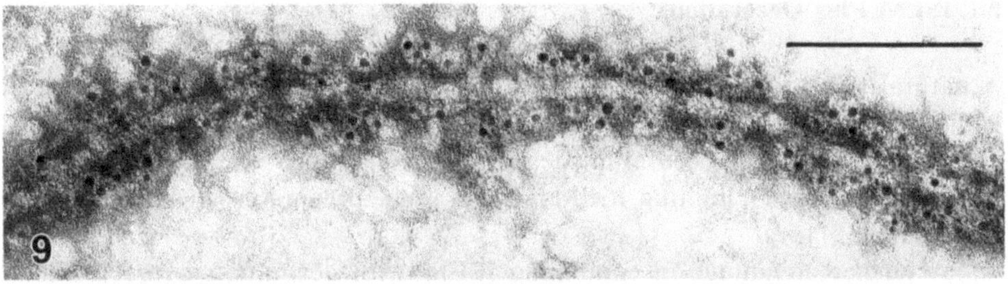
9

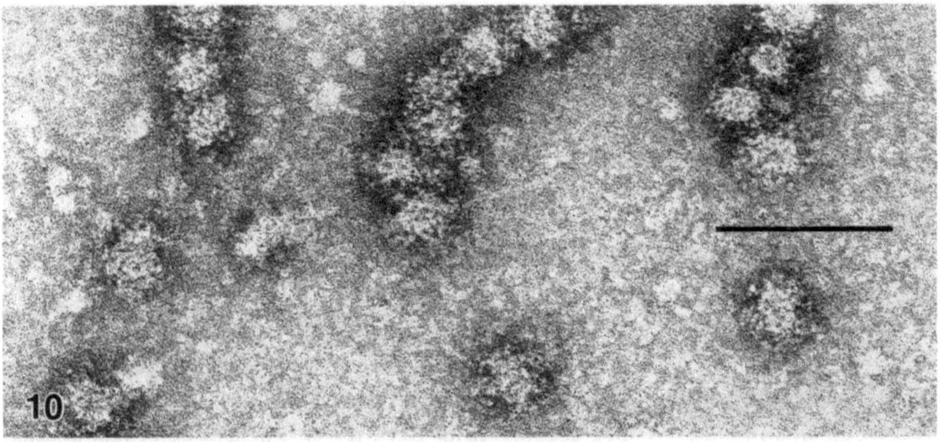
10

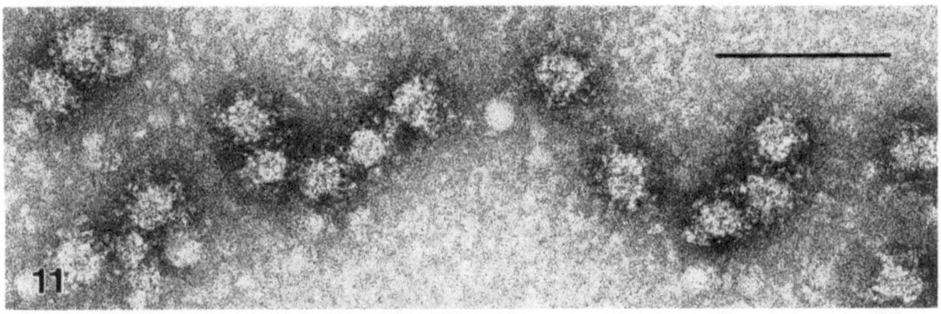
11

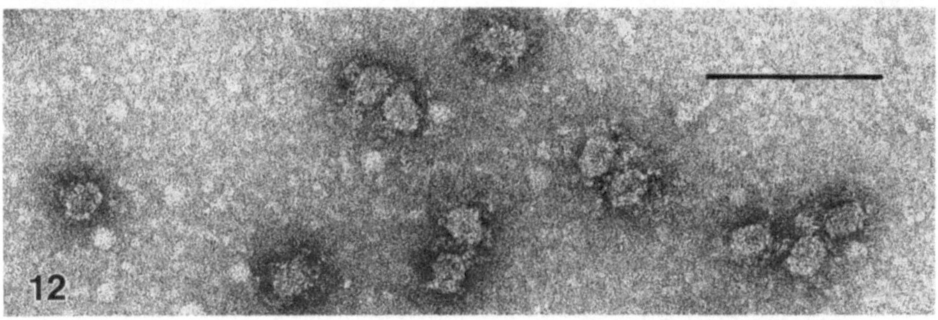
12

titration (Figs. 10–12), which can be done, from preparation to observation, easily within one hour [49]. Yet, unfortunately, the number of workers using decoration quantitatively are few whereas those using it as a yes/no measure of relationship are many.

5.3 Double Decoration

If the first decorating antibody (say, rabbit anti-plum pox virus) is followed by a second (this could be sheep anti-rabbit IgG), and the complex is negatively stained, then the particles, originally thin, pale and difficult to see at low magnification, become thickened and blackened so that they are easily noted, even at only 5000 × enlargement [30]. The stain penetrates among the attached antibodies and gives the virus particle a highly contrasted coat. The method is useful for rapid diagnosis, especially by unskilled electron microscope operators or those working with an old or low-performance instrument.

5.4 Gold Labeling

As noted by Cristoforo Colombo, "Gold is the most exquisite of things. Whoever possesses gold can acquire all that he desires in the world. Truly, for gold he can gain entrance for his soul into paradise". While electron microscopists might express themselves differently, we can see what he means. Some examples of decoration enhanced by the use of gold-labeled antibodies can be cited [5, 7, 18, 19, 25, 40, 42, 60, 72, 73].

Generally, the approach has been to decorate the antigen with the primary antibody (say, rabbit anti-virus) and follow this with either a second, gold-labeled antibody (for example, 5 nm gold-goat anti-rabbit) (Fig. 9), or with gold-labeled protein A. Louro and Lesemann [42] noted that gold labeling could carry the decoration titer four twofold dilution steps higher than was possible without the gold, although in our laboratory we would claim an increase of only two twofold steps.

Apart from the increase in sensitivity, however, the gold label has the great advantage of being easily and exactly identifiable, whereas what constitutes a trace

◄

Fig. 9. A plum pox potyvirus particle adsorbed from infected apricot leaf sap (without ISEM) after extraction in buffer containing 2% polyvinylpyrrolidone (M_r 40 000) as protectant. The particles were decorated with a 1/2000 dilution of antiserum, followed by gold labeling using a 1/100 dilution of goat anti-rabbit lgG-5nm gold (Janssen). Bar = 100 nm

Figs. 10–12. Preparation of OuMV particles trapped by ISEM as in Fig. 5, then decorated with different dilutions of the same antiserum: 1/32 (**10**), 1/256 (**11**) and 1/512 (**12**). The decoration titer of this antiserum was 1/1024. Bar = 100 nm

of unlabeled decorating antibody could be a matter of opinion, and certainly becomes so if the preparation and electron microscopy are not of a high standard. Where particular, localized epitopes are being identified, as, for example, those exposed at one end of TMV rods, only a very few antibody units (perhaps one) can attach to the site, and this is an ideal situation in which to exploit gold labeling [18]. GLAD (gold-labeled antibody decoration) has been effective in screening monoclonal antibody-secreting clones for antibodies to plum pox virus [25].

6 Mycoplasma-Like Organisms

Antibodies against MLO's, both monoclonal and polyclonal, are now becoming available, and methods are being developed to exploit such antibodies in taxonomy and diagnostics. Corn stunt spiroplasma has been detected by ISEM [16], and since then a number of MLO's have been revealed by this method [see 68]. We have used a

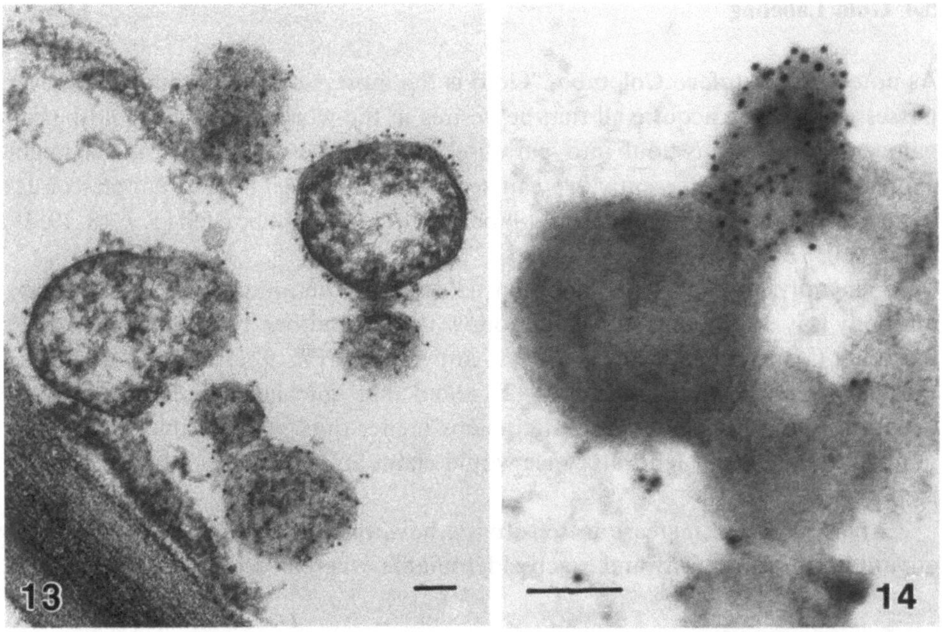

Figs. 13 and **14.** Gold labeling (5 nm gold-goat antimouse) of primula yellows MLO in thin sections and in vitro with a monoclonal antibody from M.F. Clark. **13** shows pre-embedding labeling of glutaraldehyde-fixed phloem tissue, followed by osmium fixation and classical Epon embedding. **14** shows a partially purified MLO preparation osmotically lyzed, trapped by ISEM, and decorated with gold. A highly labeled fragment (presumably MLO tissue) is seen together with unlabeled (presumed host) materials. 'Healthy' preparations gave no labeling above the rather low background. Bars = 100 nm

monoclonal from M. F. Clark [10] to label primula yellows MLO in thin sections (Fig. 13) using a pre-embedding labeling technique (R. G. Milne and R. Lenzi unpubl.) and have also labeled MLO fragments trapped by ISEM from partially purified primula yellows preparations (R. G. Milne and R. Lenzi unpubl.; Fig. 14).

Gold labeling of MLO's using embedding and sectioning methods is discussed elsewhere in this book. Our attempt to trap and label MLO fragments directly on grids was successful (Fig. 14), but the method requires further development, and we had less success in trapping intact MLO bodies. We need to be able to trap morphologically recognizable MLO's, not only fragments, and of course gold-label them convincingly. The technique should then become very useful, as it will be simple and not take more than an hour or two to perform.

7 Conclusion

This review has left many topics unexamined, but has attempted to touch on the major ones, and to discuss some areas where improvement is possible or progress is being made. My overall impression, however, is that immunoelectron microscopy of in vitro preparations (as opposed to thin sectioning) needs a new stimulus. It comprises a very useful, and indeed often used, collection of techniques, but they have become routine. I hope that this conference and similar ones will, as often happens, make contacts and produce stimuli that lead to something new.

Acknowledgment. I thank Vera Masenga and Riccardo Lenzi for technical assistance with some of the work described, and Dr. Michael Clark for MLO-infected plants and anti-MLO antibodies.

References

1. Almeida JD, Waterson AP (1969) The morphology of virus-antibody interaction. Adv Virus Res 15: 307–338
2. Anderson FA, Stanley WM (1941) A study by means of the electron microscope of the reaction between tobacco mosaic virus and its antiserum. J. Biol Chem 139: 339–344
3. Ball EM (1971) Leaf-dip serology. Methods Virol 5: 445–450
4. Ball EM, Brakke MK (1968) Leaf-dip serology for electron microscopic identification of plant viruses. Virology 36: 152–155
5. Beesley JE, Betts MP (1987) Colloidal gold probes for the identification of virus particles: an appraisal. Micron Microsc Acta 18: 299–305
6. Beier H, Shepherd RJ (1978) Serologically specific electron microscopy in the quantitative measurement of two isometric viruses. Phytopathology 68: 533–538
7. Bingren Wu, Mahoney J, Chernesky M (1989) Comparison of three protein A-gold immune electron microscopy methods for detecting rotaviruses. J Virol Methods 25: 109–118
8. Boccardo G, Milne RG, Luisoni E, Lisa V, Accotto GP (1985) Three seedborne cryptic viruses containing double-stranded RNA isolated from white clover. Virology 147: 29–40

9. Boccardo G, Lisa V, Luisoni E, Milne RG (1987) Cryptic plant viruses. Adv Virus Res 32: 171–214
10. Clark MF, Morton A, Buss SL (1989) Preparation of mycoplasma immunogens from plants and a comparison of polyclonal and monoclonal antibodies made against primula yellows MLO-associated antigens. Ann Appl Biol 114: 111–124
11. Cohen J, Loebenstein G, Milne RG (1982) Effect of pH and other conditions on immunosorbent electron microscopy of several plant viruses. J Virol Method 4: 323–330
12. Conti M, Milne RG, Luisoni E, Boccardo G (1980) A closterovirus from a stem-pitting diseased grapevine. Phytopathology 70: 394–399
13. Del Vecchio VG, Dixon C, Lemke PA (1979) Immune electron microscopy of virus-like particles of *Agaricus bisporus.* Exp Mycol 2: 138–144
14. Derrick KS (1973) Quantitative assay for plant viruses using serologically specific electron microscopy. Virology 56: 652–653
15. Derrick KS (1978) Double-stranded RNA is present in extracts of tobacco plants infected with tobacco mosaic virus. Science 199: 538–539
16. Derrick KS, Brlansky RH (1976) Assay for viruses and mycoplasmas using serologically specific electron microscopy. Phytopathology 66: 815–820
17. Doane FW, Anderson N (1977) Electron and immunoelectron microscopic procedures for diagnosis of viral infections. In: Kurstak E, Kurstak C (eds) Comparative diagnosis of viral diseases Vol 11. Academic Press, New York, pp 505
18. Dore I, Weiss E, Altschuh D, van Regenmortal MHV (1988) Visualization by elecrtron microscopy of the location of tobacco mosaic virus epitopes reacting with monoclonal antibodies in enzyme immunoassay. Virology 162: 279–289
19. El Ghorr AA, Snodgrass DR, Scott FMM (1988) Evaluation of an immunogold electron microscopy technique for detecting bovine coronavirus. J Virol Methods 19: 215–224
20. Francki RIB, Randles JW, Chambers TC, Wilson SB (1966) Some properties of purified cucumber mosaic virus (Q strain). Virology 28: 729–741
21. French RC, Price MA, Derrick KS (1982) Circular double-stranded RNA in potato spindle tuber viroid-infected tomatoes. Nature Lond 295: 259–260
22. Fukuda M, Okada Y, Otsuki Y, Takebe I (1980) The site of initiation of rod assembly on the RNA of a tobacco and a cowpea strain of tobacco mosaic virus. Virology 101: 493–502
23. Gough KH, Shukla DD (1980) Further studies on the use of protein A in immune electron microscopy for detecting virus particles. J Gen Virol 51: 415–419
24. Hamilton RI, Nichols C (1978) Serological methods for detection of pea seed-borne mosaic virus in leaves and seeds of *Pisum sativum.* Phytopathology 68: 539–543
25. Himmler G, Brix U, Steinkelner H, Laimer M, Mattanovich D, Katinger HWD (1988) Early screening for anti-plum pox virus monoclonal antibodies with different epitope specificities by means of gold-labelled immunosorbent electron microscopy. J Virol Methods 22: 351–358
26. Horne RW (1967) Electron microscopy of isolated virus particles and their components. Methods Virol 3: 521–574
27. Horne RW, Wildy P (1963) Virus structure revealed by negative staining. Adv Virus Res 10: 101–170
28. Hummeler K, Anderson TF, Brown R (1962) Identification of poliovirus particles of different antigenicity by specific agglutination as seen in the electron microscope. Virology 16: 84–90
29. Katz D, Kohn A (1984) Immunosorbent electron microscopy for detection of viruses. Adv Virus Res 29: 169–194
30. Kerlan C, Mille B, Dunez J (1981) Immunosorbent electron microscopy for detecting apple chlorotic leafspot and plum pox viruses. Phytopathology 71: 400–404
31. Kleczkowski A (1961) Serological behaviour of tobacco mosaic virus and of its protein fragments. Immunology 4: 130–141
32. Koenig R (1988) Serology and Immunochemistry. In: Milne RG (ed) The filamentous plant viruses. Plenum, New York, pp 111

33. Koenig R, Tremaine JH, Shepard F (1978) In situ degradation of the protein chain of potato virus X at the N- and C-termini. J Gen Virol 38: 329–337
34. Kojima M, Tyng-guang Chou, Shikata E (1978) Rapid diagnosis of potato leaf roll virus by immune electron microscopy. Ann Phytopathol Soc Jpn 44: 585–590
35. Lafferty KJ, Oertelis S (1963) The interaction between virus and antibody III. Examination of virus-antibody complexes with the electron microscope. Virology 21: 91–99
36. Lange L (1986) The practical application of new developments in test procedures for the detection of viruses in seed. In: Jones RAC, Torrance L (eds) Developments and Applications in Virus Testing. Association of Applied Biologists, Wellesbourne, Warwick, pp 269
37. Langenberg WG (1989) Rapid antigenic modification of wheat streak mosaic virus in vitro is prevented in glutaraldehyde-fixed tissue. J Gen Virol 70: 969–973
38. Lessmann DE, Paul HL (1980) Conditions for the use of protein A in combination with the Derrick method of immuno electron microscopy. Acta Hortic 110: 119–128
39. Lesemann DE, Bozarth RF, Koenig R (1980) The trapping of tymovirus particles on electron microscope grids by adsorption and serological binding. J Gen Virol 48: 257–264
40. Lin NS (1984) Gold-IgG complexes improve the detection and identification of viruses in leaf dip preparations. Methods Virol 8: 181–190
41. Lisa V, Milne RG, Accotto GP, Boccardo G, Caciagli P (1988) Ourmia melon virus, a virus from Iran with novel properties. Ann Appl Biol 112: 291–302
42. Louro D, Lesemann DE (1984) Use of protein A-gold complex for specific labelling of antibodies bound to plant viruses I. Viral antigens in suspensions. J Virol Methods 9: 107–122
43. Luisoni E, Milne RG, Accotto GP, Boccardo G (1987) Cryptic viruses in hop trefoil (*Medicago lupulina*) and their relationship to other cryptic viruses in legumes. Intervirology 28: 144–156
44. Luisoni E, Milne RG, Boccardo G (1975) The maize rough dwarf virion II. Serological analysis. Virology 68: 86–96
45. Luisoni E, Milne RG, Roggero P (1982) Diagnosis of rice ragged stunt virus by enzyme-linked immunosorbent assay and immunosorbent electron microscopy. Plant Dis 66: 929–932
46. Milne RG (1980) Some observations and experiments on immunosorbent electron microscopy of plant viruses. Acta Hortic 110: 129–135
47. Milne RG (1984) Electron microscopy for the identification of plant viruses in in vitro preparations. Methods Virol 7: 87–120
48. Milne RG (1986) New developments in electron microscope serology and their possible applications. In: Jones RAC, Torrance L (eds) Developments and applications in virus testing. Association of Applied Biologists, Wellesbourne, Warwick, pp 179
49. Milne RG (1988) Quantitative use of the electron microscope decoration technique for plant virus diagnostics. Acta Hortic 234: 321–329
50. Milne RG, Lesemann DE (1978) An immunoelectron microscopic investigation of oat sterile dwarf and related viruses. Virology 90: 299–304
51. Milne RG, Lesemann DE (1984) Immunosorbent electron microscopy in plant virus studies. Methods Virol 8: 85–101
52. Milne RG, Luisoni E (1975) Rapid high-resolution immune electron microscopy of plant viruses. Virology 68: 270–274
53. Milne RG, Luisoni E (1977) Rapid immune electron microscopy of virus preparations. Methods Virol 6: 265–281
54. Milne RG, Masenga V, Conti M (1986) Serological relationships between planthopper-vectored cereal rhabdoviruses. Intervirology 25: 83–87
55. Milne RG, Masenga V, Lovisolo O (1980) Viruses associated with white bryony (*Bryonia cretica* L.) mosaic in northern Italy. Phytopathol Mediterr 19: 115–120
56. Murant AF, Raccah B, Pirone TP (1988) Transmission by vectors. In: Milne RG (ed) The filamentous plant viruses. Plenum, New York, pp 237

57. Nicolaïeff A, van Regenmortel MHV (1980) Specificity of trapping of plant viruses on antibody-coated electron microscope grids. Ann Virol (Inst Pasteur) 131E: 95–110
58. Otsuki Y, Takebe I (1978) Production of mixedly coated particles in tobacco mesophyll protoplasts doubly infected by strains of tobacco mosaic virus. Virology 84: 162–171
59. Paliwal YC (1977) Rapid diagnosis of barley yellow dwarf virus in plants using serologically specific electron microscopy. Phytopathology Z 89: 25–36
60. Pares RD, Whitecross MI (1982) Gold-labelled antibody decoration (GLAD) in the diagnosis of plant viruses by immunoelectron microscopy. J Immunol Methods 51: 23–28
61. Roberts IM (1986a) Immunoelectron microscopy of extracts of virus-infected plants. In: Harris JR, Horne RW (eds) Electron microscopy of proteins, Vol 5, Viral structure. Academic Press, New York, pp 292
62. Roberts IM (1986b) Practical aspects of handling, preparing and staining samples containing plant virus particles for electron microscopy. In: Jones RAC, Torrance L (eds) Developments and applications in virus testing, Association of Applied Biologists, Wellesbourne, Warwick, pp 213
63. Roberts IM, Harrison BD (1979) Detection of potato leafroll and potato mop-top viruses by immunosorbent electron microscopy. Ann Appl Biol 93: 289–297
64. Schramm G, Friedrich-Freksa N (1941) Die Präcipitinreaktion des Tabakmosaikvirus mit Kaninchen und Schweineantiserum. Z Physiol Chem 270: 233–246
65. Shukla DD, Gough KH (1979) The use of protein *A* from *Staphylococcus aureus* in immune electron microscopy for detecting plant virus particles. J Gen Virol 45: 533–536
66. Shukla DD, Ward CW (1989) Structure of potyvirus coat proteins and its application in the taxonomy of the potyvirus group. Adv Virus Res 36: 273–314
67. Shukla DD, Jilka J, Tosic M, Ford RE (1989) A novel approach to the serology of potyviruses involving affinity-purified polyclonal antibodies directed towards virus-specific N-termini of coat proteins. J Gen Virol 70: 13–23
68. Sinha RC (1988) Serological detection of mycoplasmalike organisms from plants affected with yellows diseases. In: Hiruki C (ed) Tree mycoplasmas and mycoplasma diseases. University of Alberta Press, Edmonton, pp 143
69. Stein A, Salomon R, Cohen J, Loebenstein G (1986) Detection and characterization of bean yellow mosaic virus in corms of *Gladiolus grandiflorus*. Ann Appl Biol 109: 147–154
70. Tosi M, Anderson DL (1973) Antigenic properties of bacteriophage ϕ29 structural proteins. J Virol 12: 1548–1559
71. Valentine RC (1961) Contrast enhancement in the electron microscopy of viruses. Adv Virus Res 8: 287–318
72. Van Lent JWM, Verduin BJM (1985) Specific labelling of antibodies bound to plant viruses in mixed suspensions. Neth J Plant Pathol 91: 205–213
73. Van Lent JWM, Verduin BJM (1986) Detection of viral protein and particles in thin sections of infected plant tissue using immunogold labelling. In: Jones RAC, Torrance L (eds) Developments and applications in virus testing. Association of Applied Biologists, Wellesbourne, Warwick, pp 193
74. van Regenmortel MHV (1982) Serology and Immunochemistry of Plant Viruses, Academic Press, New York
75. van Regenmortel MHV, Nicolaïeff A, Burckard J (1980) Detection of a wide spectrum of virus strains by indirect ELISA and serological trapping electron microscopy (STREM). Acta Hortic 110: 107–115
76. Watson DH, Le Bouvier GL, Tomlinson JA, Walkey DG (1966) Electron microscopy and antigen precipitates extracted from gel diffusion plates. Immunology 10: 305
77. Williams RC (1954) Electron microscopy of viruses. Adv Virus Res 2: 183–239
78. Yanagida M, Ahmad-Zadeh C (1970) Determination of gene product position in bacteriophage T4 by specific antibody association. J Mol Biol 51: 411–421

Chapter 8 Giovanni P. Martelli

Cytochemistry of Virus-Infected Plant Cells

1 Introduction

Most plant viruses induce cytological changes in the host cells that can readily be observed with the light or electron microscope in appropriately chosen and processed tissue samples.

The products of virus multiplication are either a consequence of the direct expression of the viral genome (i.e., coat protein, replicative forms of nucleic acids, nonstructural proteins, assembled virions), or the result of deranged cell metabolism, or of modification of normal cell constituents and/or organelles. These products, regardless of their origin, are located intracellularly, and often accumulate in a specific form at preferential sites, giving rise to a variety of cytopathic structures known as "inclusion bodies". These are veritable "signatures" of the eliciting virus, quite useful for identification and classification of pathogens and disease diagnosis [8, 15, 17, 26, 34, 35, 42–44, 55].

Light microscopy of virus-infected cells has given an insight into the morphology and interacellular localization of inclusion bodies, whereas electron microscopy has provided a wealth of information on their fine structure and organization. However, neither light nor electron microscopy per se allows the identification and determination of the nature of virus-related products, nor of any change in the chemical constitution of normal or structurally modified cell constituents and/or organelles caused by viral infections. This can be approached through cytochemical microscopy, whereby the chemical nature (protein, Lipid, nucleic acid) of virus-related products can be determined and their intracellular distribution assessed at the gross (light microscopy) or fine structural (electron microscopy) level. Cytochemistry thus provides a better understanding of the virus-host relationships and represents a powerful means for correlating structure with function.

Strictly speaking, cytochemistry is based on the use of reagents (e.g., dyes, enzymes) which, by combining with, or modifying, specific substrates, produce

Dipartmento di Protezione delle Piente delle Molattie, Università degli Studi and Centro di Studio del CNR sui Virus e le Virosi delle Colture Mediterranee, I-Bari, Italy

detectable reactions by which the substrates are identified. However, certain treatments of tissue samples during fixation and embedding, or after thin sectioning, can also change the appearance of the specimen because of differential extraction or electron staining of materials, thus providing evidence of their presence.

Cytochemical methods employed in the study of biological materials have been exhaustively covered in books [19, 60] to which the reader is referred for detailed information. This presentation simply aims at illustrating the current status of cytochemical applications in plant virology with the help of selected examples.

2 Light Microscopy

Although this paper should primarily be centered on cytochemical electron microscopy, it seems appropriate to devote a section to cytochemical optical microscopy, a technique that, even in the recent past, has provided most valuable information on the nature of cytopathic inclusions.

Different types of materials, either in a fresh or fixed status, are suitable for light microscope examination: (1) epidermal strips, (2) semi-thin (1 to 2 μm) sections of tissues embedded in resins (Epon, methyl-butyl-methacrylate, glycol methacrylate), (3) thick (15 to 20 μm or more) sections of tissues, usually embedded in paraffin or paraplast, (4) whole mounts of decolorized leaf blade pieces.

Likewise, different stains can be used according to whether the investigator aims at visualizing inclusion bodies, to distinguish them from normal cell organelles, or at establishing their gross chemical composition. In the first instance, broad-spectrum aspecific dyes like phloxin, toluidine blue, azure A, and calcomine orange-luxol brilliant green are quite suitable. These have given satisfactory and consistent results with an impressive array of viruses belonging to different taxonomic groups [8]. For cytochemical tests a variety of dyes is available, each capable of recognizing and specifically staining given groups of substances: (a) mercuric bromophenol blue and ninhydrin-Schiff for total protein, (b) alkaline fast green for histones; (c) Sudan black B for lipids; (d) azure B for the basophilic reaction given by RNA; (e) Feulgen for DNA; (f) methyl green-pyronin for RNA and DNA, and the fluorochromes acridine orange for RNA, and aniline blue for callose.

In some cases, differential staining was coupled with enzymic treatment to cross-check the occurrence of specific compounds. For instance, inclusions of beet yellows closterovirus (BYV) [9], tobacco rattle tobravirus [20], celery mosaic potyvirus (CeMV) [10], broad bean wilt fabavirus (BBWV) [50], and red clover vein mosaic carlavirus (RCVMV) [29], that stained positive for RNA and proteins, lost their staining properties after exposure to RNase and/or proteolytic enzymes. Treatment with 1,3-glucan hydrolase eliminated toulidine blue fluorescence due to callose in cells locally and systemically infected by tobacco mosaic tobamovirus (TMV) [59], whereas the bright green fluorescence of Fiji disease reovirus (FDV) viroplasms was lost after exposure to pancreatic ribonuclease in low-, but not in high-salt media, indicating that the RNA of the inclusions was double-stranded [22].

Epidermal strips constitute an excellent material for detecting intracellular inclusions. They have also been successfully subjected to differential staining for the establishment of the chemical composition of cytopathic structures elicited by viruses belonging to different taxonomic groups, i.e., dahlia mosaic [48] and cauliflower mosaic [37], caulimoviruses (CaMV); CeMV [10], watermelon mosaic 2 [56], papaya ringspot W [41], and turnip mosaic [39] potyviruses; BBWV [31,50]; RCVMV [29]. In all instances, the inclusions were shown to contain substantial amounts of proteins and RNA. In addition, those of BBWV contained lipids, and those of CaMV DNA.

Semi-thin sections have been used for monitoring the distribution in the host tissues of BYV and potato Y potyvirus [16,37], but also for the chemical identification of intranuclear "satellite bodies" induced by beet mosaic potyvirus (BMV) [40], and for assessing the quantitative distribution of chromatin and nucleohistones in modified nuclei of cells infected by eggplant mottled dwarf rhabdovirus (EMDV) [51].

Cytochemical identification of protein and double-stranded RNA in FDV viroplasms [18,22] was done on thick sections, whereas decolorized leaf pieces were employed essentially for the detection and distribution of starch, callose, and other barrier substances (e.g., lignin, suberin) in localized infections by potato X potexvirus (PVX) [1,46], tomato bushy stunt (TBSV) and Cymbidium ringspot (CyRSV) tombusviruses [47,52], TMV [59], beet necrotic yellow vein furovirus [53], and olive latent ringspot nepovirus [13]. The outcome of most of these studies, in which the pattern of distribution and accumulation of barrier substances (especially callose) in local lesions was followed on a time-course basis, indicated that cell invasion by different viruses preceded callose formation. Hence, it was concluded that callose deposits are unlikely to account for restriction of cell-to-cell movement of viruses in hypersensitive tissue reactions (see [36] for review).

3 Electron Microscopy

3.1 Differential Use of Fixatives

The purpose of fixation is to stabilize the cell structure so as to minimize changes that may occur, with respect to the in vivo condition, during processing for electron microscopic observation.

It is now customary to double fix tissue samples with aldehydes and osmium tetroxide. However, before introduction of aldehydes, osmium tetroxide and potassium permanganate were used alone, both with unsatisfactory results.

Potassium permanganate is a strong oxidant which is reduced to manganese oxide, a brown precipitate that confers a marked contrast upon the structures reacting with it. Membranes are fixed extremely well, acquire a high electron density, and become outstanding. Although a permanganate-fixing schedule was developed that allows a relatively good preservation of ground cytoplasm and virus particles [61], the severe loss of cellular and virus-related components that is so often registered [57] has discouraged its use, except for special purposes. For instance, permanganate fixation found useful applications with enveloped viruses, as exemplified by the neat visualization of intracellular tomato spotted wilt virus [32] and EMDV particles, and of the connection of the outer envelope of EMDV virions with the nuclear membrane [49]. Recently, the "wall" (Fig. 1A) surrounding "particles" of the putative agent of fig mosaic was shown to be a membranous envelope (Fig. 1B) in permanganate-fixed samples (unpubl. information).

Osmium tetroxide reacts strongly with unsaturated lipid molecules which are cross-linked and insolubilized. Since osmium acts both as a fixative and a stain for lipids, enhancing their electron opacity, cytomembranes are especially well preserved. When post fixation with osmium is omitted, the lipid moiety retains its electron transparency and is partly extracted by organic solvents because it is not insolubilized by simple aldehyde fixation. It ensues that cytomembranes become almost invisible, and organelles with extensive membrane systems (e.g., chloroplasts, mitochondria, dictyosomes) show a negative image, i.e., electron-lucent lamellae surrounded by a dense stroma. Witholding osmium postfixation was useful for: (a) localizing lipid components in thin-sectioned particles of EMDV, thus giving a better insight into their structural organization [38, 49]; (b) confirming the occurrence of lipids at the periphery of the particles of the presumed agent of fig mosaic (unpubl. information).

Tannic acid increases the contrast and preservation of cell structures, especially if applied together with ordinary aldehyde and osmium fixatives. The use of tannic acid, however, did not improve the preservation nor did it change the outward appearance of TBSV-induced multivesicular bodies (MVB) [4].

3.2 Differential Use of Electron Stains

Uranyl ions bind to several groups, such as carboxyl and phosphate, thus producing a generalized increase in contrast without much specificity. However, in absence of osmium, plant cells become heavily stained with uranium salts as a likely consequence of increased electron opacity of substances having good affinity for

---

Fig. 1. A A group of fig mosaic-associated "particles" in the cytoplasm of a fig cell double-fixed with glutaraldehyde-osmium tetroxide. The particles are surrounded by a wall-like structure (*arrowheads*) and exhibit a granular internal content. **B** Appearance of the same particles fixed in potassium permanganate. Details of the internal structure have disappeared but it is clearly shown that the peripheral "wall" is in fact a membrane. Bar = 100 nm

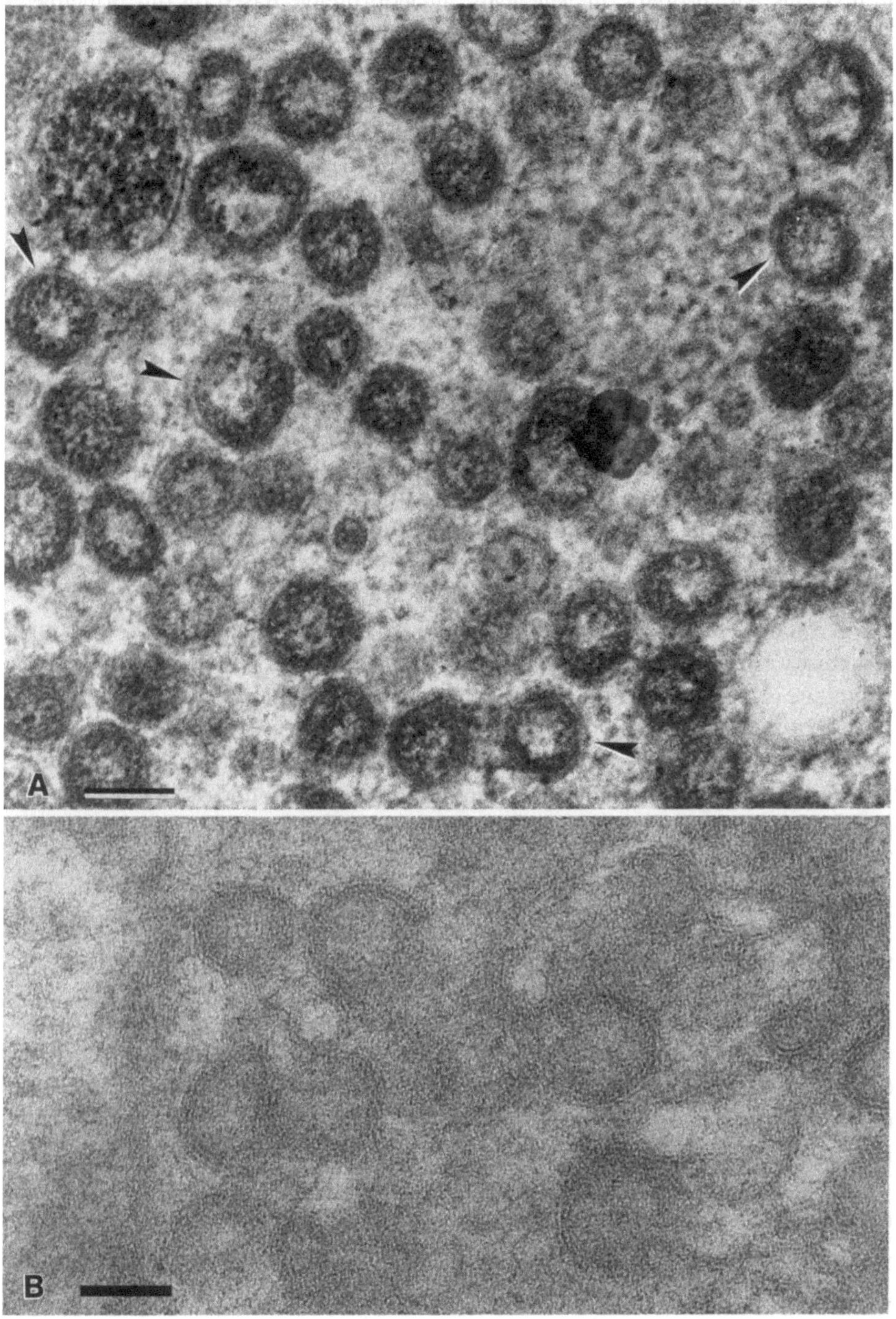
A
B

uranyl ions, e.g., proteins, nucleic acids, nucleoproteins. Based on this principle, a technique was developed (the uranyl soak method) in which osmium postfixation is omitted and tissues are dehydrated in ethanol saturated with uranyl acetate [27]. Certain cell regions and structures are overstained, and isometric viruses are more readily seen because ribosomes are lightly stained compared with virus particles. The application of this method to cells infected by different plant viruses yielded valuable information on the structural organization of EMDV particles, showed the uniform constitution and proteinaceous nature of cylindrical inclusions of potyviruses, and the finer details of CaMV viroplasms [38].

Alcoholic phosphotungstic acid is a stain with strong affinity for proteins. Its use on TBSV-infected cells greatly enhanced the electron opacity of the proteinaceous matrix of MVB and of viral coat protein accumulations in the ground cytoplasm [4].

Treatment with phosphotungstic acid-chromic acid increases selectively the electron density of the plasma membrane, and so does alkaline bismuth to the bounding membrane of chloroplasts. The application of these selective stains to TBSV-infected cells showed that the membranes of MVB were much less densely stained than the plasma membranes or plastidial envelopes, suggesting a different genesis for the membranous components of MVB [4].

A staining procedure for differentiating between RNA- and DNA-containing structures in conventionally fixed tissues was developed by Bernhard [5]. This technique is based on the use of a chelating agent (EDTA) which, when applied to thin sections, decolorizes deoxyribonucleoproteins (DNP) but not ribonucleoproteins (RNP). EDTA does not extract DNA or histones but removes uranyl ions specifically from DPN, without eliminating the stain bound to RPN [5]. This method was successfully used to demonstrate the DPN nature of the intranuclear "fibrillar rings" evoked by bean golden mosaic germinivirus [30].

3.3 Differential Use of Enzyme Digestion

Enzyme digestion tests can be performed before or after embedding of tissues. Accordingly, epidermal strips or thin tissue slices are fixed with an aldehyde, exposed to the desired enzymes, osmicated, dehydrated and embedded, or tissues embedded in epoxy or water-soluble resins (e.g., glycol methacrylate) are exposed to the enzyme, stained and observed.

Proteins and nucleic acids are the two main groups of virus-related products that have been identified by enzyme digestion in thin-sectioned cells.

3.3.1 Nucleic Acids

Intracellular nucleic acid molecules, when they are not complexed with proteins, appear as discrete networks of tiny fibrils contained within organelles or, more often,

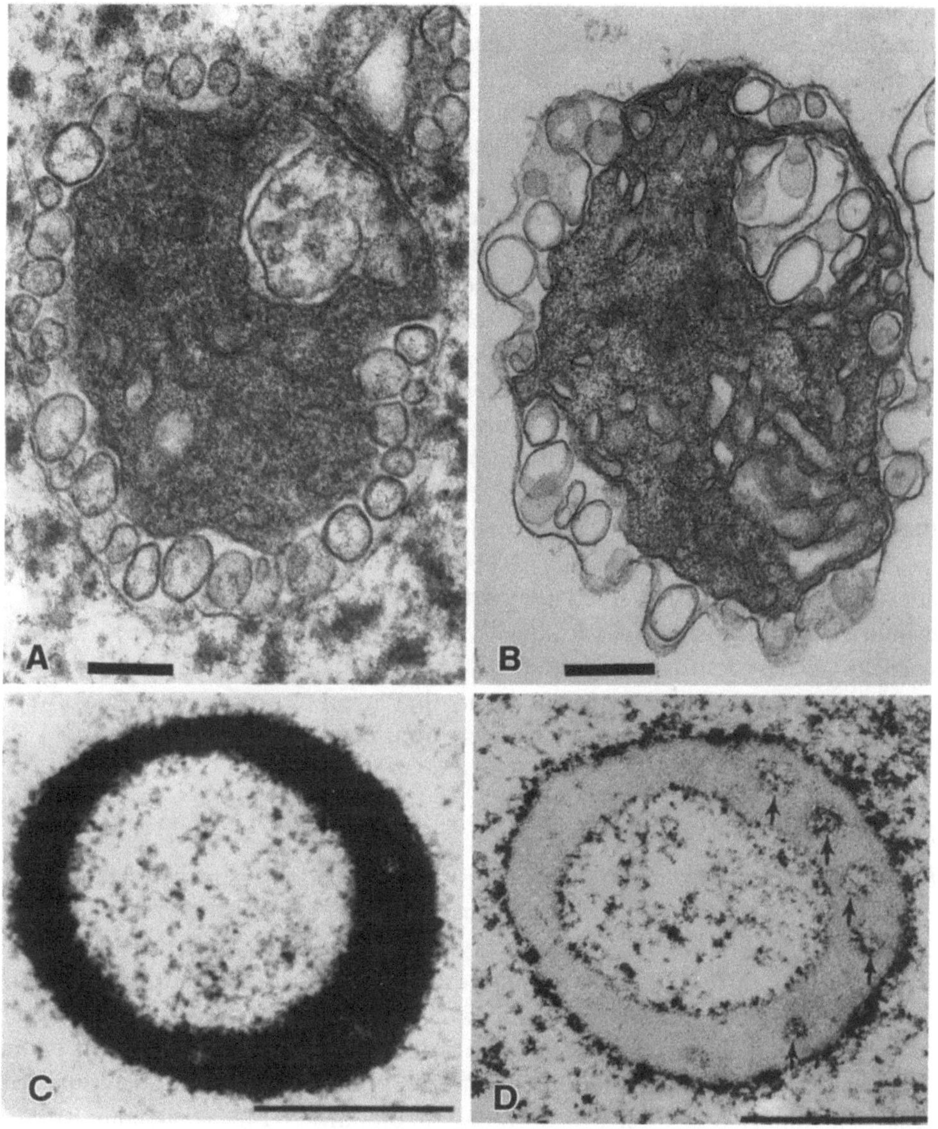

Fig. 2. A, B Effect of RNase treatment in low salt medium on the fibrils contained in the peripheral vesicles of carnation Italian ringspot virus multivesicular bodies. The fibrils, which are clearly discernible in the untreated sample (**A**), are no longer visible in the sample exposed to the enzyme (**B**). Bars = 200 nm. **C, D** Appearance of intranuclear fibrillar rings induced by bean golden mosaic virus in controls (**C**) and in EDTA-treated samples (**D**) EDTA has removed the stain, indicating that the fibrillar ring is made up of deoxyribonucleoproteins. Bars = 500 nm. [30]

various kinds of membranous vesicles. A technique that permits localization of RNA and determines whether it occurs as a single-stranded (ss) or double-stranded (ds) form has been developed [21]. It exploits the differential ability of pancreatic ribonuclease to digest ssRNA regardless of the salt concentration of the incubating medium, whereas dsRNA is susceptible to the enzyme only under low salt concentration.

This method was successfully used for the identification of dsRNA in vesicles of vesiculate-vacuolate inclusions of cowpea mosaic comovirus [21], in tonoplast-associated vesicles of cells infected by cucumber mosaic and tomato aspermy cucumoviruses [23], and in the MVB elicited by CyRSV and carnation Italian ringspot tombusvirus (CIRV) [14, 54] (Fig. 2A, B). In all cases the consensus was that the dsRNA represented replicative forms of viral RNA, and that the cytopathic structures containing it were most likely sites of viral RNA replication.

The same technique has been used for improving the intracellular visualization of virus particles containing ssRNA because RNA in ribosomes is digested whereas that of virions is not [24].

RNase treatment of cells infected by bean yellow mosaic (BYMV) [62], tobacco etch (TEV) [58], or BMV [28] potyviruses, did not change the appearance of cylindrical inclusions, not did it attack visibly the core of thin-sectioned EMDV particles [49]. By contrast, RNase digestion modified the look of maize rough dwarf reovirus (MRDV)-induced viroplasms, in that immature particles, which were plentiful in the controls, almost completely disappeared from treated sections [3].

3.3.2 Proteins

Several proteolytic enzymes (papain, trypsin, subtilisin, pepsin, pronase) have been used on virus-infected cells. It was the sensitivity to proteolytic enzymes shown by cylindrical inclusions and intranuclear crystals of several potyviruses (BMV, TEV, BMV) (Fig. 3C, D) that provided ultimate evidence that these structures were indeed made up of proteins [28, 58, 62, 64]. Likewise, it was a successful enzymic digestion that demonstrated the proteinaceous nature of the thin laminate components of the complex PVX inclisions [57] and of the "dense granules" found in TBSV-infected cells [4]. Finally, the extensive degradation of CaMV (Fig. 3A, B) and BBWV viroplasms following enzymic attack [37, 50] established beyond doubt the presence of proteins in these inclusions, a possibility that had been suggested by less conclusive cytochemical tests. Proteinaceous material was shown to constitute the

Fig. 3. A Aspect of a caulimovirus viroplasm in a normally processed cell. **B** Cauliflower mosaic virus-induced viroplasm after 1 h protease treatment of an epidermal strip prior to embedding. The viroplasm matrix has been extensively digested. Bars = 200 nm. [37] **C** Cylindrical inclusion (PW) of tobacco etch virus totally digested after 4 h exposure to subtilisin of thin sections embedded in Araldite. **D** Potyvirus-induced cylindrical inclusion (PW) in a nontreated cell. Bars = 200 nm. [58]

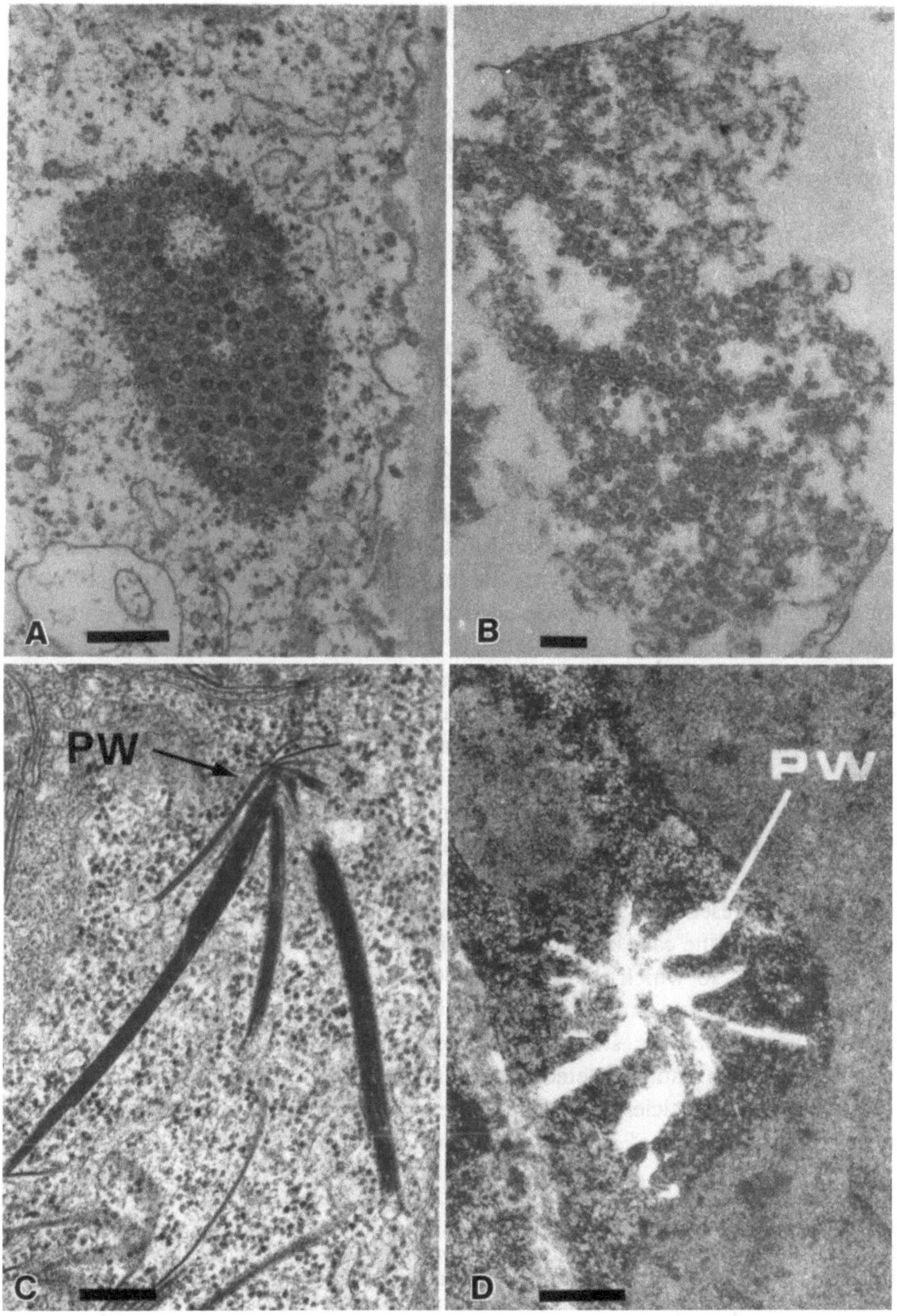
A
B
PW
C
PW
D

virus-containing tubules and much of the viroplasms of MRDV by means of protease digestion tests [3].

However, enzymatic treatments may not always be successful, as shown by the failure to digest cytoplasmic tubules associated with particles of olive latent virus 1 [7], or the matrix of TBSV-induced MVB [4], both of which had the outward appearance and staining properties of proteins.

3.4 Localization of Enzymes

Although techniques for the electron microscopic localization of some 30 different enzymes are available [19, 60], very few of these have been used in plant virology. The scanty list of examples given in a recent review [43] has not grown much over the last few years.

Enzyme cytochemistry requires careful fixation with polymerizing fixatives such as aldehydes (formaldehyde and glutaldehyde being the best) to preserve enzyme activity and prevent enzyme diffusion during tissue manipulation. Methods for intracellular localization rely on the formation, at the site of enzyme activity, of an electron-opaque reaction product detectable by direct observation under the electron microscope.

Diaminobenzidine (DAB) is one of the most used substrates for cytochemical enzyme localization. Following oxidation, DAB polymerizes, becoming insoluble. This polymer, upon exposure to osmium tetroxide, forms a product ("osmium black") that is intensely electron-opaque. Similar black reaction products are yielded by cerium chloride precipitation in presence of glycolate oxidase, or by the ferricyanide-reduction reaction of glycolate oxidase.

With plant viruses, enzyme cytochemistry has been used for: (1) correlating intracellular presence of enzymes with the level of enzymatic activity of infected samples. This is the case of leaf tissues infected by pea enation mosaic and alfalfa mosaic viruses, to which the DAB test was applied to find an explanation for increased peroxidase and catalase activity [11, 12], and (2) identifying enzymic activity in cytoplasmic structures to demonstrate possible relationships with organelles. In this instance, detection of acid phosphatase in secondary cytoplasmic vacuoles, where particles of MRDV were apparently being digested, successfully demonstrated lysosomal activity [2]. Furthermore, the identification of catalase

Fig. 4. Identification of enzyme activity in organelles and cytopathic structures. **A** A normal-looking mitochondrion and **B** a mitochondrion undergoing transformation into a multivesicular body (MVB), both treated with DAB, and showing positive reaction for cytochrome oxidase (*dark deposits*) at the level of the cristae. Bars = 200 nm. **C** A MVB of cymbidium ringspot virus in a control sample. The granular matrix (*M*) suggests a peroxisomal origin of the MVB. **D, E** Evidence that MVB originate from modified peroxisomes: dark staining reaction of the matrix following DAB test (positive for catalase) (**D**), and cerium precipitation (positive for glycolate oxidase) (**E**). Bars = 200 nm. (**D** and **E** [54])

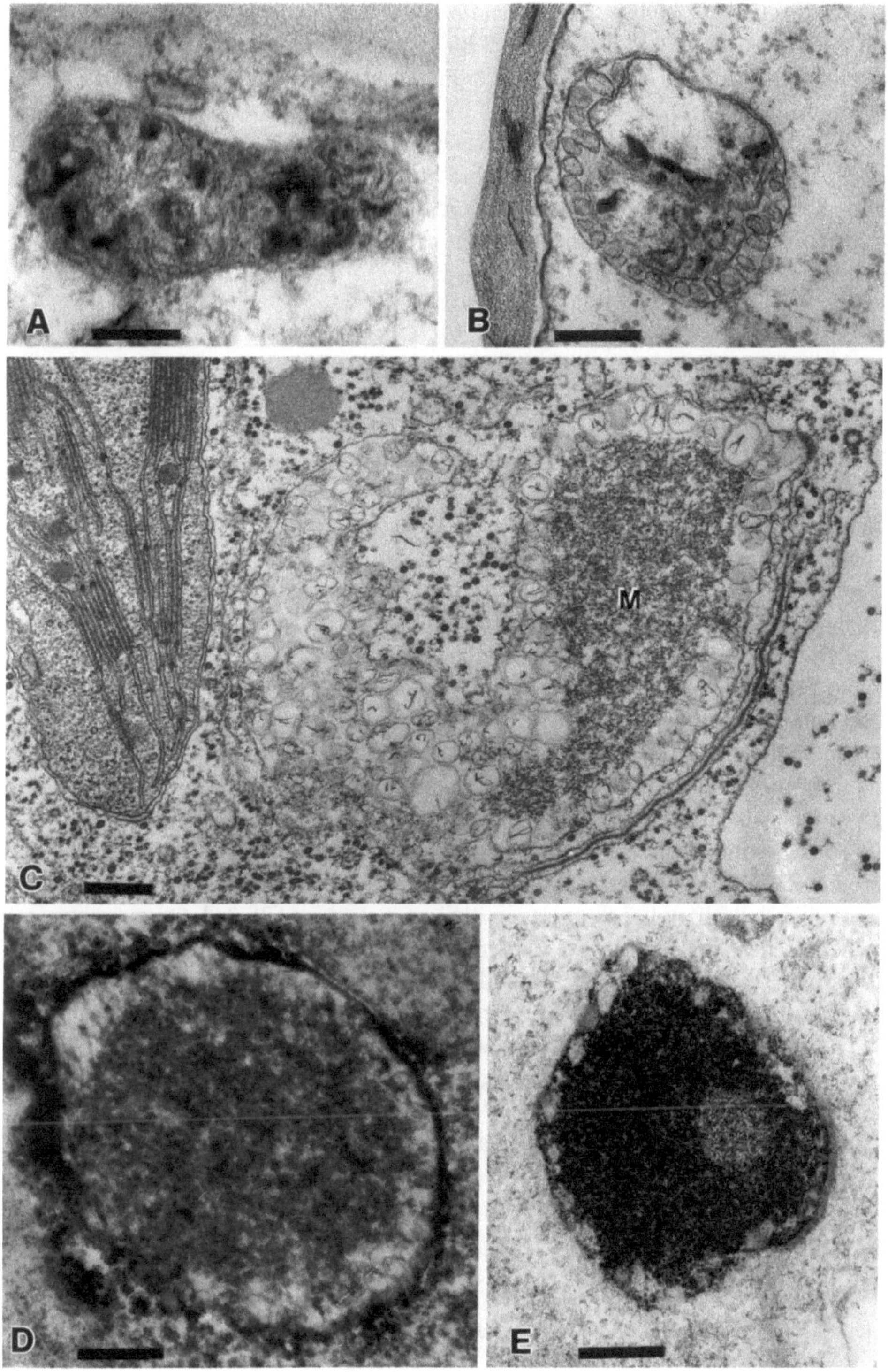
A
B
M
C
D
E

(DAB reaction) (Fig. 4D) and glycolate oxidase (cerium precipitation) (Fig. 4E) in MVB of CyRSV (Fig. 4C) and TBSV, provided ultimate evidence of their peroxisomal origin [4,45,54]. Detection of cytochrome oxidase activity (DAB reaction) (Fig. 4A, B) in MVB elicited by CIRV [14] and in "vesiculated bodies" associated with a phloem-limited isometric virus of the grapevine [6], constituted convincing proof that these cytopathic structures derive from deranged mitochondria. Cytochrome oxidase activity was also demonstrated cytochemically in hypertrophied, but still recognizable, mitochondria of cells infected by an unidentified apple virus [63].

4 Concluding Remarks

There is no doubt that cytochemical light and electron microscopy have afforded results of primary importance because of the detailed information on the intracellular localization and chemical nature of different virus-related products. This has conferred physiological significance upon some of the cytopathic structures associated with viral infections and has demonstrated the direct involvement of cell organelles in the main phases of virus replication.

The major drawback of traditional electron microscopic cytochemistry is that by its techniques (differential use of fixatives and electron stains, enzyme digestion) the chemical composition of intracellular structures can be established but not the origin of their components, i.e., whether a nucleic acid is viral or not; or which of the several proteins known to be encoded and expressed by a viral genome occurs in a given inclusion body.

It is becoming increasingly evident that functionally important nonstructural proteins coded, say, for example, by potyviruses and caulimoviruses, are differently located in the cell and accumulate in distinct inclusions [25,34]. On cytochemical evidence, these inclusions were known to be composed essentially of proteins, but the specific identification of the single constituting protein species could not be made, unless the inclusions were extracted, purified, and properly analyzed [26].

Improved immunolabeling techniques have now greatly facilitated the intracellular identification of specific viral gene products. These represent a tool whereby remarkable advances are expected in the near future in the unraveling of intracellular events that underline plant virus multiplication processes.

References

1. Allison AV, Shalla TA (1974) The ultrastructure of local lesions induced by potato virus X: a sequence of cytological events in the course of infection. Phytopathology 64: 784–793

2. Appiano A, Lovisolo O (1979) Ultrastructure of maize roots infected with maize rough dwarf virus and presence of virus particles in vacuoles with lysosomal activity. Microbiologica 2: 37–50
3. Bassi M, Favali A (1972) Electron microscopy of maize rough dwarf virus assembly sites in maize. Cytochemical and autoradiographic observations. J Gen Virol 16: 153–160
4. Bassi M, Barbieri N, Appiano A, D'Agostino G (1986) Origin and function of tomato bushy stunt virus-induced inclusion bodies. J Ultrastruct Mol Struct Res 96: 194–203
5. Bernhard W (1969) A new staining procedure for electron microscopical cytology. J Ultrastruct Res 27: 250–256
6. Castellano MA, Martelli GP (1984) Ultrastructure and nature of vesiculated bodies associated with isometric virus-like particles in diseased grapevines. J Ultrastruct Res 89: 56–64
7. Castellano MA, Di Franco A, Martelli GP (1987) Electron microscopy of two olive viruses in host tissues. J Submicrosc Cytol 19: 495–508
8. Christie RG, Edwardson JR (1977) Light and electron microscopy of plant virus inclusions. Fla Agric Exp Stn Monogr 9: 150
9. Cronshaw J, Hoefert LL, Esau K (1966) Ultrastructural features of *Beta* leaves infected with beet yellows virus. J Cell Biol 31: 429–443
10. Da Cruz ND, Medina DM, Kitajima EW, Costa AS, Landin CC (1972) Citoquimica de inclusoes intranucleares associadas ao virus do mosaico amarelo do salsao. Bragantia 31: 217–226
11. De Vecchi L, Conti GG (1975) Peroxisomes in leaves of *Nicotiana sylvestris* infected with alfalfa mosaic virus. Isr J Bot 24: 71–79
12. De Zoeten GA, Gaard G, Diez FB (1973) Localization of some heme-containing enzymes in healthy and pea enation mosaic virus-infected plants. Physiol Plant Pathol 3: 159–166
13. Di Franco A, Martelli GP, Russo M (1983) An ultrastructural study of olive latent ringspot virus in *Gomphrena globosa*. J Submicrosc Cytol 15: 539–548
14. Di Franco A, Russo M, Martelli GP (1984) Ultrastructure and origin of cytoplasmic multivesicular bodies induced by carnation Italian ringspot virus. J Gen Virol 65: 1233–1237
15. Edwardson JR, Christie RG (1978) Use of virus-induced inclusions in classification and diagnosis. Annu Rev Phytopathol 16: 31–55
16. Esau K, Hoefert LL (1971) Cytology of beet yellows virus infections in *Tetragonia*. Parenchyma cells in infected leaves. Protoplasma 72: 255–273
17. Francki RIB, Milne RG, Hatta T (1985) An atlas of plant viruses, Vol 1 and 2. CRC Press, Boca Raton
18. Giannotti J, Monsarat P (1968) Etude biologique et histochimiques des tumeurs foliaires des cannes àsucre atteintes de la maladie de Fidji. Ann Epiphyt 19: 707–712
19. Hall JL (ed) (1978) Electron microscopy and cytochemistry of plant cells. Elsevier/North Holland Biomedical Press, Amsterdam
20. Harrison BD, Stefanac Z, Roberts IM (1969) Role of mitochondria in the formation of X-bodies in cells of *Nicotiana clevelandii* infected with tobacco rattle virus. J Gen Virol 6: 126–140
21. Hatta T, Francki RIB (1978) Enzyme cytochemical identification of single-stranded and double-stranded RNAs in virus-infected plant and insect cells. Virology 88: 105–117
22. Hatta T, Francki RIB (1981a) Development and cytopathology of virus-induced galls on leaves of sugarcane infected with Fiji disease virus. Physiol Plant Pathol 19: 337–346
23. Hatta T, Francki RIB (1981b) Cytopathic structures associated with tonoplast of plant cells infected with cucumber mosaic and tomato aspermy viruses. J Gen Virol 53: 343–346
24. Hatta T, Francki RIB (1981c) Identification of polyhedral small virus particles in thin sections of plant cells by an enzyme cytochemical technique. J Ultrastruct Res 74: 116–129
25. Hiebert E, Dougherty WG (1988) Organization and expression of viral genomes. In: Milne RG (ed) The plant viruses, Vol 4. Plenum, New York

26. Hiebert E, Purciful DE, Christie RG (1984) Purification and immunological analysis of plant virus inclusion bodies. In: Maramorosch K, Koprowski H (eds) Methods Virol, Vol 8. Academic Press, New York
27. Hills GJ, Plaskitt A (1968) A protein stain for electron microscopy of small isometric plant virus particles. J Ultrastruct Res 25: 323–329
28. Hoefert LL (1969) Proteinaceous and virus-like inclusions in cells infected with beet mosaic virus. Virology 37: 498–501
29. Kahn MA, Maxwell DP, Maxwell MD (1977) Light microscopic cytochemistry and ultrastructure of red clover vein mosaic virus-induced inclusions. Virology 78: 173–182
30. Kim KS, Shock TL, Goodman RM (1978) Infection of *Phaseolus vulgaris* by bean golden mosaic virus: ultrastructural aspects. Virology 89: 22–33
31. Kistah AA, Russo M, Tolba MA, Martelli GP (1978) A strain of broad bean wilt virus isolated from pea in Egypt. Phytopathol Mediterr 17: 157–164
32. Kitajima EW (1965) Electron microscopy of vira-cabeca virus (Brazilian tomato spotted wilt virus) within host cells. Virology 26: 98–99
33. Kitajima EW, Camargo IJB, Costa AS (1968) Intranuclear crystals and cytoplasmic membranous inclusions associated with infection by two Brazilian strains of potato virus Y. J Electron Microsc 17: 144–153
34. Lesemann DE (1988a) Recent aspectes of the use of cytopathology in virus identification. Acta Hortic 234: 289–298
35. Lesemann DE (1988b) Cytopathology. In: Milne RG (ed) The Plant Viruses, Vol 4. Plenum, New York
36. Martelli GP (1980) Ultrastructural aspects of possible defence reactions in virus-infected plant cells. Microbiologica 3: 369–391
37. Martelli GP, Castellano MA (1971a) Light and electron microscopy of the intracellular inclusions of cauliflower mosaic virus. J Gen Virol 13: 133–140
38. Martelli GP, Castellano MA (1971b) A brief account of the use of the uranyl soak method for the visualization of some viruses in plant tissues. Phytopathol Mediterr 10: 78–81
39. Martelli GP, Lal SB (1985) Ultrastructural observations on virus-diseased plants of the Kabul area. Phytopathol Mediterr 24: 228–233
40. Martelli GP, Russo M (1969) Nuclear changes in mesophyll cells of *Gomphrena globosa* L associated with infection by beet mosaic virus. Virology 38: 297–308
41. Martelli GP, Russo M (1976) Unusual cytoplasmic inclusions induced by watermelon mosaic virus. Virology 72: 352–362
42. Martelli GP, Russo M (1977) Plant virus inclusion bodies. Adv Virus Res 21: 175–226
43. Martelli GP, Russo M (1984) Use of thin sectioning for the visualization and identification of plant viruses. In: Maramorosch K, Koprowski H (eds) Methods Virol, Vol 8. Academic Press, New York
44. Martelli GP, Russo M (1985) Virus-host relationships. Symptomatological and ultrastructural aspects In: Francki RIB (ed) The plant viruses, Vol 1. Plenum, New York
45. Martelli GP, Di Franco A., Russo M (1984) The origin of multivesicular bodies in tomato bushy stunt-infected *Gomphrena globosa* plants. J Ultrastruct Res 88: 275–281
46. Pennazio S, Redolfi P, Sapetti C (1981) Callose formation and permeability changes during the partly localized reaction of *Gomphrena globosa* to potato virus X. Phytopathol Z 100: 172–181
47. Pennazio S, D'Agostino G, Appiano A, Redolfi P (1978) Ultrastructure and histochemistry of resistant tissues surrounding lesions of tomato bushy stunt virus in *Gomphrena globosa* leaves. Physiol Plant Pathol 13: 165–171
48. Robb SM (1964) Location, structure and cytochemical staining reactions of the inclusion bodies found in *Dahlia variabilis* infected with dahlia mosaic virus. Virology 23: 141–144
49. Russo M, Martelli GP (1973) A study of the structure of eggplant mottled dwarf virus. Virology 52: 39–48
50. Russo M, Martelli GP (1975a) Light and electron microscopy of broad bean wilt virus intracellular inclusions. J Submicrosc Cytol 7: 335–347
51. Russo M, Martelli GP (1975b) Some cytochemical reactions of nuclei infected with eggplant mottled dwarf virus. Phytopathol Z 83: 97–102

52. Russo M, Martelli GP (1981) The fine structure of Cymbidium ringspot virus in host tissues. II. Light and electron microscopy of localized infections. J Ultrastruct Res 77: 105–118
53. Russo M, Martelli GP, Di Franco A (1981) The fine structure of local lesions of beet necrotic yellow vein virus in *Chenopodium amaranticolor*. Physiol Plant Pathol 19: 237–242
54. Russo M, Di Franco A, Martelli GP (1983) The fine structure of Cymbidium ringspot virus in host tissues. III. Role of peroxisomes in the genesis of multivesicular bodies. J Ultrastruct Res 82: 52–63
55. Russo M, Di Franco A, Martelli GP (1987) Cytopathology in the classification of tombusviruses. Intervirology 28: 134–143
56. Russo M, Martelli GP, Vovlas C, Ragozzino A (1979) Comparative studies on Mediterranean isolates of watermelon mosaic virus. Phytopathol Mediterr 18: 94–101
57. Shalla TA, Shepard JF (1972) The structure and antigenic analysis of amorphous inclusion bodies induced by potato virus X. Virology 49: 654–667
58. Shepard JF (1968) Electron microscopy of subtilisin-treated tobacco etch virus nuclear and cytoplasmic inclusions. Virology 36: 20–29
59. Shimomura T (1979) Stimulation of callose synthesis in the leaves of Samsun NN tobacco showing systemic acquired resistance to tobacco mosaic virus. Ann Phytopathol Soc Jpn 45: 299–304
60. Vaughn KC (ed) (1987) Handbook of plant cytochemistry, Vol 1 and 2. CRC, Boca Raton
61. Warmke HF, Edwardson JR (1966) Use of potassium permanganate as a fixative for virus particles in plant tissues. Virology 28: 693–700
62. Weintraub M, Ragetli HWJ (1968) Intracellular characterization of bean yellow mosaic virus-induced inclusions by differential enzyme digestion. J Cell Biol 38: 316–328
63. Weintraub M, Schroeder B (1979) Cytochrome oxidase activity in hypertrophied mitochondria of virus-infected leaf cells. Phytomorphology 29: 273–285
64. Weintraub M, Ragetli HWJ, Veto M (1969) The use of glycol methacrylate for the study of the ultrastructure of virus-infected leaf cells. J Ultrastruct Res 26: 197–215

Chapter 9 Jan W. M. van Lent and Benedictus J. M. Verduin

Immunolabeling of Viral Antigens in Infected Cells

1 Introduction

Since its introduction in 1971 [2], labeling of antigens with gold particles to facilitate the detection of these antigens in thin sections of infected tissue by electron microscopy has proven to be a powerful technique. Also in plant pathology and especially in plant virology [4], this detection system has been applied frequently, enabling studies which were difficult or impossible before the gold era. Both in the field of diagnostics and research, scientists have profited from the gold label.

Two aspects of the infection cycle were our choice for immunolabeling studies in situ with a bromovirus: (a) the possible compartimentalization of the different steps in the infection cycle; (b) the translocation of the viral genome.

The bromoviruses are icosahedral plant viruses whose genomes are divided among three separately encapsidated, messenger RNAs designated RNA-1, RNA-2 and RNA-3 (ca. 3.2, 2.8, and 2.2 kb, respectively). Each of these genomic RNA's serves as a mRNA for a separate noncapsid protein (for cowpea chlorotic mottle virus (CCMV) with respective molecular masses of 110, 94, and 33 kDa). In addition, the 3′ portion of CCMV RNA-3 encodes the coat protein (M_r 20 189), which is translated from a 0.9-kb subgenomic RNA (RNA-4). The protein coat of each particle consists of 180 of these protein subunits arranged in a shell with icosahedral symmetry.

In the infection cycle of CCMV, six different steps can be distinguished: penetration, dissociation/disassembly/uncoating, translation, replication, association/assembly and translocation/release. Our present knowledge of these steps is the following. Penetration is most likely to occur through wounds in the cell wall and plasma membrane, caused by insects or mechanical wounding with an abrasive, which results in depositing a "modified" virus particle into the cytoplasm [5, 7]. Subsequently, the virus particle binds to ribosomes and in the process of translation of the genome the particle is disassembled [6, 8]. This cotranslational

Department of Virology, Agricultural University, Binnenhaven 11, NL-6709 PD Wageningen, The Netherlands

disassembly has been demonstrated for tobacco mosaic virus, bromoviruses and southern bean mosaic virus [17]. Virus-specific proteins are synthesized and form, together with host proteins, a complex with the genome. This complex initiates and regulates the replication of the nucleic acid. The newly synthesized nucleic acid molecules or even new subgenomic messenger RNA's function as templates for the production of coat protein. Both newly synthesized nucleic acid and coat protein assemble into new virus particles. These particles, or in some cases an earlier nucleoprotein complex, are then transported from cell-to-cell and over long

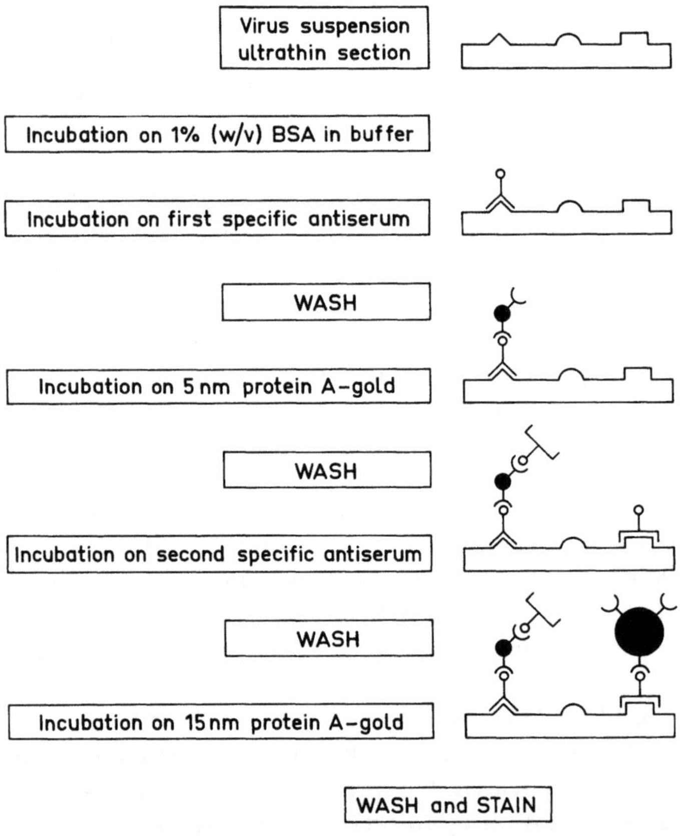

Fig. 1. Schematic representation of immunogold double labeling of two antigens. All incubations were done at room temperature. Virus suspensions or ultrathin sections on grids are incubated with 1% (w/v) bovine serum albumin in phosphate buffered saline (BSA-PBS) for 1 h. Subsequently, the grids were incubated with a 1/100 dilution of first antiserum in PBS-BSA for 1 h. Grids are washed with 30 drops of PBS and incubated with 5–7 nm protein A gold for 1 h. Again grids are washed with 30 drops of PBS. Incubation with a 1/100 dilution of the second specific antiserum in PBS-BSA was done for 1 h. Wash grids with 30 drops of PBS and incubate with 15–16 nm protein A-gold for 1 h. Wash with 30 drops of PBS, followed by 10 drops of distilled water and stain with 1% (w/v) uranyl acetate in water [9]. Air dry and view in the electron microscope

distances through the vascular tissue. A virus-specific protein may regulate this process by modifying the plasmodesmata.

This paper describes the immunogold localization of CCMV coat protein antigen and nucleic acid in infected cowpea protoplasts and plants, in an attempt to present evidence for the above described steps in the infection cycle of CCMV.

2 Labeling

The method of immunogold double labeling is schematically given in Fig. 1. The different sized protein A-gold preparations can be easily made [9, 10] or obtained commercially. For single labeling we use 7-nm gold particles routinely. When several antisera against different antigens are used, complex formation with protein A-gold is preferred over adsorption of different gammaglobulins to gold particles. To standardize the protein A-gold complexes and to control its quality upon storage

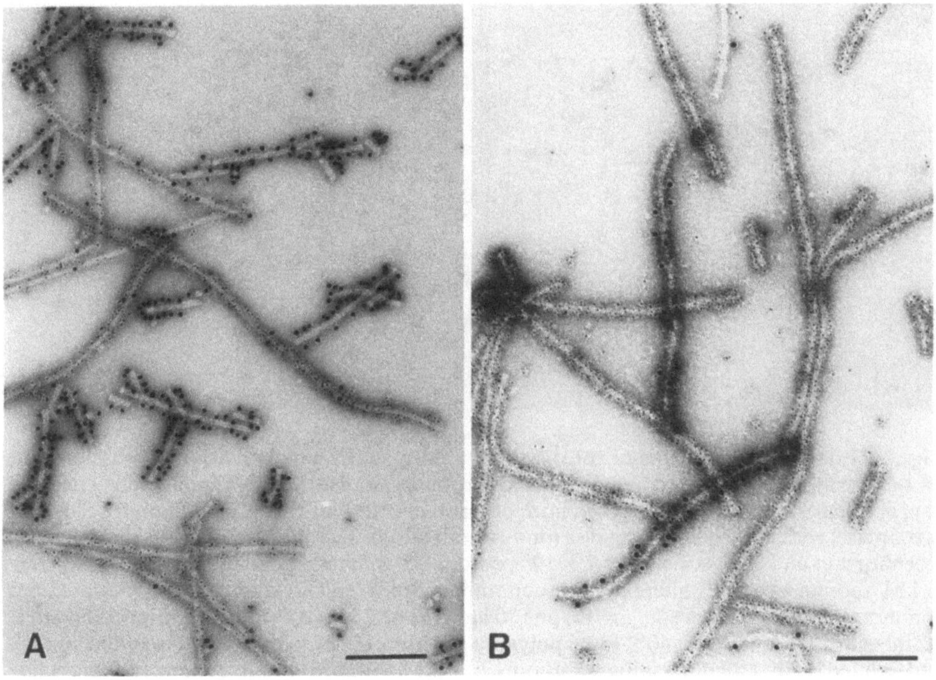

Fig. 2. Immunogold double labeling of mixtures of rodlike and filamentous plant viruses in suspension. **A** Tobacco mosaic virus (16 nm gold) and a carlavirus (7 nm gold). **B** Two potyviruses: blackeye cowpea mosaic virus (16 nm gold) and potato virus Y^N (7 nm gold). Labeling as described in Fig. 1. Different particles with similar morphology can easily be distinguished with immunogold double labeling. Bar = 200 nm

at 4 °C, we regularly label purified virus suspensions and count the number of gold particles per virus particle. Figure 2 shows the double labeling of mixed virus suspensions with either different (A) or identical morphology (B). Next to discrimination of morphologically identical particles it also indicates that the size of the gold particle influences the number of protein A-gold particles that can bind to a virus particle. Due to steric hindrance, the 16-nm gold particles bind less frequently than the 7-nm particle.

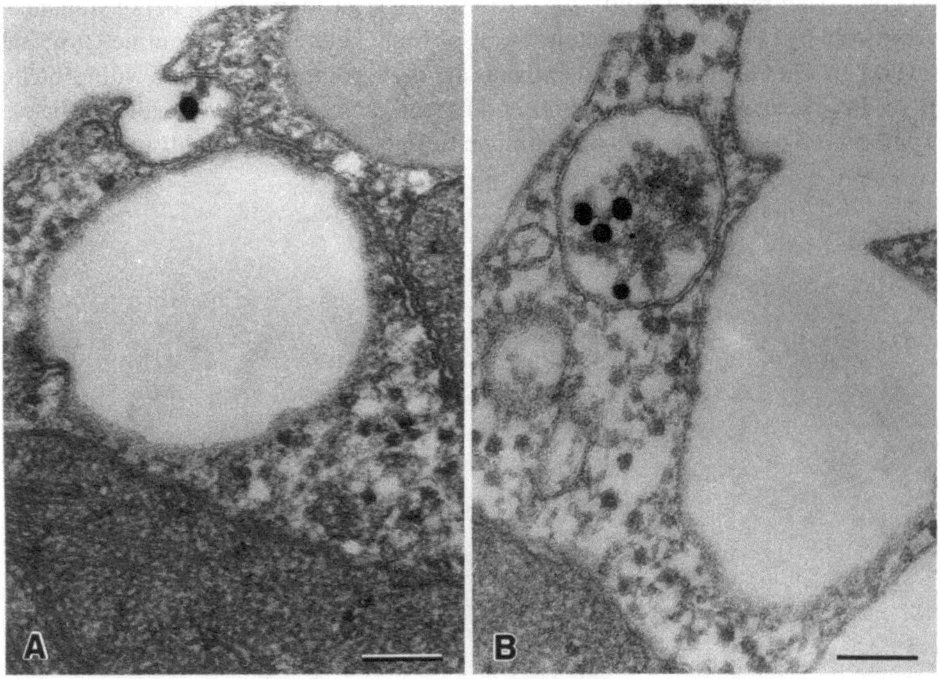

Fig. 3. Endocytosis by cowpea protoplasts of gold particles coated with cowpea chlorotic mottle virus (CCMV) coat protein. Viral coat protein was adsorbed to 25 nm gold particles in a procedure similar to the production of protein A gold [9]. Inoculation of cowpea protoplasts with coated gold particles (or virus or RNA) was done essentially as described by Roenhorst et al. [5, 7]. A pellet of 2×10^2 protoplasts was resuspended in 25 μl virus buffer (0.1 M sodium acetate buffer, pH 5.0 containing 1 mM EDTA and 1 mM sodium azide) containing coated gold ($A_{530\ nm} = 13$) or 100 μg virus or 25 μg RNA. Immediately thereafter, 0.2 ml of 3 mM $CaCl_2$ with 40% (w/v) polyethylene glycol (PEG, M_r 6000) was added, mixed, and diluted with 1.8 ml washing solution [2.5 mM 2(N-morpholino)ethane sulfonic acid-KOH (MES), pH 5.6, 0.6 M mannitol]. After incubation for 30 min at 25 °C, the protoplasts were sedimented and washed three times in washing solution at room temperature. Then the inoculated protoplasts were resuspended in culture medium and incubated as described [7]. Embedding and sectioning of antigen in protoplasts was done as described [10] **A** A CCMV-antigen-coated gold particle present in an invagination of the plasma membrane. **B** Four CCMV-antigen-coated gold particles in a closed vesicle. Bar = 200 nm

3 Initial Interaction

In attempts to elucidate the initial interactions between virus particles and plant cells, we inoculated cowpea protoplasts with 25-nm gold particles coated with CCMV-coat protein. Gold particles were adsorbed with isolated and purified coat protein [15], as described for protein A. After inoculation and incubation, protoplasts were fixed with formaldehyde/glutaraldehyde, embedded in Lowicryl or London Resin White, and sectioned as described [10]. It is known that protoplasts after isolation from leaf tissue show invagination of the plasma membrane. This occurs even more frequently if the protoplasts are treated with polyethylene glycol in the standard inoculation procedure. Coat protein-coated gold particles are observed in the cavities of the invaginating plasma membrane (Fig. 3A) and they can be found in closed vesicles (Fig. 3B). Like virus particles, they remain in vesicles and are not released in the cytoplasm [7]. Therefore we concluded that uptake of virus particles by this type of endocytosis hardly ever contributes to infection. Mechanical wounding of the (cell wall and) plasma membrane or transient discontinuities of the membrane are more likely ways of entry. The observations of cytoplasmic extrusions after inoculation of tobacco leaves with several plant viruses [3] indicate that this mechanism is also applicable to plants.

4 Site of Coat Protein Synthesis/Virus Assembly

When cowpea protoplasts were inoculated with CCMV particles and subsequently incubated for 9 h under continuous light, several protoplasts could be found with gold label in both the vesicles and cytoplasm (Fig. 4). The label in the vesicles was attributed to the viral inoculum and the label in the cytoplasm to newly synthesized coat protein or virus particles. Final proof of protein synthesis rather than label bound to inoculum was obtained by inoculation with glutaraldehyde-fixed virus particles or with RNA [14]. With fixed virus, label was almost exclusively found in vesicles and with RNA, high concentrations of label were found in the cytoplasm 9 h after inoculation (Fig. 5). With RNA, as early as 6 h after inoculation, antigen could be detected in Lowicryl-embedded protoplasts. At t = 12 h after inoculation CCMV coat protein antigen could be detected in the entire cytoplasm. Cytoplasmic areas with higher labeling densities, possible infection sites, were still present. Quantification of labeling density in complete sections of three infected and three mock-inoculated protoplasts at 12 h was done for the different organelles and the cytoplasm. It showed specific labeling of the cytoplasm with ca. 65 protein A-gold particles per μm^2 compared to 4 for the control. With time, the labeling density of the cytoplasm increased up to more than 200 at 24 and 40 h after inoculation.

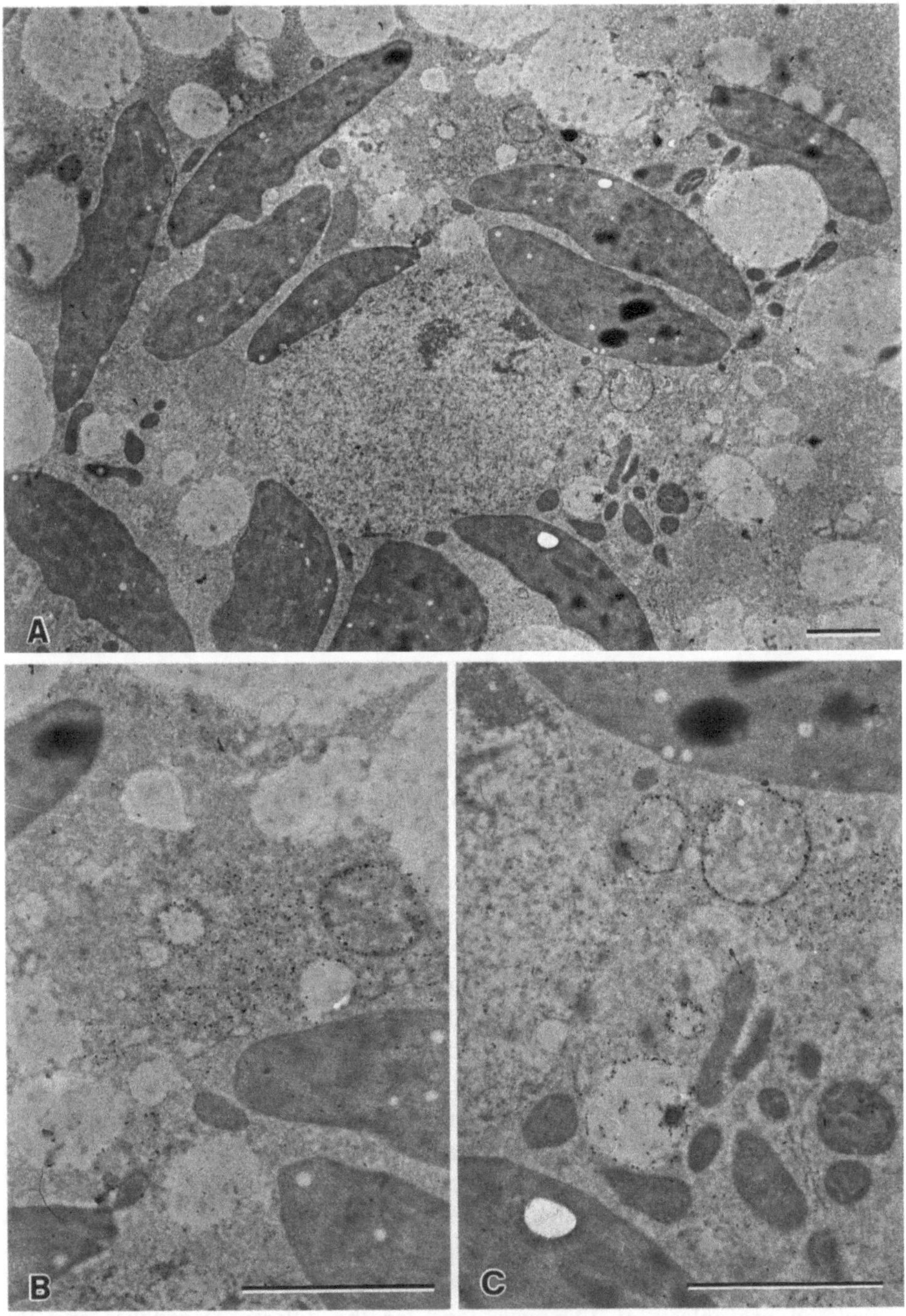

Fig. 4. Specific immunogold detection of cowpea chlorotic mottle virus antigen in cowpea protoplasts 9 h after inoculation with virus particles. **A** View of an inoculated and infected cell with nucleus, chloroplasts and vesicles. **B** and **C** Details of **A** showing specific labeling of inoculum virus particles in vesicles and newly synthesized antigen in cytoplasm. Bar = 200 nm

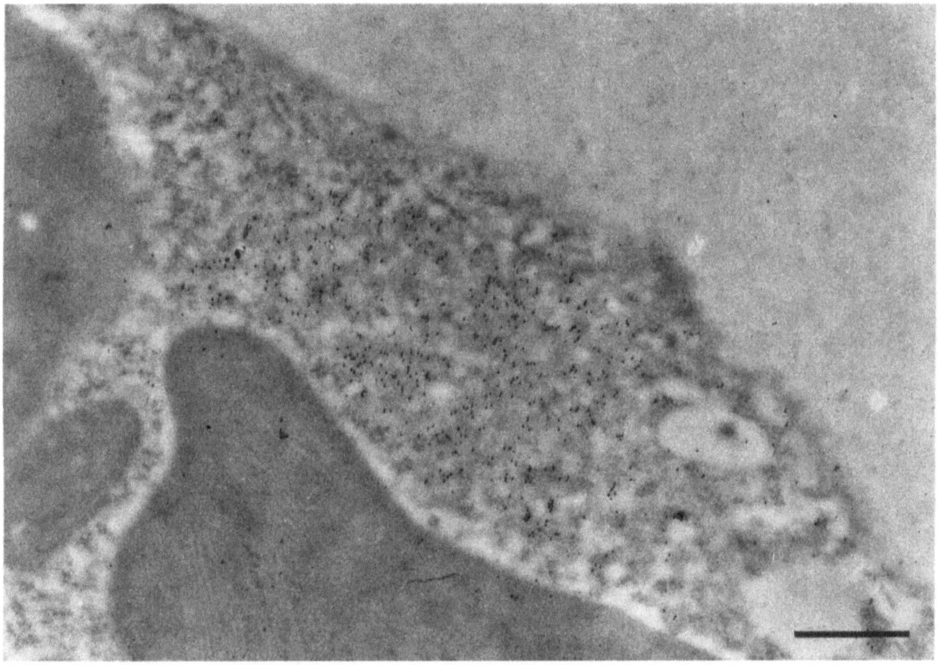

Fig. 5. Specific immunogold detection of cowpea chloroptic mottle virus antigen in cowpea protoplasts 9 h after inoculation with RNA. Label and, because of inoculation with RNA, newly synthesized antigen is only found in the cytoplasm. Bar = 0.5 μm

5 Translocation of CCMV Genome

After this localization in the cell we were interested in the spread of the virus. Therefore we developed the silver-enhancement technique [11, 12, 13, 16] to view a large number of cells at low magnification in the electron microscope and to be able to see label at the light microscope level. Figure 6 shows specific immunogold labeling with silver enhancement of CCMV coat protein antigen in infected petiolules of cowpea. At this low magnification infected and uninfected cells are easily distinguished. Tissue sections cytochemically stained with gold and silver were examined with epi-illumination and polarized light in a Leitz Orthoplan microscope (Fig. 7).

Spread of the virus was studied in systemically infected cowpea plants. Infection was caused by systemic inoculation using the differential temperature inoculation technique as described [1]. Briefly, the primary cowpea leaves, inoculated with CCMV, were kept in the light at 27–30 °C, and the secondary leaves were kept in the dark. After 3 days the plants were transferred to a growth cabinet at 25 °C and continuous light. The transfer was defined as time zero (t = 0) of systemic inoculation.

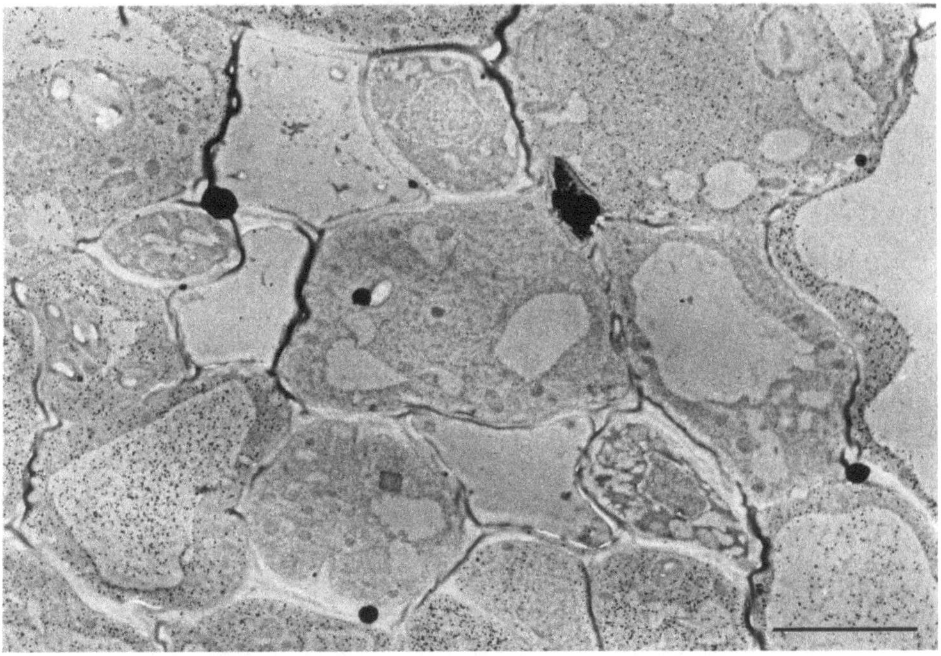

Fig. 6. Specific immunogold labeling with silver enhancement of cowpea chlorotic mottle virus in phloem of infected petiolules of cowpea. Infected and uninfected cells are easily distinguishable at lower magnification to monitor the spread of virus infection through plant tissue. Bar = 2 μm

Up to t = 6 h after systemic inoculation none or only small quantities of virus were detected in the secondary leaves by enzyme-linked immunosorbent assay (ELISA). Virus concentration increased rapidly between 6 and 24 h, indicating some synchrony of virus multiplication. CCMV coat protein antigen was detected by immunogold-silver staining and epipolarization microscopy in semi-thin sections of aldehyde fixed and London Resin White-embedded samples of the stems, petioles, petiolules, and leaves of systemically inoculated plants. At t = 0 the antigen could be detected only in the phloem and bundle sheath of stems and petioles, but not in the petiolules and leaves. In the petiolule, coat protein antigen was first detected in phloem parenchyma cells at 3 h after systemic inoculation. Figure 7A and B shows some labeled phloem parenchyma cells at 6 h. From t = 3 h onwards the antigen was detected progressively in the neighboring tissues, first in the bundle sheath, followed by the cortex and finally at t = 24 h also in the procambium and xylem (Fig. 7C, D). Similar patterns were observed in the main veins of secondary leaves and in leaf blade tissues, where coat protein antigen was first detected at t = 3 h in the phloem of minor veins and subsequently in the mesophyl and epidermal cells. CCMV is apparently transported through the phloem, following the routes of assimilate

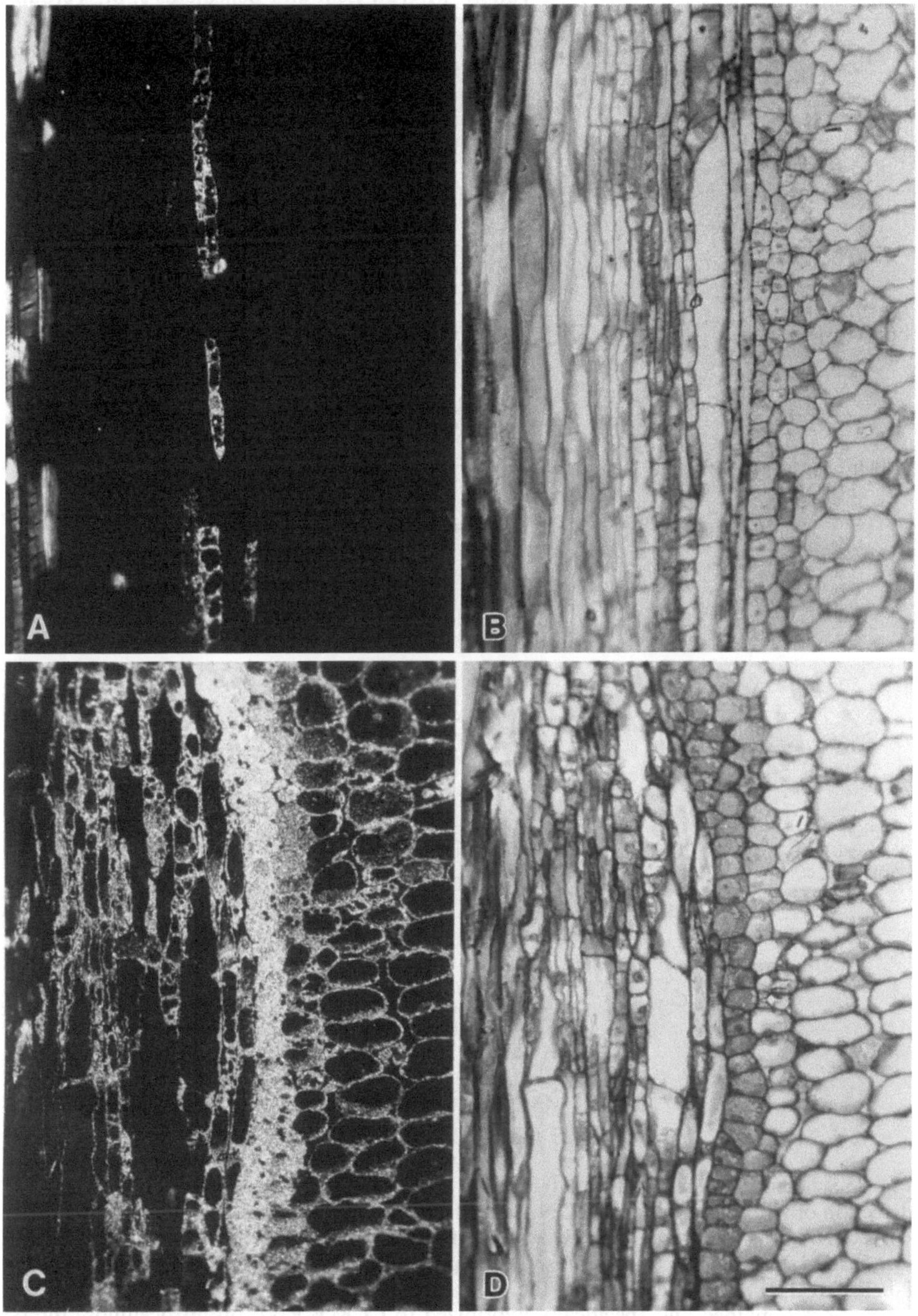

Fig. 7. Immnogold-silver staining of cowpea chlorotic mottle virus (CCMV) coat protein in systemically infected petiolules of cowpea plants. Semi-thin sections of London Resin White-embedded petiolules were immunolabeled with anti-CCMV-coat protein, further incubated with 7 nm gold particles adsorbed to protein A and finally processed in a physical developer. Label was visualized with epipolarization (**A, C**) and bright field transillumination (**B, D**) microscopy at t = 6 (**A, B**) and 24 h (**C, D**) after systemic inoculation. Bar = 20 μm

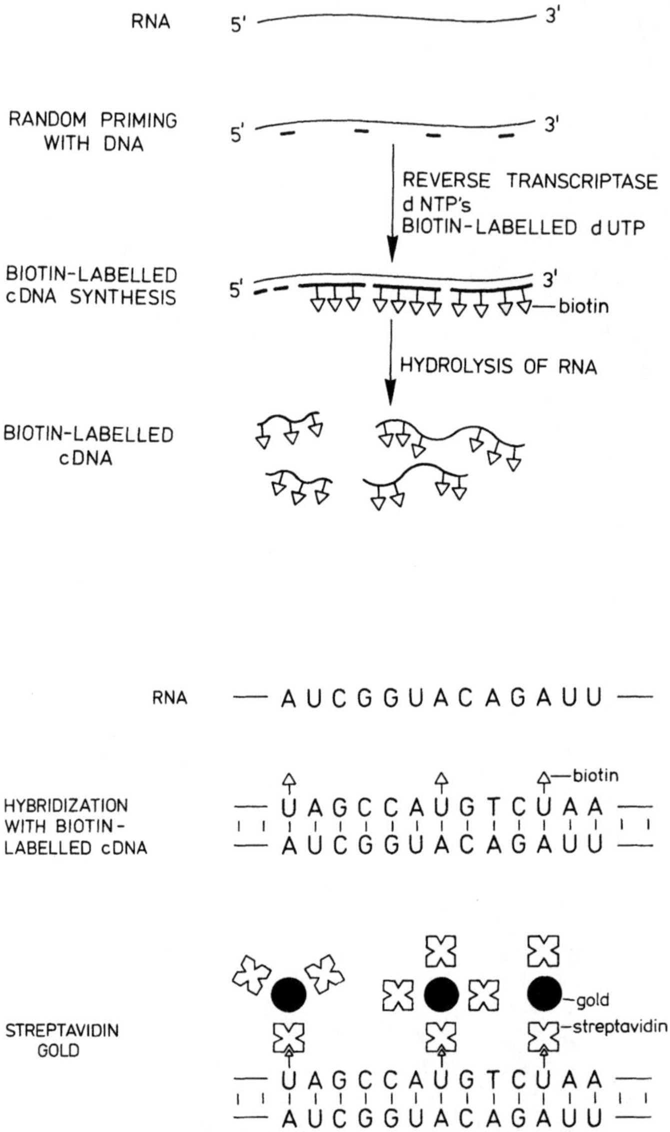

Fig. 8. Schematic representation of random synthesis of biotin-labeled cDNA (**A**) and detection of viral RNA with this cDNA and streptavidin gold (**B**). Cowpea chlorotic mottle virus RNA was randomly primed with calf thymus DNA fragments and cDNA was synthesized with reverse transcriptase in the presence of deoxyribonucleotides and biotin-labeled dUTP. After hydrolysis of the RNA biotin-labeled DNA fragments of 150–200 nucleotides remained and specifically hybridized to CCMV-RNA on nitrocellulose filters. Fixation, embedding, and sectioning of these filters was done prior to incubation with streptavidin gold. This label revealed the presence of biotin if CCMV-RNA was present on the filter

transport. No viral antigen was detected in mature sieve elements or companion cells at any time after inoculation, indicating that the virus is possibly transported in a form other than whole particles or its concentration is very low.

6 Detection of RNA

The viral genome may be transported over short or long distances in the form of free nucleic acid, or as nucleoprotein particle containing only a few coat protein subunits per genome. Therefore we tested in situ hybridization and immunogold labeling to detect and localize CCMV RNA. Figure 8 shows a schematic representation of random synthesis of biotin-labeled DNA complementary to CCMV-RNA (cDNA, A) and detection of viral RNA after hybridization with this cDNA and subsequent detection of these hybrids with streptavidin gold (B). Similar sections of London Resin White-embedded petiolules 24 h after systemic inoculation with CCMV were treated with antiserum and protein A-gold to label the coat protein antigen (Fig. 9A)

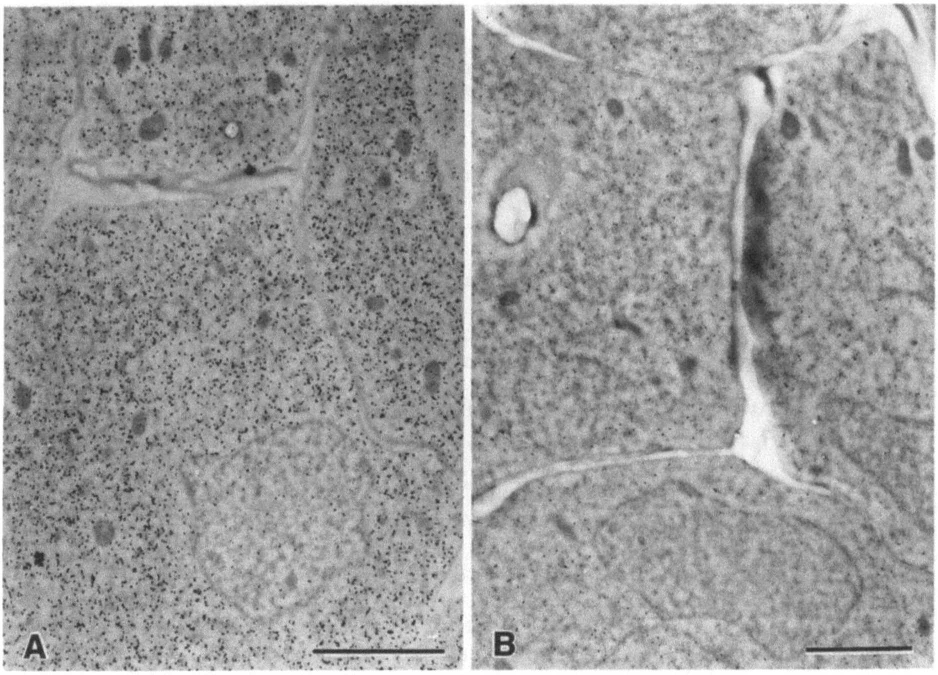

Fig. 9. Streptavidin gold labeling (**B**) of biotin-labeled DNA complementary to cowpea chlorotic mottle virus (CCMV) hybridized to thin sections of London Resin White-embedded petiolules 24 h after systemic inoculation with CCMV and compared to protein A-gold labeling with anti-CCMV-coat protein (**A**). Bar = 1 μm

or hybridized with biotin-labeled cDNA, followed by streptavidin gold (Fig. 9B). Both sections were enhanced with silver. Although there is specific labeling of CCMV-RNA in the sections the concentration of label is much lower than the labeling of coat protein antigen. Both probe preparation and hybrid detection need further improvement. Using riboprobe and antiserum against biotin we are trying to improve sensitivity.

Future research will focus on localization of the nonstructural proteins of CCMV in both protoplasts and plants. Antisera against synthetic peptides of the three nonstructural proteins have been raised and show specific activity in ELISA tests. The results so far obtained with brome mosaic virus (BMV) 1a and 2a protein localization in the plasma membrane of infected barley protoplasts is puzzling (Lawrence R. Griffing, Texas, pers. commun.) and immunogold localization has to be repeated for BMV in both protoplasts and plants including CCMV controls.

Acknowledgments. We should like to thank Hanke Bloksma for excellent technical assistance and the Netherlands Foundation for Fundamental Biological Research, which is subsidized by the Netherlands Organization for Scientific Research, for financial support.

References

1. Dawson WO, Schlegel DA (1976) Synchronization of cowpea chlorotic mottle virus replication in cowpeas. Intervirology 7: 284–291
2. Faulk WP, Taylor GM (1971) An immunocolloid method for the electron microscope. Immunochemistry 8: 1081–1083
3. Laidlaw WMR (1987) A new method for mechanical virus transmission and factors affecting its sensitivity. EPPO Bull 17: 81–89
4. Patterson S, Verduin BJM (1987) Immunogold labelling in animal and plant virology. Arch Virol 97: 1–26
5. Roenhorst JW, Verduin BJM (1987) Initial interactions between virus and protoplasts. Acta Bot Neerl 36: 109
6. Roenhorst JW, Verduin BJM (1988) Isolation and characterization of virus-ribosome complexes from cell-free translation systems supplemented with cowpea chlorotic mottle virus particles. Acta Bot Neerl 37: 539
7. Roenhorst JW, Van Lent JWM, Verduin BJM (1988) Binding of cowpea chlorotic mottle virus to cowpea protoplasts and relation of binding to virus entry and infection. Virology 164: 91–98
8. Roenhorst JW, Verduin BJM, Goldbach RW (1989) Virus-ribosome complexes from cell-free translation systems supplemented with cowpea chlorotic mottle virus particles. Virology 168: 138–146
9. Van Lent JWM, Verduin BJM (1985) Specific gold-labeling of antibodies bound to plant viruses in mixed suspensions. Neth J Plant Pathol 91: 205–213
10. Van Lent JWM, Verduin BJM (1986) Detection of viral protein and particles in thin sections of infected plant tissue using immunogold labelling. In: Jones RAC, Torrance L (eds) Developments in applied biology. 1. Developments and applications in virus testing. Association of Applied Biologists, Wellesbourne, UK, pp 193–211
11. Van Lent JWM, Verduin BJM (1987) Ultastructural localization of cowpea chlorotic mottle virus in cowpea plants and protoplasts by immunogold labelling. Proc R Microsc Soc 22: 84

12. Van Lent JWM, Verduin BJM (1987) Detection of viral antigen in semi-thin sections of plant tissue by immunogold-silver staining and light microscopy. Neth J Plant Pathol 93: 261–272
13. Van Lent JWM, Verduin BJM (1987) In situ detection of the translocation of virus in plants. Acta Bot Neerl 36: 327
14. Van Lent JWM (1988) Localization of viral antigens in leaf protoplasts and plants by immunogold labelling. PhD Thesis Agricultural University, Wageningen. pp 122
15. Verduin BJM (1974) The preparation of CCMV-protein in connection with its association into a spherical particle. FEBS Lett 45: 50–54
16. Verduin BJM, Van Lent JWM (1987) In situ detection of the translocation of virus in plants. Abstracts Seventh International Congress of Virology, Edmonton, Canada, pp 82
17. Wilson TMA (1985) Nucleocapsid disassembly and early gene expression by positive-strand RNA viruses. J Gen Virol 66: 1201–1207

Chapter 10 EL DESOUKY AMMAR

Mechanisms of Plant Virus Transmission by Homopteran Insects

1 Introduction

The great majority of plant viruses and all plant pathogenic mollicutes (*Spiroplasma* and mycoplasma-like organisms) depend on insects and other arthropods for their natural spread. About 76% of the insect vectors of plant viruses belong to the order Homoptera [15], mainly aphids (Aphididae), leafhoppers (Cicadellidae), planthoppers (Delphacidae), whiteflies (Aleyrodidiae), and mealybugs (Pseudococcidae). These insects have piercing and sucking mouthparts which are well suited for acquisition and inoculation of viruses and other "internal" plant pathogens during probing, feeding, and salivation activities on their host plants [6, 7].

In addition to its significant role in the discovery and characterization of plant viruses and mollicutes, electron microscopy (EM) has contributed much to our understanding of the mechanisms involved in insect transmission and vector specificity of these pathogens. Some of these mechanisms are now generally well understood (e.g., with propagative and circulative viruses), whereas others are much less clear (e.g., with nonpersistent and semi-persistent viruses). Several reviews have dealt with pathogen-vector relationships of homopteran insects [14, 15, 22, 26, 29, 34]. Here, I will first outline the main characteristics of various types or modes of virus transmission by these vectors, and the role of EM in the elucidation of transmission mechanisms in each case. Second, special problems involved in studying virus-vector relationships by EM, and suggested ways to overcome these problems, will be discussed.

2 The Role of EM in the Elucidation of Transmission Mechanisms

2.1 Propagative Transmission

Propagatively transmitted (*propagative*) plant viruses are those which are known to multiply in their vectors. After being ingested from infected plants, these viruses pass

Department of Economic Entomology, Faculty of Agriculture, Cairo University, Giza, Egypt through the vector's gut wall into the hemolymph. Following a latent period, during which virus multiplication and transport takes place in several tissues of the vector, virus particles are discharged with the salivary secretions during feeding and salivation into the host plants. Propagative viruses require longer latent periods (days-weeks) compared to circulative viruses (hours); they persist in their vectors usually for life, and some are vertically (transovarially) transmitted through the female to subsequent generations [26, 34]. At least 48 plant viruses belonging to four taxonomic groups are propagative in their homopteran vectors: 3 marafiviruses transmitted by leafhoppers, 5 tenuiviruses by planthoppers, 13 reoviruses by either leafhoppers or planthoppers, and 27 rhabdoviruses by aphids, leafhoppers, or planthoppers [26, 34].

Multiplication of these viruses in their vectors has been proven by several methods including transovarial transmission, serial injection, quantitative serology, and transmission EM (TEM). Additionally, TEM has played a major role in studying the multiplication sites and cytopathology in vectors of several reoviruses, rhabdoviruses and marafiviruses [14, 26, 34]. Large accumulations of reovirus-like particles (icosahedral, 65–75 nm in diameter) were found in the cytoplasm of fat body, malpighian tubules, hypodermis, tracheoblasts, muscle, mycetome, gut epithelium, hemocytes, and all lobes of the salivary glands of *Agallia constricta*, the most efficient vector of wound tumor virus, whereas fewer virus accumulations were found in the inefficient vector *Agalliopsis novella* [14, 32]. With another reovirus, rice dwarf virus (RDV), high concentrations of virus particles were found in almost all organs of the vector *Nephotettix cincticeps* [32]. Intracytoplasmic viroplasms, which are similar to the "virus factories" containing mature and immature progeny virions in cells infected with vertebrate reoviruses, occurred in insect and plant cells infected with plant reoviruses [14, 32, 33]. The viroplasm consists of electron-dense granular or fine fibrillar material and has no surrounding membranes. Shikata [32] indicated that the viroplasms caused by viruses belonging to the planthopper-borne subgroup of the reoviruses (Fijiviruses) occupy large areas of infected cells, are amorphous, and spread within the cytoplasm; whereas those caused by the leafhopper-borne subgroup (phytoreoviruses) are smaller and are more or less spherical. With maize rough dwarf virus and other Fijiviruses, immature smaller particles (probably inner capsids) are embedded within the viroplasms, whereas the larger—apparently complete—virions accumulate at the periphery of the viroplasm [32]. Autoradiography and immuno-EM with ferritin labeling indicated that the viroplasms formed in vector cell monolayers of *N. cincticeps* infected with RDV are the sites where the viral RNAs and proteins accumulate [14].

Electron microscopy has also provided the only evidence so far obtained on a possible mechanism for transovarial transmission of a plant virus. Nasu [25] reported that RDV particles were found to enter oocytes of the vector from mycetocytes of the ovarioles. Virions in these mycetocytes appear to be selectively attached to the surface membrane of the L-symbiotes present in the cytoplasm of the mycetocytes, which migrate to the adjacent oocytes. Other mechanisms for infecting

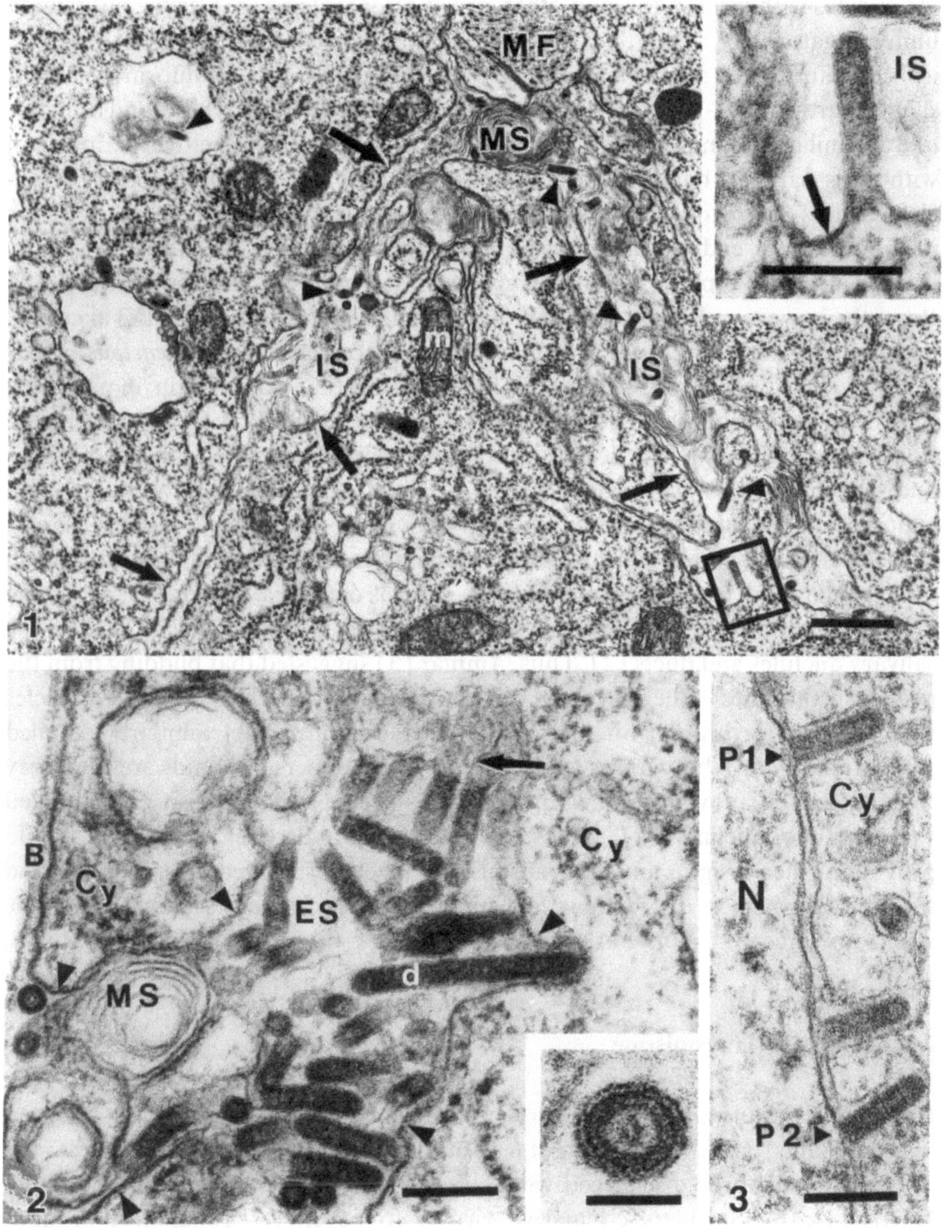

Figs. 1–3. Assembly sites of rhabdovirus particles in tissues of the planthopper *Peregrinus maidis* infected with maize mosaic virus. **1, 2** Particles budding through plasma membranes (*arrows* and *arrowheads*) in cells of the principal salivary gland, and accumulating in intercellular (*IS*) or extracellular (*ES*) spaces. *Boxed area* in **1** is enlarged in the *inset*; in **2** the *inset* shows cross-section of a virus particle. **3** Particles budding through the inner nuclear membrane (at *P1* and *P2*) in a brain cell. *B* basal lamina; *Cy* cytoplasm; *d* double-length particle; *MF* myofibrils; *MS* multilaminar structures; *N* nucleus. Bars: **1** 500 nm and 200 nm (*inset*); **2** 200 nm and 50 nm (*inset*); **3** 200 nm [5]

the oocytes with virus in this and other virus-vector systems are feasible but remain uninvestigated.

Ultrastructural studies on several plant rhabdoviruses (bacilliform or bullet-shaped particles, 45–100 × 100–430 nm) suggested that most assemble and accumulate at similar sites in plant and vector cells [17]. These sites are either perinuclear, with virus particles budding on nuclear membranes and accumulating in perinuclear spaces, or cytoplasmic with virus particles budding through cytoplasmic membranes (e.g., endoplasmic reticulum). In a third category, particles apparently budded from cytoplasmic viroplasms into surrounding membranes [11, 32]. Recently, however, ultrastructural studies on maize mosaic virus (MMV) in cells of maize plants and in various organs of the planthopper vector *Peregrinus maidis* (Figs. 1–3) have revealed differences in budding sites that may explain, in part, how this rhabdovirus is transmitted by its vector. In plant cells and in most of the vector tissues examined, MMV particles budded from nuclear and cytoplasmic membranes and accumulated in perinuclear spaces or in dilated cisternae connected with endoplasmic reticulum. However, in secretory cells of the principal salivary glands, MMV particles budded mainly from the plasma membranes and accumulated in intercellular and extracellular spaces of various acini [3, 5]. These spaces are apparently connected with extracellular vacuoles and canaliculi that lead to the salivary ductules and ducts [2]. Thus, Ammar [3] suggested that budding from the plasma membranes in the salivary glands allows efficient discharge of virus particles with the saliva, through an intercellular-extracellular route. Similarly detailed studies on other rhabdoviruses transmitted by leafhoppers or aphids are necessary to compare the above results with other virus-vector systems. Also, EM coupled with immuno-labeling [31], vector cell lines [21], and other techniques should provide further insights on the transmission mechanisms, vector specificity, and transmission barriers of propagative plant viruses in their vectors, as has been done with some mosquito-borne arboviruses [13].

2.2 Circulative Transmission

Circulatively transmitted (*circulative*) viruses are ingested by their vectors from infected plants, pass through the gut wall into the hemolymph, and then pass to the salivary glands to be discharged with salivary secretions of the vector. Although Harris [14, 15] considers circulative viruses to be either propagative or nonpropagative in their vectors, both Sylvester [34] and Nault and Ammar [26] use the term circulative only for nonpropagative viruses that require a latent period in their vectors. Under the latter definition, circulative viruses include the aphid-borne pea enation mosaic virus (PEMV) and luteoviruses, in addition to the geminiviruses transmitted by leafhoppers, whiteflies, aphids, or treehoppers. However, EM studies on transmission mechanisms have been done only on PEMV and the barley yellow dwarf virus (BYDV) complex.

With PEMV, viruslike particles were found in the gut epithelium, connective tissue, fat body, and the accessory—but not the principal—salivary glands of vector aphids [14]. Based on a series of EM studies on BYDV, Gildow [12] proposed a route for this and other luteoviruses in their aphid vectors, which also revealed some possible mechanisms for vector specificity. Ultrastructural studies, supported by ferritin labeling, on the vector *Rhopalosiphum padi* indicated that BYDV particles were specifically associated with the apical plasmalemma of the hindgut. Most particles were observed individually in shallow pits and in coated vesicles near the plasmalemma. The results suggested that receptor-mediated endocytosis in the hindgut may be involved in luteovirus acquisition by aphid vectors. Furthermore, in the salivary glands (SG), BYDV particles have been observed associated only with the simple accessory SG, rather than the larger and more complex principal SG. It was suggested that BYDV particles suspended in the hemolymph diffuse through the fibrous basal lamina of the accessory SG and come into contact with the basal plasmalemma. Virus-specific receptors may then bind the virus to the membrane, and the virus may be endocytosed within invaginations which bud off the plasmalemma to form tubular vesicles. Virions in these vesicles are apparently packaged in a specific manner in coated vesicles for transport to the apical plasmalemma and exocytosed into the salivary canal. Earlier work had shown that vector-specific transmission of BYDV isolates occurred even when purified virus preparations were injected into the hemocoel [30]. Thus, at least in some luteovirus-aphid combinations the site preventing transmission of nonvectored isolates is not associated with the alimentary canal, but rather with specific receptors in the accessory SG [12]. However, it is not known whether the above findings are true for other circulative viruses, particularly the geminiviruses, for which no detailed TEM studies in their vectors have been reported. Among this group are several economically important viruses, e.g., beet curly top, maize streak, and tomato yellow leaf curl viruses [15].

2.3 Semi-Persistent Transmission

Semi-persistently transmitted (*semi-persistent*) viruses have no latent period in their vectors, cannot be recovered from vector hemolymph, and are lost during ecdysis of the vector. Nonpersistent viruses share these characteristics but persist in their vectors only for a few seconds to 1 h, whereas semi-persistent viruses are retained by their vectors for some hours to a few days [24, 26]. TEM has been used to study the retention sites of three semi-persistent viruses (with isometric particles, ca. 30 nm diameter) in their aphid or leafhopper vectors. Murant et al. [23] found viruslike particles (VLP) in the foregut of the aphid *Cavariella aegopodii* carrying anthriscus yellows virus (AYV) alone or with parsnip yellow fleck virus (PYFV). The VLP, thought to be those of AYV, which acts as a helper for aphid transmission of PYFV, were associated with a 15–20 μm-long portion of the cuticular lining of the pharynx

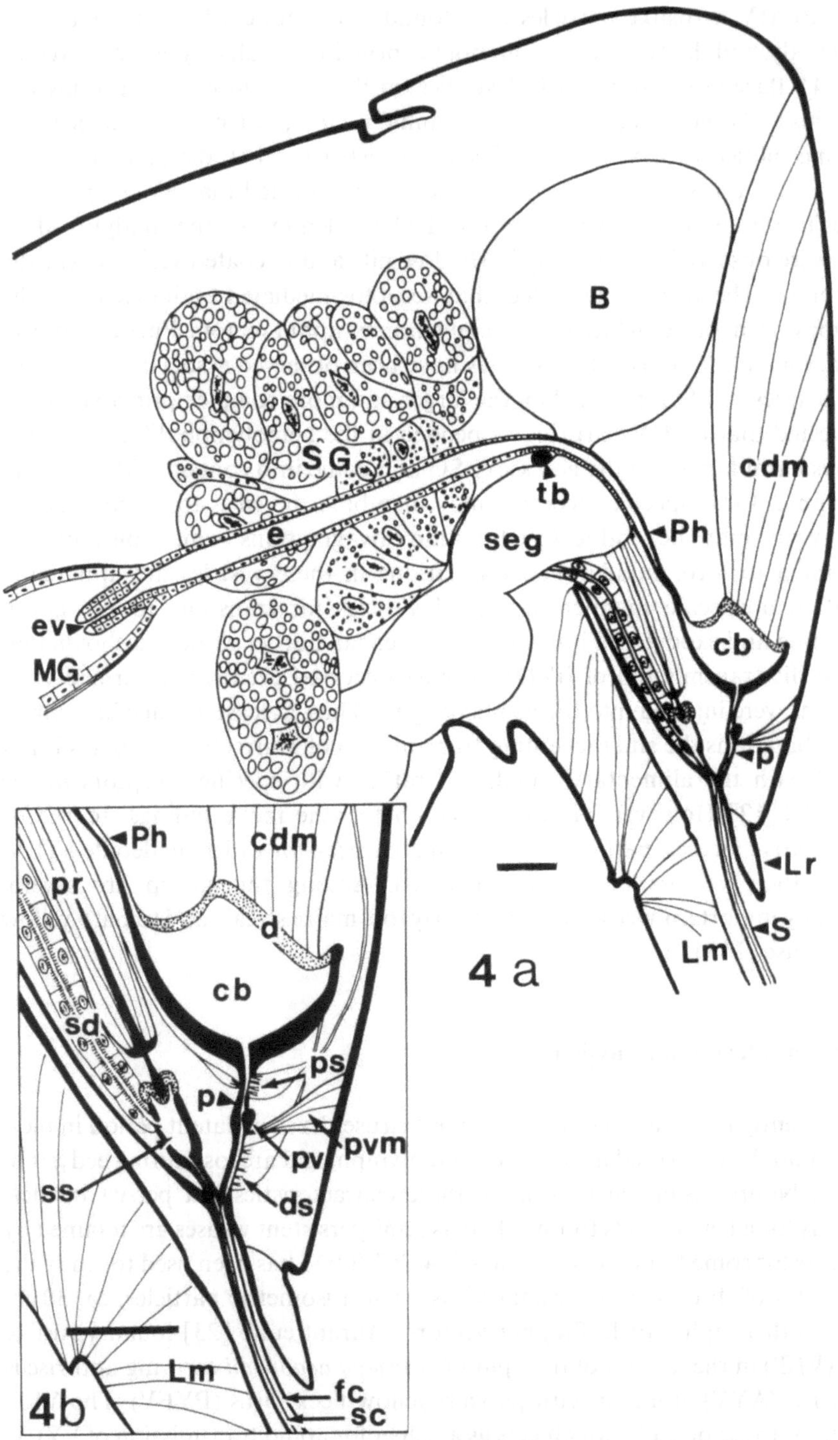
B
SG
cdm
tb
e
seg
Ph
ev
MG
cb
p
Lr
S
Lm
4 a
Ph
cdm
pr
d
cb
sd
ps
p
pvm
pv
ss
ds
Lm
fc
sc
4b

where it passes over the tentorial bar. These VLP were surrounded by densely staining material which was embedded in a matrix of lightly staining material.

A second study [9, 10] was on maize chlorotic dwarf virus (MCDV) in its leafhopper vector *Graminella nigrifrons* (Fig. 4). Numerous retention sites were reported in virus-exposed vector leafhoppers but not in nonvirus-exposed vector controls or virus-exposed nonvector leafhoppers (*Dalbulus maidis*). However, in a recent study by Ammar, Nault and others [1, 26, 27 and unpubl.], MCDV-like particles were found embedded in a dense matrix (Figs. 5–8) attached to the cuticular lining of parts of the foregut and maxillary food canal in the vector *G. nigrifrons* and in the nonvector leafhopper *D. maidis*, but not in the nonvector delphacid planthopper *P. maidis*, either immediately or 4–5 h after removal of these insects from infected plants. These results suggest that attachment of virus to the foregut cuticle may be only partly responsible for vector specificity of MCDV. Other factors that could be involved are: specificity of detachment of virus from its retention sites, the presence of virus inhibitors in the salivary secretions of nonvectors or virus activators or protectants in vectors, or the involvement of helper components. Apart from the last possibility, evidence for which was provided by Hunt et al. [16], the other options remain largely uninvestigated. Additionally, no TEM studies on the vectors have been reported for rice tungro spherical or bacilliform viruses, which have a helper-dependent relationship in their leafhopper vectors [26], or the closteroviruses (with filamentous particles), many of which are transmitted by aphids [24].

2.4 Nonpersistent Transmission

The great majority of aphid-borne viruses, and more than half the viruses vectored by insects, are nonpersistently transmitted. These (*nonpersistent*) viruses are acquired by the vector in a few seconds to a few minutes of feeding on infected plants; the vector can immediately inoculate them into healthy plants, but rarely remains able to do so for longer than an hour following acquisition [24].

The role of EM in the elucidation of transmission mechanism(s) of these viruses has been limited so far, probably because: (a) the retention period in the vector is very brief, and (b) most of these viruses have narrow filamentous particles (10–12 nm

◀

Fig. 4 a, b. Drawings made from light micrographs of semi-thin sections of the head and prothorax of the leafhopper vector *Graminella nigrifrons*, showing sagittal view of the anterior alimentary canal and salivary system. **b** Details of the precibarium and salivary syringe areas. *B* brain; *cb* cibarium (sucking pump); *cdm* cibarial dilator muscle; *d* cibarial diaphragm; *ds* distal precibarial sensilla; *e* esophagus; *ev* esophageal valve; *fc* food canal; *Lm* labium; *Lr* labrum; *MG* midgut; *p* precibarium; *Ph* pharynx; *pr* piston retractor muscle; *ps* proximal precibarial sensilla; *pv* precibarial valve; *pvm* precibarial valve muscle; *S* stylets; *sc* salivary canal; *sd* salivary duct; *seg* subesophageal ganglion; *Sg* salivary gland; *ss* salivary syringe; *tb* tentorial bar. Bars = 50 μm. [26]

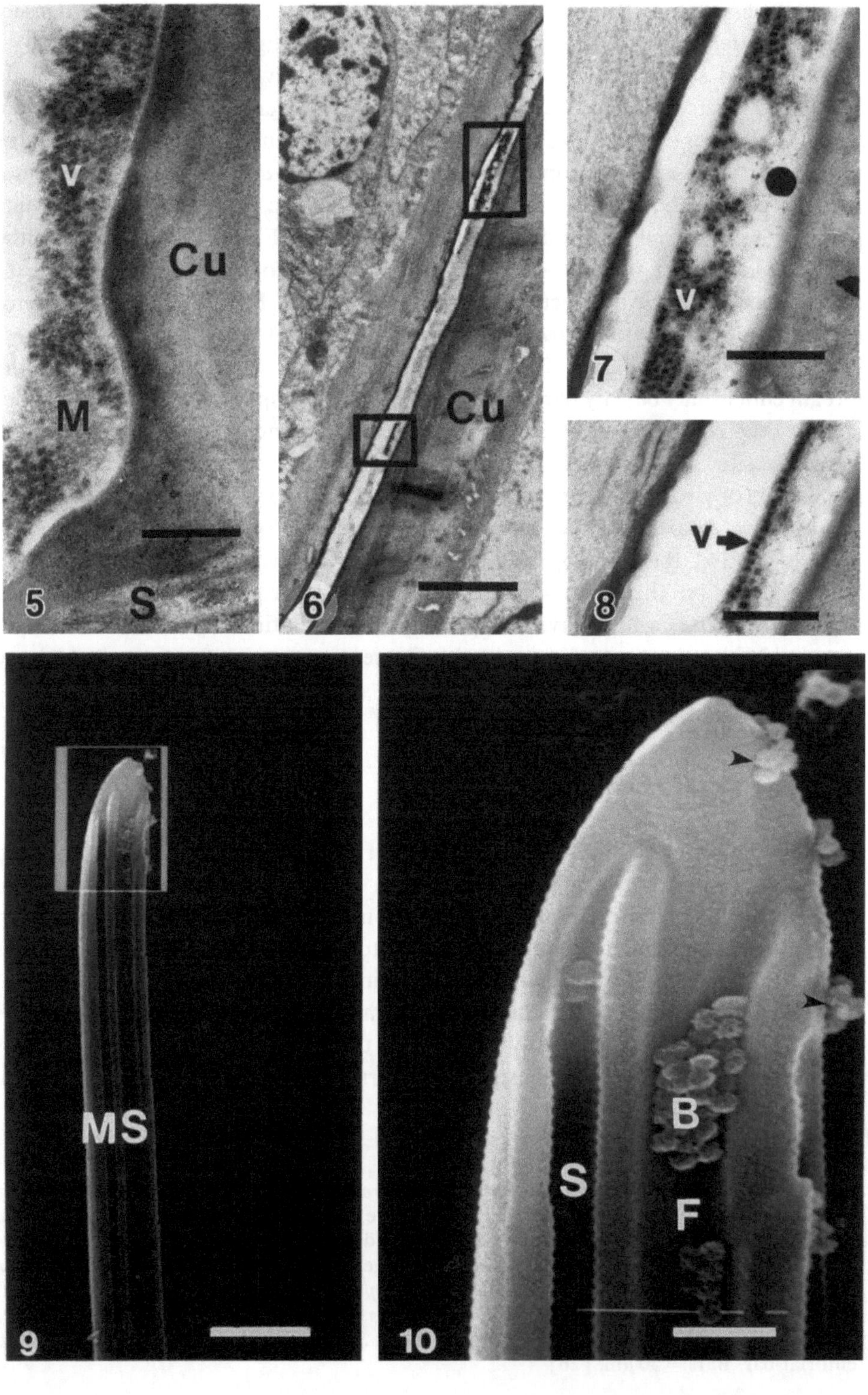
v
Cu
M
5
S
6
Cu
7
v
8
v
9
MS
10
B
S
F

in diameter) which are difficult to detect in the vector by EM. The only report of a filamentous virus found in the vector by EM was that by Taylor and Robertson [35], who found very few potyvirus-like particles associated with the distal 20 μm of the maxillae of *Myzus persicae* carrying tobacco etch virus. Lim et al. [18], on the other hand, used labeling with antibody-sensitized latex to detect a potyvirus on aphid stylets by scanning EM. These studies, in addition to those using virus-inhibition treatments (reviewed in [24]), suggested that nonpersistent viruses are retained on the aphid stylets; this is consistent with the term stylet-borne previously used for these viruses. More recently, Berger and Pirone [8] used light-microscope autoradiography to detect retention sites of ^{125}I-labeled virions (or antigens) of two potyviruses. In the presence of helper component (HC), label was associated with the maxillary stylets, the foregut and the midgut, whereas in the absence of HC, label accumulated only in the midgut. These results implied a binding mechanism for the HC and suggested that nonpersistent viruses are retained both in the aphid stylets and in the foregut.

Additionally, TEM is playing an increasing role in studying the feeding behavior of aphids in relation to nonpersistent transmission of viruses. Until recently, studies on superficial feeding probes were conducted by viewing aphids or their stylet paths by eye, hand lens, or light microscopy [20]. These studies led to the idea that transmission of nonpersistent viruses is an *inter-epidermal* event. However, using EM, Lopez-Abella and Bradley [19] demonstrated that stylet paths sometimes enter epidermal cells during brief probes. Later, it was found that only aphids that made probes in which the stylets came into contact with the epidermal cytoplasm acquired and transmitted nonpersistent viruses [20]. Fasted aphids, which are more efficient in transmission than nonfasted ones, make one or more brief probes (15–59 s) within the epidermis, whereas nonfasted aphids generally make longer probes that penetrate past the epidermis to the phloem.

Much remains to be done to understand the transmission mechanisms and vector specificity of nonpersistent and semipersistent viruses. Electron microscopy, in conjunction with immunolabeling and other techniques, will be essential for revealing aspects of these mechanisms, particularly the specific retention sites in the

Figs. 5–8. Apparent retention sites of the semi-persistently transmitted maize chlorotic dwarf virus in the foregut of the leafhopper vector *Graminella nigrifrons* (cf Fig. 4). Viruslike particles (*v*) embedded in a dense matrix (*M*) attached to the cuticular lining of the precibarium. **6–8** matrix-embedded particles (*v*) in the pharynx; *boxed areas* in **6** are enlarged in **7** and **8**. *Cu* cuticle; *S* precibarial sensillum. Bars: **5**, **7** and **8** 0.3 μm; **6** 2 μm. [26]

Figs. 9 and **10.** Scanning electron micrographs of an inner surface of the distal part of a maxillary stylet (*MS*) of the vector leafhopper *G. nigrifrons*, showing the food canal (*F*) and salivary canal (*S*) *Boxed area* in **9** is enlarged in **10**. The spherical bodies (*B* and *arrowheads*) contaminating the stylet are brochosomes, an excretory product usually found on the cuticle of leafhoppers [1]. Bars: **9** 8 μm, **10** 1.6 μm. (Ammar E.D. and Styer W.E. unpubl.)

vector, and the events that take place at these sites between viruses, helper components—if present—and the insect cuticle (or binding sites), during the acquisition, retention, and inoculation phases of transmission.

3 Studies on Insect Structures Involved in Transmission

Scanning and transmission EM, in addition to light microscopy, have been used in studying insect organs, tissues and structures involved in the transmission of viruses or mollicutes. In several cases, these studies have been crucial in understanding transmission mechanisms of these pathogens. I have already mentioned that studying the ultrastructure of the salivary glands of *P. maidis* [2] has led to a proposed mechanism for the multiplication, accumulation and transport of maize mosaic virus in its vector [3]. Similarly, studying the ultrastructure of aphid salivary glands has led to a model for vector specificity and transmission mechanism of luteoviruses [12]. Other studies on the structure of homopteran stylets, alimentary canal, and salivary systems [1, 6, 28 and (Figs. 4, 9, 10)] are important in understanding certain aspects of virus transmission by these insects.

Recently, in a series of studies by scannining and transmission EM, Backus, McLean and others (reviewed in [6] and [7]) have elucidated the sensory mechanisms associated with the feeding behavior of leafhoppers and aphids. These studies revealed the presence and structure of the precibarial valve and precibarial sensilla (Fig. 4) and led to a proposed theory for the feeding mechanism of homopteran insects that might explain several aspects of nonpersistent and semi-persistent transmission of plant viruses [6, 7, 20, 26]. Future studies, including light and electron microscopy, electronic monitoring, radiolabeling, and infectivity assays, should be useful in following the feeding behavior of homopteran insects in relation to virus acquisition and inoculation.

4 Special Problems in Studying Virus-Vector Relations by EM

4.1 Fixation and Embedding

One of the most important factors affecting the success or failure of fixation and embedding in these studies is whether *insect bodies, parts or organs* are being processed. With smaller homoptera, e.g., whiteflies and some aphids, whole insects may be processed, preferably after severing their legs and wings, or making some punctures in the cuticle to help penetration of fixatives and embedding media. Advantages of this method are convenience (no need to dissect tiny organs or parts), and conservation of the integrity and relationships of various organs (e.g., Figs. 4.2–

4.4 in [12], for sections of whole aphids). However, with larger homoptera, e.g., leafhoppers and planthoppers, penetration of fixatives and/or embedding media into whole insects is usually poor [9]; Ammar unpub.). In such cases, preliminary fixation of whole insects in 2–3% glutaraldehyde, followed by cutting the insect transversely into two or three parts before further processing, may improve fixation and embedding [23]. Alternatively, insect organs or parts needed for a particular study may be dissected out in buffer or in glutaraldehyde, following anesthetization of the insect with CO_2 [1, 2]. However, considerable experience is required to dissect such tiny organs or parts, to identify them, and to understand their structural relationship with other parts. A case in point is dissection of the foregut of the leafhopper *G. nigrifrons*, which apparently led to erroneous drawings and nomenclature of various parts (Fig. 3b in [9]), as compared to results obtained by sectioning of the whole head and prothorax of the same species (Fig. 4).

Another problem usually encountered in studying virus vectors by EM is separation of the insect cuticle from the embedding medium during sectioning and/or viewing. In most cases, this problem may be overcome by one or a combination of the following: (a) using a harder mixture of a low-viscosity embedding medium, (b) infiltration under negative pressure [20], (c) using a diamond or a sharper glass knife for sectioning, (d) using coated, fine (300–400 mesh) grids rather than uncoated or coarser grids, and (e) gradual exposure of sections to the electron beam during viewing.

To process a large number of samples, particularly of small insects or insect parts, capped porous teflon (or perforated stainless steel) capsules, fitted to an automatically rotated processor may be used (R. E. Whitmoyer pers. commun.). This minimizes the handling of such tiny specimens, and facilitates their transfer through various reagents undisturbed and hopefully undamaged. This procedure made possible the fixation and embedding of dissected intact salivary glands of the planthopper *P. maidis*, that consist of several tiny lobes (acini) which otherwise might have been broken and thus difficult or impossible to identify [2].

4.2 Orientation and Sectioning

To make longitudinal or cross-sections in insects or insect parts, a certain orientation of the specimen is desired before sectioning. In my experience, the best way to achieve this is to start with flat embedding of the specimens in shallow vials or caps, followed by cutting these specimens and glueing them to stubs in the desired orientation. Unnecessary parts of the specimen may be trimmed out, and identification of the desired structures or organs is aided by examination of semi-thin sections (1–2 μm thick), stained with toluidine blue or other stains, and examined by light microscopy. Extreme care is sometimes necessary in order not to lose some of the desired parts in this process, e.g., the pharynx in leafhoppers is only a few microns wide in some areas (Fig. 4) and it might be lost if more than a few semi-thin sections were cut in these areas. Although using fine grids (300–400 mesh) provides greater stability of the specimen under the electron beam, as indicated earlier, coarser (100–200 mesh) grids coated with formvar-carbon may be more suitable for viewing larger areas. This is particularly useful if lower magnification pictures are desired for survey or identification purposes, and if accumulations of virus particles can be detected at such lower magnifications (Figs. 1, 5–8).

4.3 Identification of Virus and of Insect Structures

A knowledge of the morphology, and if possible the cytopathology, of the studied virus (or virus group) is usually necessary for successful localization of this virus in various tissues of the insect vector. Also, since insect tissues are usually more varied than those of plants, it is important to understand the morphology and ultrastructure of the insect organs and tissues involved, at least for proper identification of these parts. Similar parts from nonviruliferous control insects (unexposed to the studied virus) must be examined as thoroughly as those of viruliferous (or virus-exposed) ones. Preferably the age, sex, nutrition, and other variables of the viruliferous and control insects should be comparable, so that the cytopathology of the virus may be distinguished from normal ultrastructure of the same cell types under similar physiological conditions. Descriptions of the internal morphology of leafhoppers and planthoppers [1] and of aphids [28] are available, but they are by no means complete, particularly at the ultrastructural level. Much remains to be studied in the morphology and physiology of the homopteran salivary glands and other structures involved in transmission. As mentioned earlier, a detailed knowledge of these structures is sometimes necessary to understand transmission mechanisms and vector specificity [1–3, 12].

In studying plant viruses and other pathogens in their insect vectors, one must not forget that these vectors may harbor their own (nonplant pathogenic) viruses or other microorganisms. Several viruses, bacteria, rickettsiae, and yeasts have been reported as pathogens or symbiotes of homopteran insects [1, 22, 25, 29]. Thus, one must be reasonably sure about the identification of the plant virus in question, by thorough examination of control insects and/or by immunolabeling. Gildow [12] described two different procedures for ferritin labeling of BYDV isolates in aphids salivary glands. The first involved incubating dissected salivary glands in IgG fraction of rabbit antibodies made against BYDV, and later incubating them in ferritin-conjugated, goat-antirabbit IgG. In the second method, the tedious dissection procedure, as well as the loss of ultrastructural detail resulting from tissue incubation in buffer, were avoided. Instead, live aphids were microinjected with rabbit IgG made against BYDV, allowed to feed overnight, then injected with ferritin-conjugated, goat-antirabbit IgG. In both cases aphids were later processed for TEM as usual. However, gold labeling of thin sections, which has several advantages over ferritin labeling, and has been used successfully to detect several viruses in plant and animal tissues [31], holds much promise for detection and identification of plant or insect viruses in the tissues of their insect vectors or hosts. This is particularly necessary where viruses of similar morphology may be involved, or with viruses whose morphology makes them difficult to detect by EM in the vector, e.g., the tenuiviruses, which are associated with very thin (3–8 nm) filamentous particles [3, 26].

Acknowledgments. The author undertook the literature research and electron microscopy work included here while at the Ohio Agricultural Research and Development Center (OARDC), The Ohio State University (Wooster, OH 44691, USA), as a Visiting Professor and Research Scientist. Thanks are due to L.R. Nault (OARDC), F.E. Gildow (Pennsylvania State University, USA) and R.G. Milne (Torino, Italy) for their helpful comments on the manuscript, and to W.E. Styer, R.E. Whitmoyer, and staff of the EM Laboratory at the OARDC for their valuable help in preparing specimens for electron microscopy.

References

1. Ammar ED (1985) Internal morphology and ultrastructure of leafhoppers and planthoppers. In: Nault LR, Rodriguez JG (eds) The leafhoppers and planthoppers. Wiley, New York, p 127
2. Ammar ED (1986) Ultrastructure of the salivary glands of the planthopper *Peregrinus maidis* (Hom., Delphacidae). Int J Insect Morphol Embryol 15: 417–428
3. Ammar ED (1987) Ultrastructural studies on the planthopper, *Peregrinus maidis* (Ashmead), vector of maize mosaic and maize stripe viruses. In: Wilson MR, Nault LR (eds) Proc 2nd Int Workshop Leafhoppers Planthoppers Econo Import. Commonw. Inst Entomol, London, p 83
4. Ammar ED, Gordon DT, Nault LR (1987) Ultrastructure of maize chlorotic dwarf virus-infected maize and viruliferous leafhopper vectors. Phytopathology 77: 1743
5. Ammar ED, Nault LR (1985) Assembly and accumulation sites of maize mosaic virus in its planthopper vector. Intervirology 24: 33–41
6. Backus EA (1985) Anatomical and sensory mechanisms of planthopper and leafhopper feeding behavior. In: Nault LR, Rodrigues JG (eds) The leafhoppers and planthoppers. Wiley, New York, p 163
7. Backus EA (1988) Sensory mechanisms and behaviors which mediate hemipteran plant-feeding: a taxonomic overview. J Insect Physiol 34: 151–165
8. Berger PH, Pirone TP (1986) The effect of helper component on the uptake and localization of potyviruses in *Myzus persicae*. Virology 153: 256–261
9. Childress SA (1980) The fate of maize chlorotic dwarf virus (MCDV) in the black-faced leafhopper, *Graminella nigrifrons* (Forbes) (Homoptera, Cicadellidae). Ph.D. Thesis. Texas A & M University, Texas
10. Childress SA, Harris KF (1989) Localization of virus-like particles in the foreguts of viruliferous *Graminella nigrifrons* carrying the semi-persistent maize chlorotic dwarf virus. J Gen Virol 70: 247–251
11. Conti M, Appiano A (1973) Barley yellow striate mosaic virus and associated viroplasms in barley cells. J Gen Virol 21: 315–322
12. Gildow FE (1987) Virus-membrane interactions involved in circulative transmission of luteoviruses by aphids. In: Harris KF (ed) Current topics in vector-research, Vol IV. Springer, Berlin Heidelberg, New York Tokyo, p 93
13. Hardy JL (1988) Susceptibility and resistance of vector mosquitos. In: Monath TP (ed) The arboviruses: epidemiology and ecology, Vol I. CRC, Boca Raton, Fla, p 87

14. Harris KF (1979) Leafhoppers and aphids as biological vectors: vector-virus relationships. In: Maramorosch K, Harris KF (eds) Leafhopper vectors and plant disease agents. Academic Press, New York, p 217
15. Harris KF (1981) Arthropod and nematode vectors of plant viruses. Annu Rev Phytopathol 19: 391–426
16. Hunt RE, Nault LR, Gingery RE (1988) Evidence for infectivity of maize chlorotic dwarf virus and a helper component in its leafhopper transmission. Phytopathology 78: 499–504
17. Jackson AO, Francki RIB, Zuidema D (1987) Biology, structure and replication of plant rhabdoviruses. In: Wagner RR (ed) The rhabdoviruses. Plenum, New York, p 427
18. Lim WL, De Zoeten GA, Hagedorn DJ (1977) Scanning electronmicroscopic evidence for attachment of a nonpersistently transmitted virus to its vector's stylets. Virology 79: 121
19. Lopez-Abella D, Bradley RHE (1969) Aphids may not acquire and transmit stylet-borne viruses while probing intercellularly. Virology 39: 338–342
20. Lopez-Abella D, Bradley RHE, Harris KF (1988) Correlation between stylet paths made during superficial probing and the ability of aphids to transmit nonpersistent viruses. Adv Dis Vector Res 5: 251–285
21. Maramorosch K (1979) Leafhopper tissue culture. In: Maramorosch K, Harris KF (eds) Leafhopper vectors and plant disease agents. Academic Press, New York, p 485
22. Maramorosch K, Harris KF (1979) Leafhopper vectors and plant disease agents. Academic Press, New York
23. Murant AF, Roberts IM, Elnagar S (1976) Association of virus-like particles with the foregut of the aphid *Cavariella aegopodii* transmitting the semipersistent viruses anthriscus yellows and parsnip yellow fleck. J Gen Virol 31: 47–57
24. Murant AF, Raccah B, Pirone TP (1988) Transmission by vectors. In: Milne RG (ed) The plant viruses, Vol IV. The filamentous viruses. Plenum, New York, p 237
25. Nasu S (1965) Electron microscopic studies on transovarial passage of rice dwarf virus. Jpn J Appl Entomol Zool 9: 225–237
26. Nault LR, Ammar ED (1989) Leafhopper and planthopper transmission of plant viruses. Annu Rev Entomol 34: 503–529
27. Nault LR, Hunt RE, Ammar ED (1987) Specificity in transmission of leafhopper-borne maize chlorotic dwarf virus. In: Proc 6th Auchen. Meeting, 7–11 Sept 1987, Turin, Italy, p 463
28. Ponsen MB (1977) Anatomy of an aphid vector: *Myzus persicae*. In: Harris KF, Maramorosch K (eds) Aphids as virus vectors. Academic Press, New York, p 63
29. Purcell AH (1982) Insect vector relationships with procaryotic plant pathogens. Annu Rev Phytopathol 20: 397–417
30. Rochow WF (1969) Biological properties of four isolates of barley yellow dwarf virus. Phytopathology 59: 1580–1589
31. Roth J (1986) Post-embedding cytochemistry with gold-labelled reagents: a review. J Microsc 143: 125–137
32. Shikata E (1979) Cytopathological changes in leafhopper vectors of plant viruses. In: Maramorosch K, Harris KF (eds) Leafhopper vectors and plant disease agents. Academic Press, New York, p 309
33. Shikata E (1979) Rice viruses and MLO's and leafhopper vectors. In: Maramarosch K, Harris KF (eds) Leafhopper vectors and plant disease agents. Academic Press, New York, p 515
34. Sylvester ES (1980) Circulative and propagative virus transmission by aphids. Annu Rev Entomol 25: 257–286
35. Taylor CE, Robertson WM (1974) Electron microscopy evidence for the association of tobacco severe etch virus with the maxillae in *Myzus persicae* (Sulz). Phytopathol Z 80: 257

Chapter 11 DIETRICH-E. LESEMANN

Specific Cytological Alterations in Virus-Infected Plant Cells

1 Introduction

The aim of this review is to give a short description of the principal cytological alterations occurring in virus-infected plant cells which may be specific for the inducing virus but are relatively independent from the host plant and its general reactions to the metabolic disturbances caused by the virus. Thus, alterations like the degradation of chloroplast membrane system in chlorotic tissues, or alterations of cytoplasm and organelle structure indicating heavy disturbance of metabolism which may result in cell death (necrosis), will not be considered here. The virus-specific effects treated represent either accumulations of different viral genome products, or alterations of the host cell membrane system which are supposed to be functionally important for virus replication and which, as a rule, can be observed in the early stages of infection. If these disturbances persist until late stages of infection, they are more and more accompanied by nonspecific pathological reactions.

Several detailed reviews on the vast amount of information on virus-induced cytological effects obtained by electron microscopy in the past 30 years or more exist already (e.g., [16, 17, 32, 37]). In view of the limited space, only some major examples will be given, whereas attention will be focused on newer lines of knowledge which have evolved in recent years, concomitantly with the rapid development of new methods of labeling antigens and nucleic acids in ultrathin sections. Thus, functional aspects can now be attributed to virus-specific cellular structures long known in morphology. Much new information has been obtained especially for viruses with positive sense single-stranded RNA [(+) ssRNA]. Therefore, this presentation will concentrate on this large category of plant viruses.

Biologische Bundesanstalt für Land- und Forstwirtschaft, Institut für Viruskrankheiten der Pflanzen, Messeweg 11/12, D-3300 Braunschweig, FRG

2 Viral Genome Products as Infection-Specific Constituents of Cells

2.1 Virions

Virus particles are the main genome product of a virus-infected cell. They have been sought, and their appearance in the cells described, in many virus-host systems. Often they form the most conspicuous cellular modification. The particles of (+) ssRNA viruses are, as a rule, located in the cytoplasmic compartment and/or the nucleoplasm. Less frequently, viruslike particles have been found in extracytoplasmic compartments, e.g., plastids and mitochondria (e.g., tobamoviruses, tombusviruses) or in the central vacuoles (many groups), see [17]. In the case of tobacco mosaic tobamo virus (TMV), such particles have been shown to be noninfectious pseudoparticles [48], in the other cases, the significance of the extracytoplasmic localization of the particles is not known. The central vacuoles can contain particles if the tonoplast is broken. Depending on their size and shape, particles of (+) ssRNA viruses may be individually seen in thin sections, like the elongated helically constructed types. Isometric viruses present more problems in their direct recognition. If the particles occur in scattered distribution and low concentration they can be detected only in special cases. Since the size of their dark-staining RNA-containing central part is in the same order of that of ribosomes, only techniques of specific ribosome digestion facilitate the reliable identification of isometric particles occurring at low concentration [26]. However, if virus particles occur at high concentration, they can be recognized either in random distribution (Fig. 1) or aggregated into compact masses or crystals. Virus crystals may be compact (Fig. 1, inset) with crystal lattices depending on the virus, but also tubular, like with some strains of broad bean wilt virus [17]. Likewise, particles of elongated viruses can be found in different arrangements, i.e., scattered, attached to membranes in monolayers (often with potyviruses), in platelike aggregates (Fig. 2) which, in addition may be stacked in series to form either compact crystals (tobamoviruses) or banded inclusions (elongated flexuous particles of potex-, carla-, closteroviruses) [35]. But also special arrangements like angled-layer or brush-like aggregates are found [17, 35]. Apart from the particle length determining virus group-specific dimensions of aggregates, also the organization of the particle aggregates may be specific for individual viruses or virus groups.

2.2 Nonstructural Proteins

In terms of genome products of (+) ssRNA viruses, the particles represent the RNA coated most frequently by only one of the several proteins encoded in the genome, i.e., the coat protein. It determines the shape and many properties of the particles. However, other proteins coded by the RNA may be as many as 7 in the case of

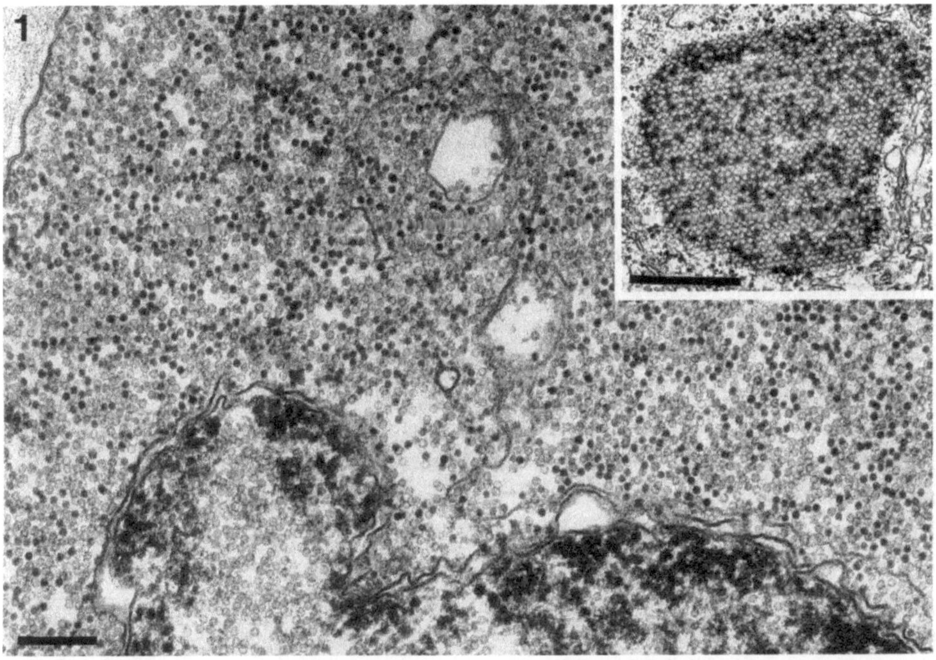

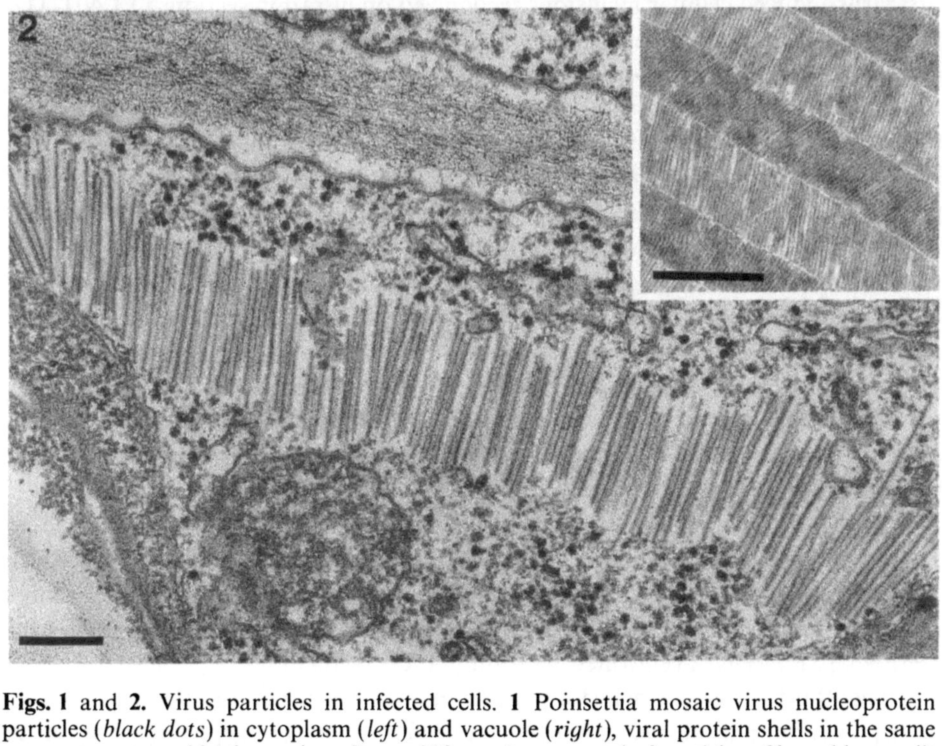

Figs. 1 and **2.** Virus particles in infected cells. **1** Poinsettia mosaic virus nucleoprotein particles (*black dots*) in cytoplasm (*left*) and vacuole (*right*), viral protein shells in the same compartments and in the nucleus. Bar = 200 nm. *Inset* crystal of particles of broad bean wilt fabavirus isolate Feng. Bar = 500 nm. **2** Plate-like aggregate of rod-shaped particles of unidentified tobamovirus from lily. Bar = 200 nm. *Inset* crystal of particles of an unidentified tobamovirus isolate Mosel. Bar = 500 nm

potyviruses, and may be produced in equal amounts as the coat protein, depending on the translation strategy of the virus. With several (+) ssRNA viruses accumulations of nonstructural proteins have been identified in infected cells in the form of large inclusions in the cytoplasmic compartment, sometimes with very specific structure.

TMW has long been known to induce X-bodies as cytoplasmic inclusions in addition to virion aggregates (Fig. 3). Recent immunolabeling work proved that the main tubular constituent of these X-bodies contain the virus-coded 126 K nonstructural protein, which is associated with RNA-replicase and protease function [29, 46, 50]. It is still unknown whether these functions are carried out in the X-bodies or whether the tubules represent only aggregates of the 126 K protein comparable to the crystalline aggregates of replicase and protease in nuclei of tobacco etch potyvirus (TEV)-infected cells (see below). X-bodies with dark staining material organized into regularly structured sheets, or into amorphous granular masses have been recorded with tobamoviruses other than TMV (Fig. 4) (D.-E. Lesemann unpubl.).

Potyviruses are one of the best-studied virus groups. They produce eight genome-coded proteins, some of which aggregate into several types of conspicuous cellular inclusions [for review 12, 14, 28]. Some of these inclusions were among the first virus-induced cellular inclusions recognized on ultrathin sections [13, 40]. The aggregates of nonstructural proteins are found in form of cylindrical inclusions, crystalline nuclear or cytoplasmic inclusions, and as amorphous cytoplasmic inclusions. Additionally, several types of other cytoplasmic and nuclear inclusions have been found whose nature is still obscure (for review see [35]).

Cylindrical inclusions (CI) are produced by all viruses assigned to the potyvirus group. The term pinwheel inclusion is also widely used for describing the typical appearance of CI in cross-section (Fig. 5). CI are composed of one virus-encoded 66–75 K protein which is aggregated to form a complicated structure made up of a central tubule to which several curved sheets are attached. The virus-specific modifications of CI morphology can be used as characters of potyviruses which are not based on the coat protein properties [14]. The function of CI proteins is as yet unknown, but supposedly they may play a role in the RNA replicating systems, since they show some sequence homology to "nucleotide binding" domains of picornaviruses [12, 23].

The crystalline nuclear inclusions (NI) induced by TEV (Fig. 6), bean yellow mosaic potyvirus (BYMV) and related viruses are composed of RNA polymerase and protease in equimolar amounts (for review see [6, 28]. Corresponding proteins are supposedly formed by all potyviruses and can be detected by serological methods [6], but NI are formed with only a few of them.

The third class of inclusions made up of nonstructural protein are the amorphous cytoplasmic inclusions (AI) induced by papaya ringspot and some other potyviruses (Fig. 7). This protein has been identified as the aphid-transmission

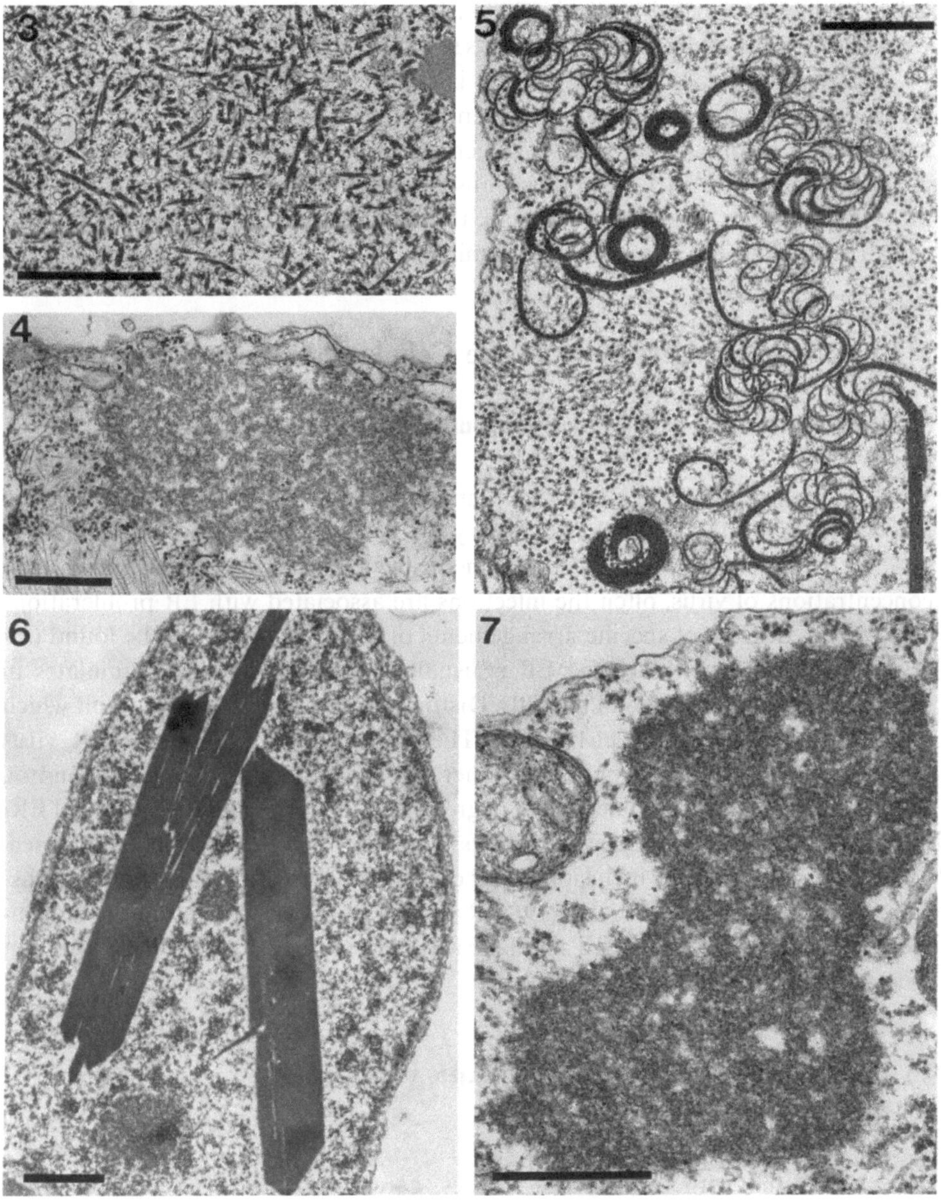

Figs. 3–7. Aggregates of virus-coded nonstructural gene products. **3** Part of X-body with tubular elements induced by tobacco mosaic virus. Bar = 5 μm. **4** X-body with granular structure induced by isolate BS of tobacco mosaic virus. Bar = 500 nm. **5** Cross-section of cylindrical (pinwheel) inclusions formed by peanut stripe potyvirus. Bar = 500 nm. **6** Nuclear inclusions induced by tobacco etch potyvirus. Bar = 1 μm. **7** Amorphous cytoplasmic inclusion induced by an isolate of watermelon mosaic 2 potyvirus, Morocco strain. Bar = 500 nm

helper component, which again is obviously produced by many, but found as AI with only a few potyviruses (for review see [28, 35]).

Many additional types of cellular inclusions have been found with other potyviruses. They might represent the genome products mentioned above in other arrangement or may be aggregates of genome products not yet found in cellular inclusions until now. Inclusions of various types are found also in association with other virus groups not mentioned here, though not as often [17, 35, 37]. In most cases their function has not been determined.

3 Virus-Specific Alterations of the Cellular Membrane System

3.1 Proliferation of Endoplasmic Reticulum-Like (ER) Membranes

Virus-infected cells very often show an enlarged cytoplasm volume and, in association with this, ER elements may be seen in greater amounts than in noninfected cells. Although ER can be inconspicuous also in cells producing high concentrations of virus, often the infections are associated with ER proliferations and various degrees of specific arrangements of the membranes can be found (for review see [35, 37]. Proliferated ER is usually not scattered but accumulates in certain, often only one, area of the cells. Distinct inclusions bodies may result which may be larger than the nucleus (Fig. 8). These inclusions contain rough ER, virus particles, and varying amounts of other cell constituents like mitochondria, microbodies, and lipid globules. Their degree of compactness and the pattern of ER-branching can be characteristic of the virus group [17, 35, 37]. Extremely condensed, crystal like membranous-inclusions (coat of mail structures) are induced by fungus-transmitted potyviruses [30], (Fig. 9). ER proliferations may represent the morphological expression of increased synthetic activities induced by virus infections, but their function is unknown.

3.2 Formation of Vesicles Containing ds-RNA Associated with Viral RNA-Replication

In cells infected with (+) ssRNA viruses, careful eximation often reveals the existence of one of two distinct types of small vesicles. Both types may contain finely fibrillar material (Fig. 11, inset), which supposedly is dsRNA, as proven in some instances (e.g., [8, 11, 15, 17, 19, 25, 33, 38, 39, 44, 49]).

3.2.1 Smooth Vesicles Accumulating in the Cytoplasm

Cytopathic structures mainly composed of densely clustered vesicles which may contain dsRNA-like fibrils were described for the groups of como-, faba- (Fig. 10)

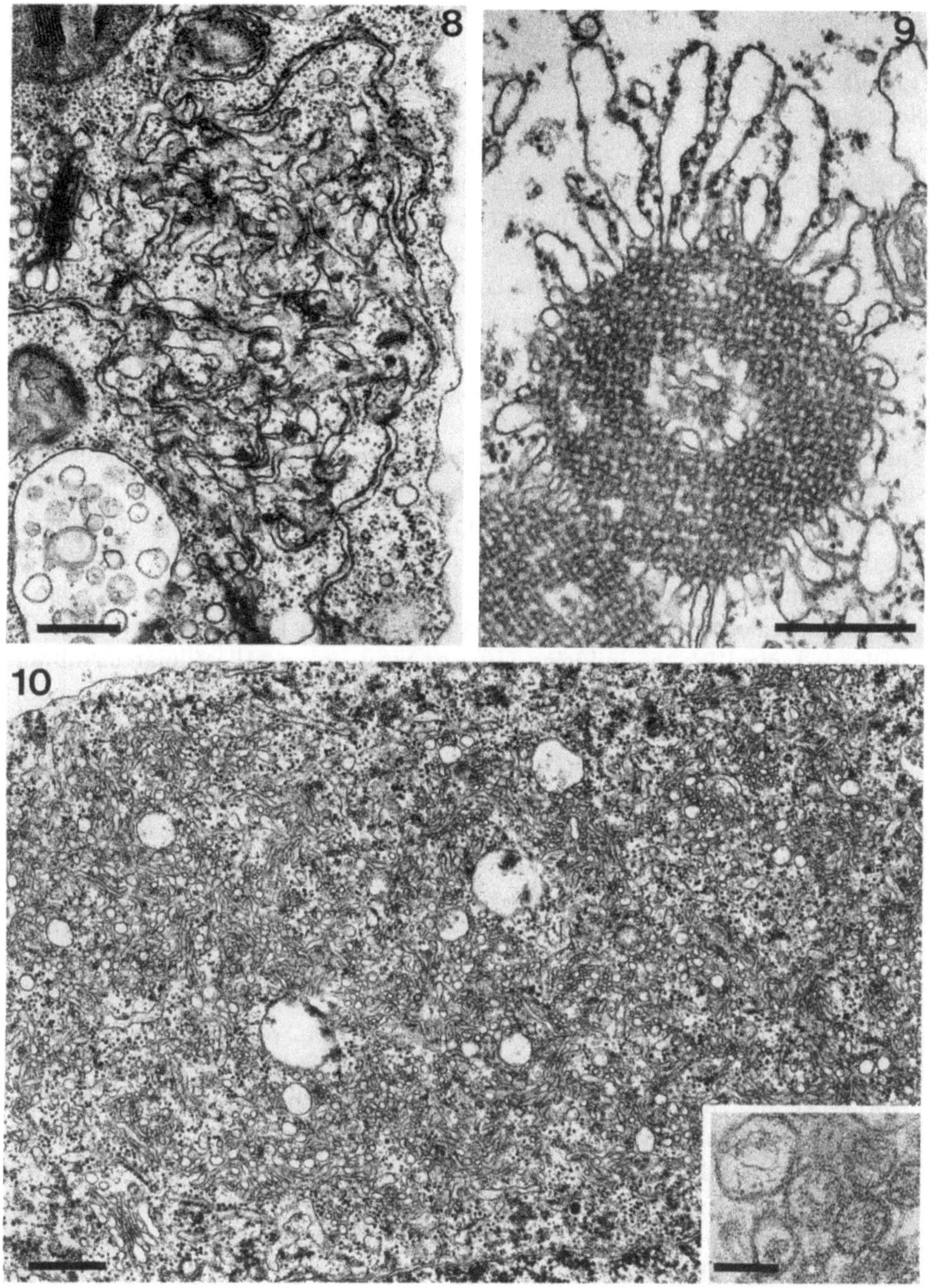

Figs. 8–10. Proliferation of endoplasmic reticulum (ER) and accumulation of free vesicles. **8** "Inclusion" composed mainly of accumulated ER induced by Indian peanut clump furovirus. Bar = 500 nm. **9** Crystal-like arrangement of ER-derived membranes induced by barley yellow mosaic virus. Bar = 500 nm. **10** Cytopathic accumulation of vesicles induced by broad bean wilt fabavirus. Bar = 5 μm. *Inset* vesicles containing dsRNA-like fibrils induced by eggplant severe mottle potyvirus. Bar = 200 nm

and nepoviruses. Also the potyvirus-induced complex cytoplasmic inclusions often contain considerable amounts of vesicles which may contain dsRNA-like material (Fig. 10, inset), (for review see [17]). The origin of these vesicles has not yet been elucidated, an origin from ER or dictyosomes seems possible. For poliovirus, a picorna virus, vesicles very similar to those of plant viruses have been shown to originate from ER membrane protrusions which finally bud off. The vesicles are shown to carry on their surface the sites of viral RNA replication [2]. In view of the genome homologies between the above-mentioned plant virus groups and picorna viruses (discussed below, [33]), a similar mechanism of vesicle formation from ER might be supposed also for these plant viruses.

3.2.2 Membrane-Associated Vesicles Formed as Cytoplasm Evaginations

These vesicles are usually only 60–100 nm in diameter and appear as flask-shaped structures attached by a thin neck to the "mother" membrane (Fig. 11). The vesicles are always formed towards the extracytoplasmic side of the respective membranes and represent, therefore, membrane-bounded evaginations of the cytoplasmic compartment. The contents of the vesicles often appears dsRNA-like (Fig. 11, inset), which has been confirmed in several cases (for review see [39]). The location of the vesicles is typical for the inducing virus or virus group. All cellular membrane systems except for dictyosomes and plasmalemma have been shown to support vesicle formation. The tonoplast seems, however, to be more active than other membranes. Excessive vesicle formation may lead to conspicuous changes in the structure of the respective organelles (e.g., Fig. 12). In double membrane-bounded organelles (plastids, mitochondria) the vesicles may be double membrane-bounded (Fig. 12), or single membrane-bounded, depending on the inducing virus. A summary of the information obtained up to now is given in Table 1.

3.2.3 Homology Between Plant and Animal (+) ssRNA Viruses

Comparisons of the genome organization of (+) ssRNA plant and animal viruses has led to the recognition to two principal groups, which have been designated picorna-like and sindbis-like, according to the similarities of plant viruses to the respective groups of animal viruses [21, 22, 23]. Table 1 allows comparison of this grouping with the distribution of the two vesicle types among the plant viruses. It turns out that all viruses assigned by Goldbach [22] to picorna-like viruses induce smooth free vesicles accumulating in the cytoplasm, whereas all plant viruses grouped with sindbis-like viruses induce membrane-associated vesicles. This observation matches cytological studies in animal viruses. Poliovirus induces free vesicles, budding off the ER and accumulating in the cytoplasm [2], whereas sindbis virus and the closely related Semliki Forest virus induce the formation of "light bulb-shaped" vesicles lining the extracyto-plasmic side of the membranes of endosomes and lysosomes [18]. Thus, these animal viruses exhibit the same principal types of

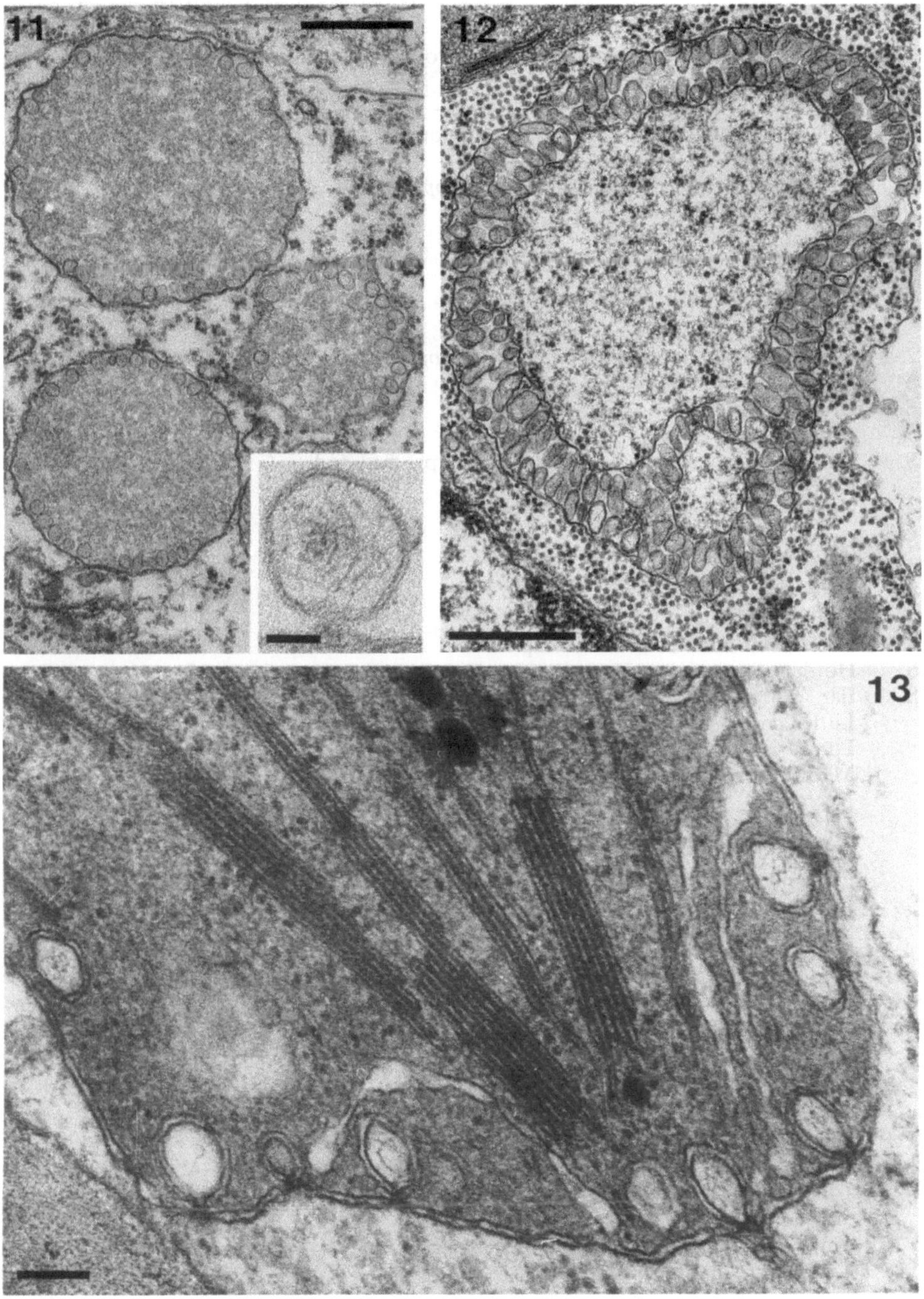

Figs. 11–13. Membrane-associated vesicles formed by cytoplasma evagination. **11** Single membrane-bounded vesicles induced at the periphery of peroxisomes by maize chlorotic mottle virus. Bar = 500 nm. *Inset* single vesicle containing dsRNA-like material induced at the tonoplast by groundnut rosette symptom-inducing agent. Bar = 50 nm. **12** Multivesicular body induced by Moroccan pepper tombusvirus, originating from a peroxisome and carrying numerous membrane-associated vesicles. Bar = 500 nm. **13** Double membrane-bounded vesicles at the chloroplast periphery induced by belladonna mottle tymovirus. Bar = 100 nm

Table 1. Two types of vesicles suspected or proven to contain viral ds-RNA induced by (+)ssRNA plant viruses. (After [36])

	Virus group or virus	Location	Reference
1. Free smooth vesicles accumulating in the cytoplasm			
P[a]	como	cytoplasm	17, 49
	diantho		17, 31
	faba		17
P	nepo		17
P	poty		17, 35
2. Membrane-associated evaginations of the cytoplasm			
S[a]	Alfalfa mosaic	tp[b]	17, 38
S	Bromo	ne	38
	Carla	tp	17, 35[c]
S	Carmo	tp, er, ne, px, mts, cpd	1, 9, 10, 27, 41, 43
	Carrot mottle	tp	17
	Clostero	u, mts	17, 35
S	Cucumo	tp, er	17, 38
	Fir (*Abies*) filament	mts	47
S	Furo	px	[c]
	Grapevine virus	mtd	4
	Groundnut rosette	tp	[c]
S	Hordei	cps	17
S	Ilar	tp	38
	Little cherry	u	42
	Luteo	er, ne, mts	17
	Maize chlorotic mottle	px	[c]
	Necro	tp, er	17[c]
	Olive latent 1	tp, ne	5
	Olive latent 2	tp, u	5
	Panicum sabulorum	mtd	20
	Pea enation mosaic	ne	7
	Pelargonium zonate spot	ne	3
	Potex	tp, er	17, 35
	Sobemo	er, ne	17
S	Tobamo	tp, mts	17
S	Tobra	mts	24
S	Tombus	px, mts, cps	39, 45
	Tymo	cpd, mtd	17, 34

[a] P = Picorna-like viruses S = sindbis-like viruses sensu [22] and [23].
[b] tp = Tonoplast; er = endoplasmic reticulum; ne = nuclear envelope.
cp = Chloroplast (d = double, s = single membrane); mt = mitochondrion (d = double, s = single membrane); px = peroxisome;
u = Unknown origin.
[c] = Lesemann, unpublished results.

vesicles involved in RNA replication as the proposed "supergroups" of plant viruses. The two vesicles types could, therefore, be the morphological expression of two major mechanisms of RNA replication used by the two "super groups" of plant and animal viruses. In both cases, an accumulation of dsRNA results which is located in the cytoplasmic compartment with sindbis-like viruses and in the extracytoplasmic compartment with picorna-like viruses.

4 Conclusions

This chapter has briefly outlined the present state of knowledge about the principal virus-specific cytoplasmic alterations which are produced by (+) ssRNA viruses in the host cells. Although many strucutral details are known, a detailed connection between structure and function can as yet be given in only a few instances. As to viral genome products and virus-induced membrane alternations, more insight can be expected in the near future, since newly developed methods for localizing proteins and nucleic acids present excellent tools for cytologists. The same will be true for plant viruses with other genome types that cause cytological effects other than those described here (for review see [17]).

References

1. Behnken GM, Dale JL (1984) Glycine mottle virus: a possible member of the tombusvirus group. Intervirology 21: 159–166
2. Bienz K, Egger D, Pasamontes L (1987) Association of polioviral proteins of the P2 genomic region with the viral replication complex and virus-induced membrane synthesis as visualized by electron microscopic immunocytochemistry and autoradiography. Virology 160: 220–226
3. Castellano MA, Martelli GP (1981) Electron microscopy of pelargonium zonate spot virus in host tissues. Phytopathol Mediterr 20: 64–71
4. Castellano MA, Martilli GP, Savino V (1983) Virus-like particles and ultrastructural modifications in the phloem of leafroll-affected grapevine. Vitis 22: 32–39
5. Castellano MA, DiFranco A, Martelli GP (1987) Electron microscopy of two olive viruses in host tissues. J Submicros Cytol 19: 495–508
6. Chang CA, Hiebert E, Purcifull DE (1988) Purification, characterization, and immunological analysis of nuclear inclusions induced by bean yellow mosaic and clover yellow vein potyviruses. Phytopathology 78: 1266–1275
7. De Zoeten GA, Gaard G, Diez FB (1972) Nuclear vesiculation associated with pea enation mosaic virus-infected plant tissues. Virology 48: 638–647
8. De Zoeten GA, Assing MA, Van Kammen A (1974) Association of cowpea mosaic virus-induced double-stranded RNA with a cytopathological structure in infected cells. Virology 59: 341–355
9. DiFranco A, Martelli GP (1987A) Comparative ultrastructural investigations on four soil-borne cucurbit viruses. J Submicrosc Cytol 19: 605–613

10. DiFranco A, Martelli GP (1987B) Some observations on the ultrastructure of galinsoga mosaic virus infections. Phytopathol Mediterr 26: 54–56
11. DiFranco A, Russo M, Martelli GP (1984) Ultrastructure and origin of cytoplasmic multivesicular bodies induced by carnation Italian ringspot virus. J Gen Virol 65: 1233–1237
12. Dougherty WG, Carrington JC (1988) Expression and function of potyviral gene products. Annu Rev Phytopathol 26: 123–143
13. Edwardson JR (1966) Electron microscopy of cytoplasmic inclusions in cells infected with rod-shapped viruses. Am J Bot 53: 359–364
14. Edwardson JR (1974) Some properties of the potato virus Y-group. Fla Agric Exp Stn Monogr Ser No 4: 398
15. Eggen R, Kaan A, Goldbach R, Van Kammen A (1988) Cowpea mosaic virus RNA replication in crude membrane fractions from infected cowpea and *Chenopodium amaranticolor*. J Gen Virol 69: 2711–2720
16. Fraenkel-Conrat H, Wagner RR (eds) (1985–1988) The viruses, subseries: The plant viruses, Vol 1, 1985; Vol 2, 1986; Vol 3 and 4, 1988. Plenum, New York
17. Francki RIB, Milne RG, Hatta I (1985) Atlas of plant viruses, Vols I and II. CRC, Boca Raton, pp 22, 284
18. Froshauer S, Kartenbeck J, Helenius A (1988) Alphavirus RNA replicase is located on the cytoplasmic surface of endosomes and lysosomes. J Cell Biol 107: 2075–2086
19. Garnier M, Candresse T, Bove JM (1986) Immunocytochemical locatization of TYMV-coded structural and nonstructural proteins by the protein A-gold technique. Virology 151: 100–109
20. Gill CC, Chong J, Caetano VE (1981) Cytopathological alterations associated with a virus-like disorder in *Panicum sabulorum*. Can J Plant Pathol 3: 129–135
21. Goldbach RW (1986) Molecular evolution of plant RNA viruses. Annu Rev Phytopathol 24: 289–310
22. Goldbach R (1987) Genome similarities between plant and animal RNA viruses. Microbiol Sci 4: 197–202
23. Goldbach R, Wellink J (1988) Evolution of plus-strand RNA viruses. Intervirology 29: 250–267
24. Harrison BD, Stefanac Z, Roberts JM (1970) Role of mitochondria in the formation of X-bodies in cells of *Nicotiana clevelandii* infected by tobacco rattle viruses. J Gen Virol 6: 127–140
25. Hatta T, Francki RIB (1978) Enzyme cytochemical identification of single-stranded and double-stranded RNAs in virus-infected plant and insect cells. Virology 88: 105–107
26. Hatta T, Francki RIB (1981) Identification of small polyhedral virus particles in thin sections of plant cells by an enzyme cytochemical technique. J Ultrastruct Res 74: 116–129
27. Hatta R, Francki RIB, Grivell GJ (1983) Particle morphology and cytopathology of galinsoga mosaic virus. J Gen Virol 64: 687–692
28. Hiebert E, Dougherty WG (1988) Organization and expression of the viral genomes. In: Milne RG (ed) The plant viruses, Vol 4. Plenum, New York, pp 159–178
29. Hills GJ, Plaskitt KA, Young ND, Dunigan DD, Watts JW, Wilson TMA, Zaitlin M (1987) Immunogold localization of the intracellular sites of structural and nonstructural tobacco mosaic virus proteins. Virology 161: 488–496
30. Huth W, Lesemann DE, Paul HL (1984) Barley yellow mosaic virus: purification, electron microscopy, serology and other properties of two types of the virus. Phytopathol Z 111: 37–54
31. Koenig R, An D, Lesemann DE, Burgermeister W (1988) Isolation of carnation ringspot virus from a canal near a sewage plant: cDNA hybridization analysis, serology and cytopathology. J Phytopathol 121: 346–356
32. Kurstak E (ed) (1981) Handbook of plant virus infections. Comparative diagnosis, Elsevier/North Holland Biomedical Press, Amsterdam, p 943

33. LaFleche D, Bove C, Dupont G, Mouches C, Astier T, Garnier M, Bove JM (1972) Site of viral RNA replication in the cells of higher plants. TYMV-RNA synthesis on the cloroplast outer membrane system. In: Proc 8th FEBS Meeting Amsterdam RNA viruses/ribosomes, Vol 27. North Holland, Amsterdam, pp 43–71
34. Lesemann DE (1977) Virus group-specific and virus specific cytological alterations induced by members of the tymovirus group. Phytopathol Z 90: 315–336
35. Lesemann DE (1988) Cytopathology. In: Milne RG (ed) The plant viruses, Vol 4. Plenum, New York, pp 179–235
36. Lesemann DE (1988) Recent aspects of the use of cytopathology in virus identification. Acta Hortic 234: 289–298
37. Martelli GP, Russo M (1977) Plant virus inclusion bodies. Adv Virus Res 21: 175–226
38. Martelli GP, Russo M (1985) Virus-host relationship. Symptomatological and ultrastructural aspects. In: Francki RIB (ed) The plant viruses, Vol 1. Plenum, New York, pp 163–205
39. Martelli GP, Gallitelli D, Russo M (1988) The tombusvirus group. In: Koenig R (ed) The plant viruses, Vol 3. Plenum, New York, pp 13–72
40. Matsui C, Yamagushi A (1964) Electron microscopy of host cells infected with tobacco etch virus. I Fine structure of leaf cells at later stages of infection. Virology 22: 40–47
41. Morris TJ, Carrington JG (1988) Carnation mottle virus and viruses with similar properties. In: Koenig R (ed) The plant viruses, Vol 3. Plenum, New York, pp 73–112
42. Raine J, Weintraub M, Schroeder B (1975) Flexuous rods and vesicles in leaf and petiole phloem of little-cherry diseased *Prunus* spp. Phytopathology 65: 1181–1186
43. Russo M, Martelli GP (1982) Ultrastructure of turnip crinkle and saguaro cactus virus-infected cells. Virology 118: 109–116
44. Russo M, DiFranco A, Martelli GP (1983) The fine structure of Cymbidium ringspot virus infections in host tissues. III Peroxisomes in the genesis of multivesicular bodies. J Ultrastruct Res 82: 52–63
45. Russo M, DiFranco A, Martelli GP (1987) Cytopathology in the identification and classification of tombusviruses. Intervirology 28: 134–143
46. Saito T, Hosokawa D, Meshi T, Okada Y (1987) Immunocytochemical localization of the 130 K and 180 K proteins (putative replicase components) of tobacco mosaic. Virology 160: 477–481
47. Schulz A, Behnke HD (1987) Feinbau und Differenzierung des Phloems von Buchen, Fichten und Tannen aus Waldschadensgebieten. Projekt Europäisches Forschungszentrum für Maßnahmen zur Luftreinhaltung, Forschungsbericht KfK-PEF 16, Kernforschungszentrum, Karlsruhe, p 95
48. Shalla TA, Petersen LJ, Giunchedi L (1975) Partial characterization of virus-like particles in chloroplasts of plants infected with the U5-strain of TMV. Virology 66: 94–105
49. Wellink J, Van Lent J, Goldbach R (1988) Detection of viral proteins in cytopathic structures in cowpea protoplasts infected with cowpea mosaic virus. J Gen Virol 69: 751–755
50. Wijdeveld MMG, Goldbach RW, Verduin BJM, Van Loon LC (1988) Association of viral 126 kDa protein-containing X-bodies with nuclei in mosaic-diseased tobacco leaves. Arch Virol 104: 225–239

Detlev Riesner[1], J. Harders[1], N. Lukacs[1], R. Gruner[1], U. J. Santore[2], G. Klotz[3], M. Robert-Nicoud[4], T. M. Jovin[4] and H. L. Sänger[5]

Chapter 12

Structure, Cellular Location, and Cytopathology of Viroids

1 Introduction

Viroids are plant pathogens distinguished from viruses by the absence of a protein coat and by their small size. They are circular, single-stranded RNA molecules consisting of a few hundred nucleotides, the smallest having about 240 and the largest about 600 nucleotides. Since viroids can have only a very limited coding capacity and as there is no experimental evidence for a viroid-coded translation product, one has to assume that viroid replication and pathogenesis depend completely on the enzyme systems of the host (for reviews see [1–5]).

To date, about 14 different viroid "species" and series of isolates of these have been found. The majority of known viroids cause diseases of economically important crops, such as potato, tomato, cucumber, avocado, citrus, and coconut. The cadang-cadang viroid of coconuts, for example, causes severe and large-scale losses in palm plantations, and hence seriously threatens the economy of countries like the Philippines. It is expected that detailed knowledge of viroid structure, replication, and interaction with the host will finally lead to control, and prevention of the losses caused to agriculture.

Since their discovery in 1971/72, viroids have been studied extensively. This report places emphasis on and summarizes findings which were obtained with the electron microscope and by fluorescence microscopy.

[1] Institut für Physikalische Biologie, Heinrich-Heine-Universität Düsseldorf, Universitätsstr. 1, D-4000 Düsseldorf, FRG
[2] Botanisches Institut, Heinrich-Heine-Universität Düsseldorf, Universitätsstr. 1, D-4000 Düsseldorf, FRG
[3] Abteilung für Virologie, Universität Ulm, Oberer Eselsberg, D-7900 Ulm, FRG
[4] Max-Planck-Institut für Biophysikalische Chemie, Am Faßberg, D-3400 Göttingen, FRG
[5] Max-Planck-Institut für Biochemie, Am Klopferspitz, D-8033 Planegg-Martinsried, FRG

2 Structure and Structural Transitions of Viroids

Electron microscopy was first applied in viroid research to elucidate the native form of viroids. Their rod-like shape became evident in these early studies [6, 7]. The native structure and structural transitions of potato spindle tuber viroid (PSTVd) were systematically investigated [8], and the parameters of its conformational transitions were determined by thermodynamic experiments; this knowledge was used to prepare viroids in different well-defined conformations for electron microscope study. In order to eliminate potential artifacts which may arise during viroid purification, and to depict the thermodynamically most stable form, purified PSTVd was heated and renatured by slowly cooling to room temperature. After the heating-cooling cycle, the molecules were rod-shaped with a length of 37 ± 6 nm (Fig. 1a); the length was slightly shorter when purified viroid RNA was imaged without a heating-cooling cycle.

Viroids were also studied after denaturation at 65 °C in the presence of 3 M urea and spreading on water at 40 °C. Since the T_m value of the main conformational transition (as determined by optical melting curves under the same conditions) is 50 °C, a distribution of different molecular conformations was expected. Indeed,

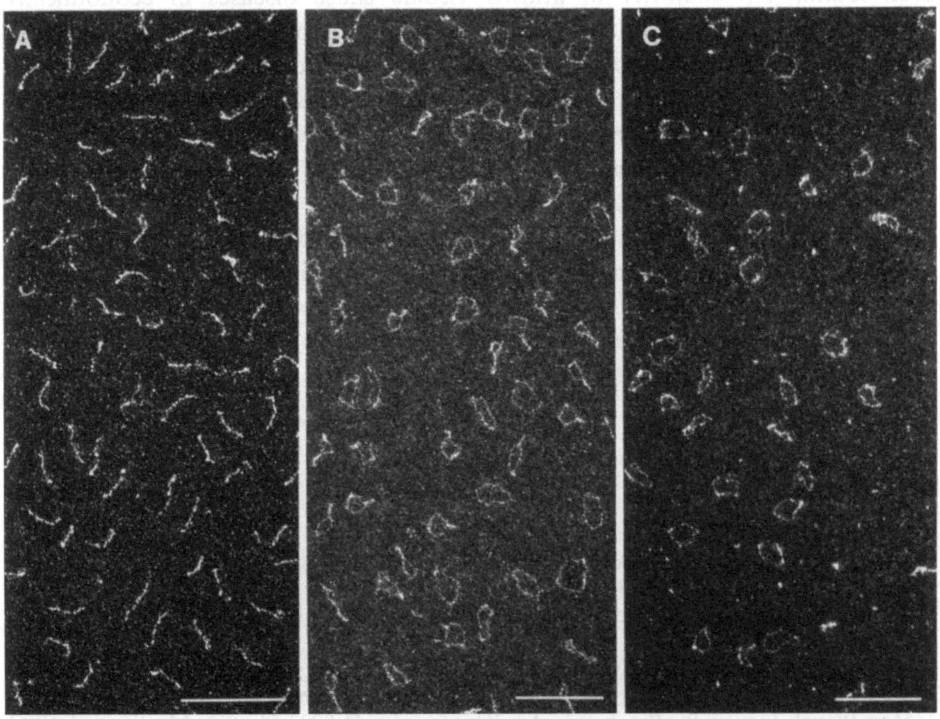

Fig. 1. Electron micrographs of PSTVd in different conformational states. **A** renatured, **B** intermediate with stable hairpin formation, **C** completely denatured circle. Bars = 100 nm. Preparation of the micrographs is described in [8], figures from [15]

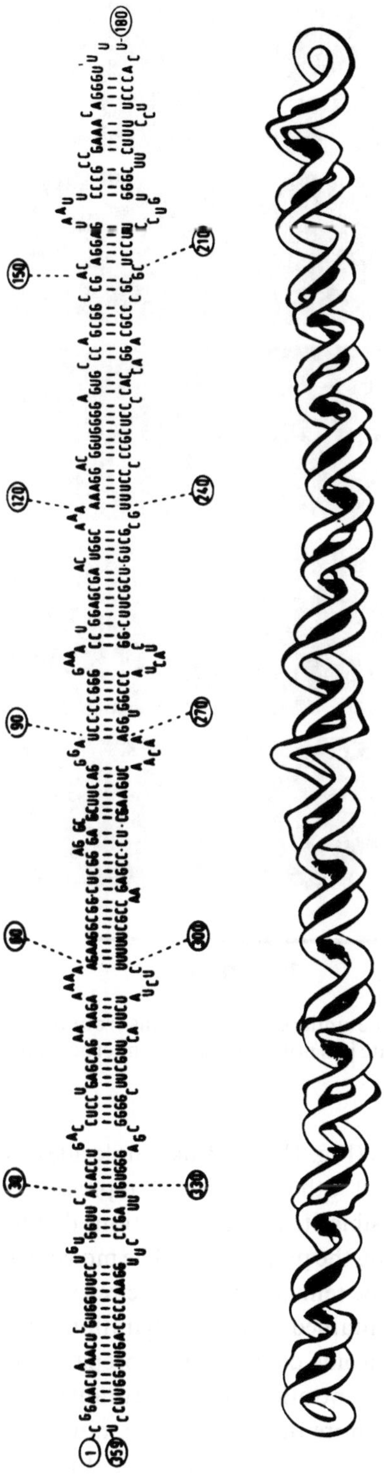

Fig. 2. Structure of PSTVd: base pairing scheme (*top*) and three-dimensional representation (*bottom*). [17]

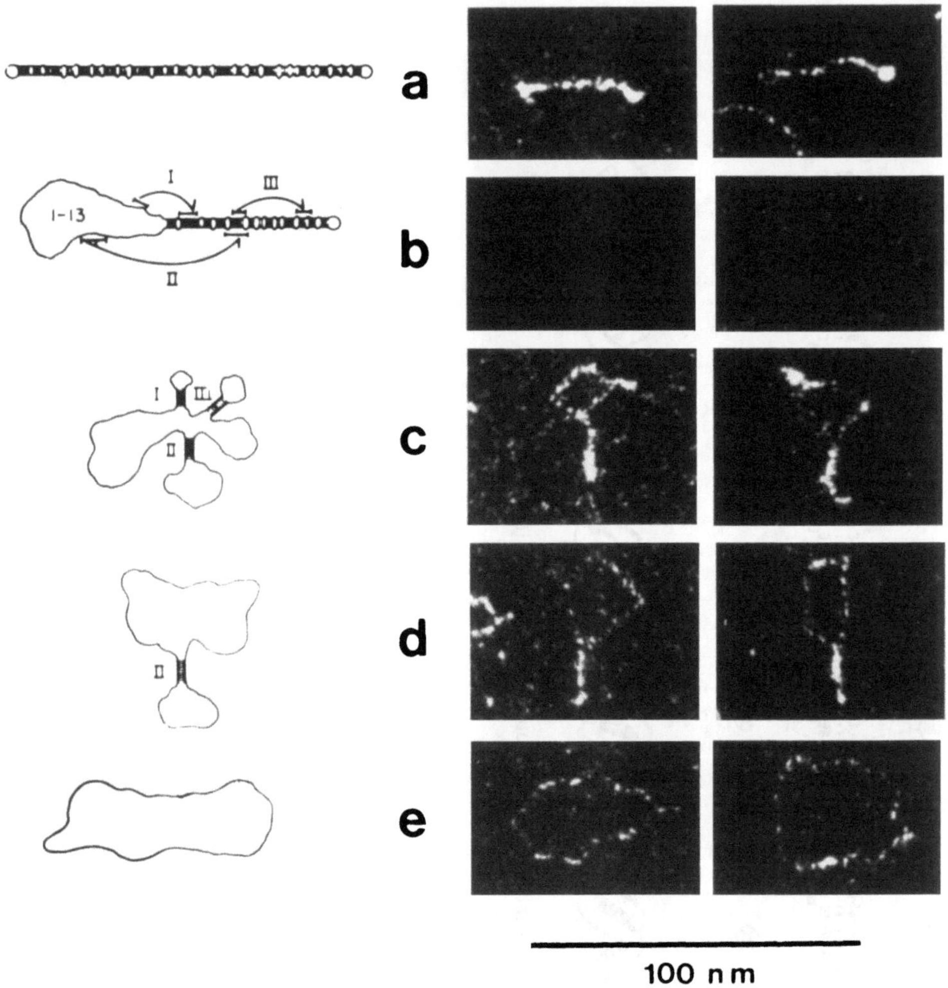

Fig. 3. Denaturing mechanism of PSTVd. The molecule undergoes transitions from **a** to **e** with increasing temperature. As **b** is a kinetic intermediate it is not observable in micrographs. [15]

the micrographs showed distinct classes of shapes (Fig. 1b). Rod-like molecules as well as completely open circles were found. All other structures represented intermediates of the helix-coil transition with possible structural alterations due to the influence of surfactant and to interactions during the preparation. The molecules could be classified as being shaped like dumbbells, tennis rackets, or circles.

Complete denaturing conditions were maintained when preparation and spreading was carried out at 70 °C. Denatured viroids could also be prepared at lower temperatures in the presence of high concentrations of denaturants like formamide or urea. Under denaturing conditions more than 80% of the molecules appeared as single-stranded circles devoid of any obvious base-pairing (Fig. 1c).

This finding demonstrated the existence of a natural circular RNA molelcule for the first time [7]. The original observation with PSTVd was subsequently confirmed for all viroids studied by this method [9, 11, 13, 14].

The mechanism of viroid structure formation was derived from experimental results and from theoretical considerations. The theoretical calculations were greatly aided by the fact that structure formation in viroids may be attributed solely to Watson-Crick and wobble base-pairing. Many single-stranded RNA's, e.g., tRNA's and rRNA's, are known to form structures not confined to two dimensions. However, viroids seem to be an exception, and there is no evidence to suggest that they possess a globular tertiary structure. It was thus possible to calculate the secondary structure using lowest free energy values [15,,16]. The secondary structure of PSTVd is illustrated in Fig. 2. The rod-like appearance of the molecule, as shown in the structure model, is supported both by direct observations in the electron microscope and by a series of physicochemical investigations and calculations (for summary see [15]).

In addition to the electron microscope studies, summarized in Fig. 1, thermodynamic and kinetic investigations have revealed a highly cooperative transition during which all base pairs of the native structure dissociate and new particularly stable helical regions evolve. As shown in Fig. 3, the molecule switches from a highly base-paired rod-shaped form to a branched secondary structure in which three stable hairpins are formed. At higher temperatures these stable hairpins in turn dissociate according to their individual thermal stabilities. At full denaturation, this completely reversible mechanism yields a circle with no internal base pairing. Recent results indicate that these conformational states—which were observed and could be thermodynamically explained nearly 10 years ago—may have relevance for viroid replication and processing [18, 19].

3 Complexes of Viroids and Polymerase

Considerable experimental evidence points to involvement of the specific nuclear DNA-dependent RNA polymerase II in viroid replication, which *in vivo* [20] is inhibited by very low concentrations of α-amanitin, a selective inhibitor of this enzyme under these conditions. Significantly, transcription studies with DNA-dependent RNA polymerase II, purified from wheat germ and tomato plants, have shown that the enzyme can synthesize full-length (−)viroid RNA copies in vitro [21]. The interaction of viroids and purified polymerase II from wheat germ has been studied quantitatively by in vitro transcription, determination of binding constants in the analytical ultracentrifuge, and by electron microscopy of viroid-polymerase complexes [22]. Figure 4 shows electron micrographs of the complexes which can be observed between polymerase II and viroid (top panel), virusoid (middle panel), and a 187 bp DNA fragment (bottom panel). The latter two nucleic acids were used as controls for the specificity of binding under the given conditions.

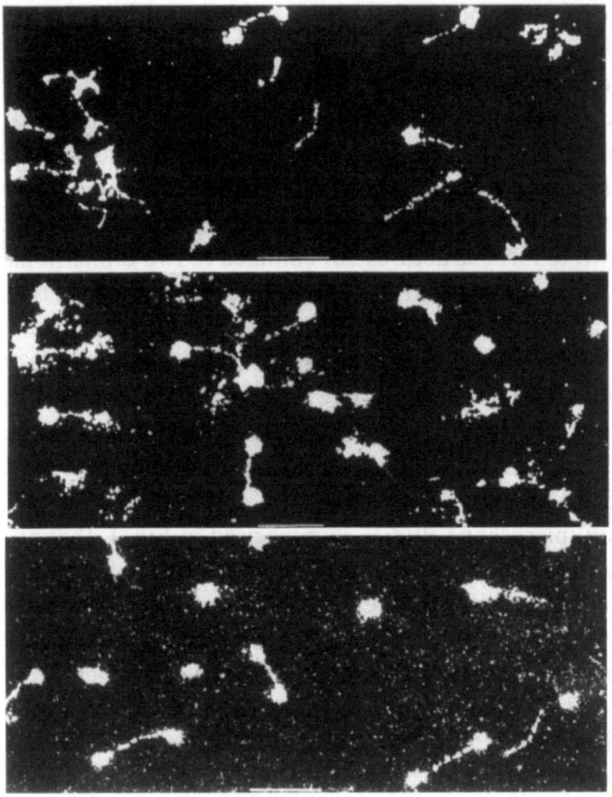

Fig. 4. Electron micrographs showing formation of complexes between DNA-dependent RNA polymerase II and PSTVd (*top panel*), VTMoV (velvet tobacco mottle virusoid, *middle panel*) and 187 bp DNA fragment (*bottom panel*). [22]

Virusoids are circular satellite RNA's which resemble viroids in size and structure but are not replicated by the enzyme. The 187-bp-long fragment of *E. coli.* DNA has the same size as PSTVd but no specific binding site for the eukaryotic RNA polymerase. The electron micrographs provided visual evidence for polymerase binding on all three nucleic acids; the binding occurred in most cases at the ends of the rod-like molecules. With the twofold molar excess of enzyme used in the experiment, both 1:1 and 2:1 enzyme to nucleic acid stoichiometries were found. From electron microscope evidence only, it was not possible to distinguish between specific and nonspecific polymerase binding, so the binding constants of the polymerase-nucleic acid complexes were determined by quantitative evaluation of the sedimentation profiles measured in the analytical ultracentrifuge. The 1:1 polymerase-viroid complex showed a binding constant of $1.9 \times 10^7\,M^{-1}$ but the affinity of the 2:1 complex was markedly lower. This suggests that (a) the 2:1 polymerase-viroid complex is not physiologically significant and (b) the binding at one of the ends of the viroid RNA must be nonspecific. The virusoid and the 187 bp

DNA fragment were bound to the polymerase with a distinctly lower affinity than viroid RNA. It can be concluded that the polymerase binding on virusoid and 187 bp DNA seen on the micrographs must be nonspecific [22].

4 Intracellular Location

Viroids are not homogeneously distributed inside the cell, but are associated with nuclei and may be protected from nucleases in complexes with cellular components [23, 24, 27].

The intracellular location of viroids was investigated recently by light microscopy at high resolution, using viroid-specific in situ hybridization with biotinylated probes and detection with rhodamine/lissamine conjugated streptavidin. The distribution of the hybridization signals was analyzed by fluorescence and confocal laser scanning microscopy. In situ hybridization was also carried out on ultrathin sections of suitably embedded material and some preliminary observations were made by electron microscopy.

4.1 Fluorescence Microscopy

For the fluorescence studies, isolated nuclei from green leaf tissue of PSTVd-infected tomato plants were attached to microscope slides and fixed with formaldehyde. Biotinylated PSTVd-specific hybridization probes were synthesized by in vitro transcription of cloned PSTVd-cDNA. For details of the hybridization and staining procedure see the original literature [25].

The results of in situ hybridization with a PSTVd (+)strand-specific probe are shown in Fig. 5. In two of the nuclei a strong rhodamine fluorescence is detectable in a limited circular area of the nucleus. Similar fluorescence was not seen in nuclei from uninfected tissue. To identify the fluorescing area all samples were double-labeled: after completing in situ hybridization, the samples were incubated with the nucleolus-specific monoclonal antibody Bv96 and with an FITC-labeled goat-anti-mouse antibody. The location of fluorescence signals of rhodamine and FITC were compared. The viroid-containing areas in Fig. 5A are unequivocally stained by the antibody (Fig. 5B) showing that the viroid-contained areas coincide with nucleoli. The result is in agreement with earlier fractionation data [27, 28]. In order to determine the percentage of nuclei infected with PSTVd, nuclei were incubated with the fluorescent dye DAPI (4,6-diamidino-2-phenylindole-hydrochloride) and the number of DAPI-stained nuclei was compared with the number of fluorescence-labeled nuclei after in situ hybridization (data not shown). In DAPI-stained nuclei the nucleoli appear as dark areas in the strongly fluorescing karyoplasm. A statistical analysis of viroid distribution was carried out for two viroid strains, namely the

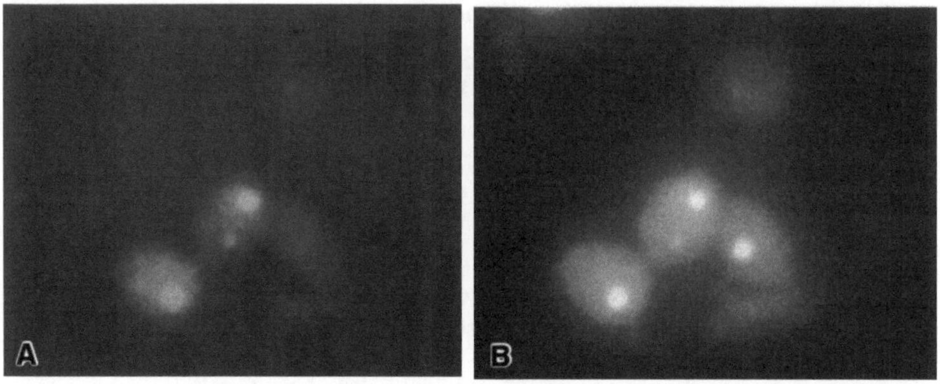

Fig. 5

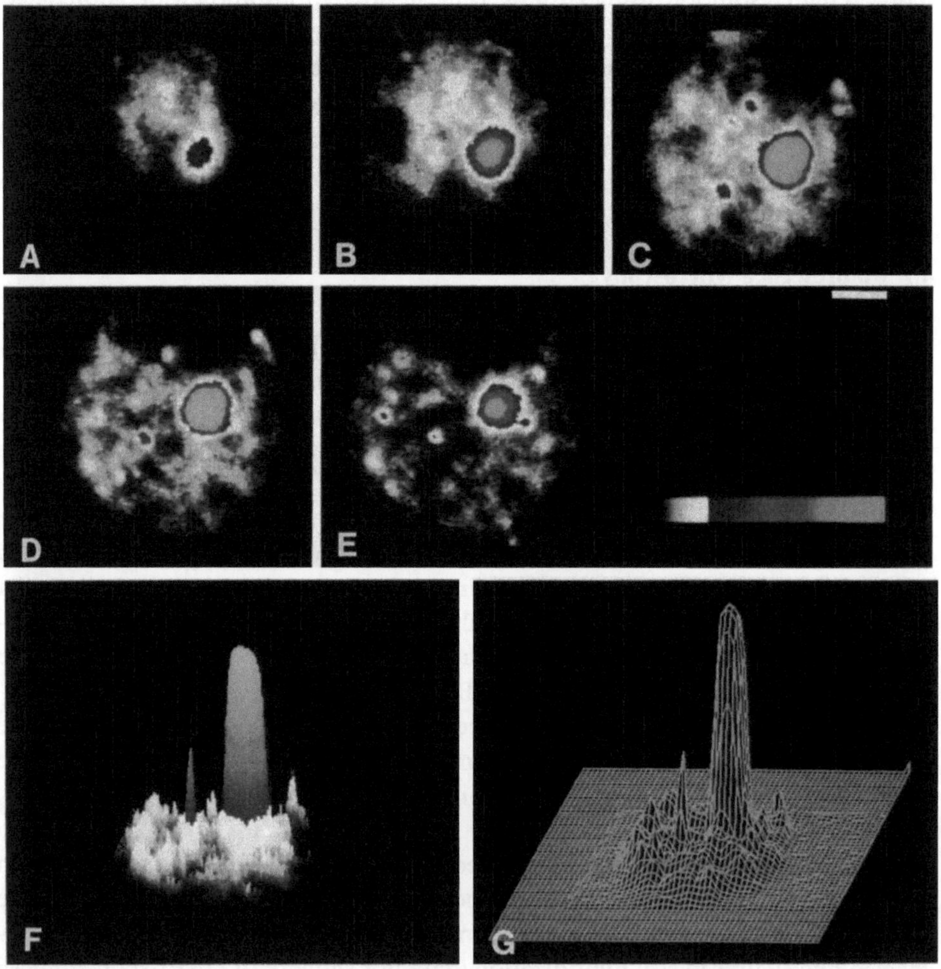

Fig. 6

"intermediate" and the "lethal" strain, which induce different symptoms. Similar results were found for both strains: 6–18% of nucleoli showed hybridization signals.

The local distribution of PSTVd-(−)strand sequences, which are replication intermediates, was analyzed by applying biotinylated PSTVd (+)strand RNA transcripts as hybridization probes. In order to exclude all unwanted (+)strand/(+)strand hybrids, the conditions applied for (+)strand/(−)strand hybridization were extremely stringent. Although much weaker signals were obtained with (+)strand-specific probes, it could be concluded that the PSTVd-(−)strand sequences are also located predominantly in the nucleoli.

4.2 Confocal Laser Scanning Microscopy (CLSM)

The observations described so far still do not give the precise location of the viroid molecules. Whether they are on the surface or within the nucleolus is very critical for discussions about function, and this question can be answered optically by confocal laser scanning microscopy. This method scans fluorescence intensity with a focal depth of around 0.4 μm and shows excellent discrimination against emissions from out-of-focus regions [29].

Figure 6 shows serial scans of a single nucleus in five different focal planes (A–E). The distance between each plane was 0.8 μm. Fluorescence intensities were transformed to colors in order of increasing intensity; they are shown in the insert and progress from grey, red, orange, to yellow. When focus is in the plane of the largest nucleolar cross-section (Fig. 6D), fluorescence intensity increases steeply from the edge of the nucleous and reaches a maximum within a 0.3-μm-wide central zone, where it remains nearly uniform. This is clearly shown in Fig. 6F, in which the distribution of fluorescence in the cross-section 6D is displayed as a three-dimensional projection with pixel grey values converted to vertical lines of proportional length. In Fig. 6G the same values are represented as a wireframe plot.

Fig. 5. Location of viroids in nuclei from PSTVd-infected tomato plants by in situ hybridization showing rhodamine/lissamine fluorescence with the viroid-(+)strand-specific probe **A**, and FITC fluorescence in Bv96-stained nucleoli **B**. Within the nuclei the viroids are labeled in the regions corresponding to the nucleoli. Bars = 10 μm. [25]

Fig. 6. Three-dimensional distribution of viroids in PSTVd-infected tomato nuclei reconstructed by in situ hybridization and evaluation of sequential images by CLSM. **A** to **E** (*from left, first row to middle, second row*) show a set of optical sections, separated by 0.8 μm, through one nucleolus. Rhodamine fluorescence intensity was measured at $\lambda > 570$ nm and is illustrated in the pseudocolor range grey, red, orange, and yellow (increasing values). **F** (*left, third row*) is a 3-D projection of the fluorescence intensity distribution from the cross-section of **D**. Here pixel gray values are convered to vertical lines of proportional length, using the same color code. **G** (*right, third row*) illustrates the same values as the 3-D plot in a wireframe. Bar = 2 μm. [25]

These observations permit the conclusion that viroid concentration is virtually constant throughout the nucleous.

4.3 Optical and Electron Microscopy of Sections

Adaptation of the in situ hybridization procedure to sections of isolated nuclei promises viroid detection with a much higher level of spatial resolution. Special fixation and embedding in the hydrophilic resin Lowicryl K4M keeps the nucleic acids accessible for hybridization with biotinylated (+)strand-specific probes. Following hybridization, the sections are incubated with fluorescence-labeled streptavidin and examined by light microscopy. Extremely weak but reproducible fluorescence signals could be observed in some of the nuclei. In all instances the viroid-specific signals were limited to the nucleoli but were spread uniformly over their entire cross-section. Examination of a number of serial sections showed that there was no higher fluorescence intensity at the periphery of nucleoli, in agreement with the results obtained with CLSM. This shows clearly that the viroids are located inside the nucleoli and not on their surface.

To date, the greatest spatial resolution of viroid location has been achieved with in situ hybridization electron microscopy. Biotinylated probes are labeled with streptavidin-gold on thin sections. Some preliminary results are illustrated in Fig. 7. The unstained section (Fig. 7A) shows a significantly higher density of gold particles (arrowhead) on the nucleolus than over the rest of the nucleus. Fig. 7B shows a nucleus that, after gold labeling, has been stained with uranyl acetate. This general stain increases the contrast and permits identification of, for example, heterochromatin, which appears rather denser than the nucleolus. A cluster of gold grains is located near the border of the nucleolus and suggests that this level of resolution may identify clustering of viroid molecules. However, one cannot dismiss the homogenous distribution that has been observed with optical instruments until the in situ hybridization procedures are shown to exclude artifacts.

The intracellular location of viroids clearly differentiates them from viruses. Most viruses accumulate and replicate in the cytoplasm, just a few in the nucleus.

Fig. 7. Electron micrographs of thin sections of nuclei from PSTVd-infected tomato plants after in situ hybridization and gold labeling. After formaldehyde fixation and embedding in Lowicryl K4M thin sections were placed on nickel grids. For in situ hybridization these were floated on drops of the relevant solution (for details see [25, 33]), rinsed with distilled water, air-dried, and examined in an EM. **A** Unstained section of a nucleus after hybridization with a (+)strand-specific probe. Note gold particles (*arrow head*) located on the nucleous. **B** Uranylacetate/lead citrate-stained section after hybridization with a (+)strand-specific probe. Gold particles are shown over the nucleous as *black points* that are difficult to recognize (*arrows*). A distinct cluster (*arrow heads*) can be seen at the edge of the nucleous, and several individual grains are located elsewhere over its matrix

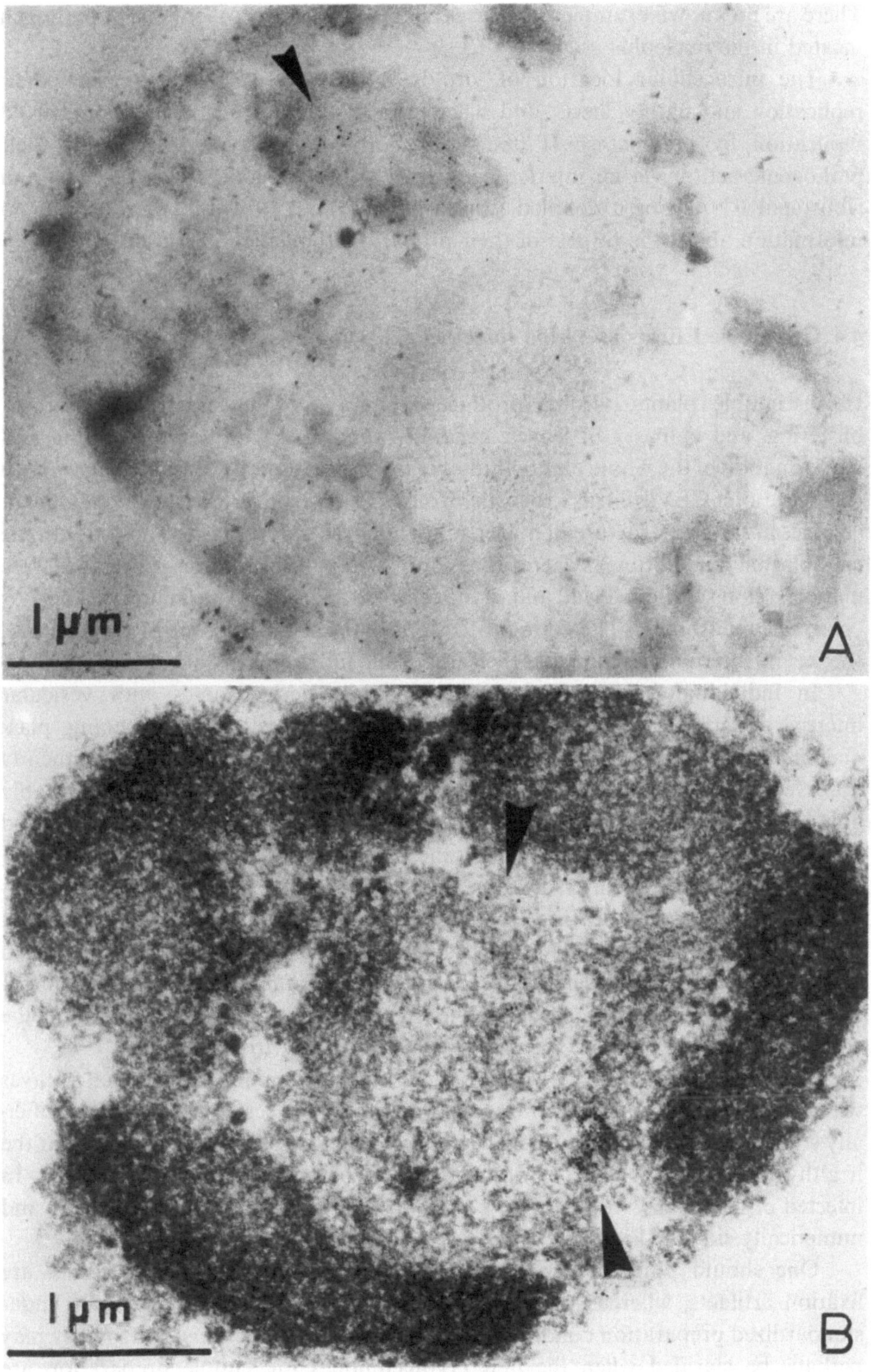
1 µm
A
1 µm
B

There are no known examples of viruses, viral RNA, or any other form of pathogen located in the nucleolus.

The intracellular location of viroids is relevant when we consider their replication and pathogenesis, and when we evaluate current ideas about viroid replication by polymerase II (see above) and recent hypothesis about their pathogenic action via an interference with 7SRNA [31], snRNA's [27, 32], and ribosomal RNA. For a detailed discussion see [25]. However, there is still no information about the targets of their primary pathogenic interaction.

4.4 Cytopathic Effects of Viroid Infection

In susceptible plants, viroids produce symptoms of mosaic, veinal necrosis, blistering, and epinasty of leaves, generally combined with severe stunting and malformation of the whole plant. Studies of the cytopathology of *Gynura aurantiaca* leaf tissue after CEVd (citrus exocortis viroid) infection did not reveal any significant changes in the fine structure of nuclei, mitochondria, chloroplasts, ribosomes in the endoplasmic membrane system. The only cytopathic changes observed were malformations of the cell wall and of the cell wall-associated "paramural bodies" or "plasmalemmasomes". These are protrusions of the cell membrane of variable size, shape, and internal structure and their origin and function has not yet been clarified.

In individual healthy *Gynura* plants, plasmalemmasomes with vesicular internal structures are found (Fig. 8A). In regions of cell wall thickening, plasmalemmasomes between cell wall and plasma membrane are quite frequently observed. In CEVd-infected *Gynura*, however, plasmalemmasomes are notably abnormal in size, shape, and internal structure (Fig. 8B) and have never been found between cell wall and plasma membrane. Moreover, malformed plasmalemmasomes occur exclusively in association with the irregular cell walls typical of CEVd-infected and symptom-bearing *Gynura* leaf tissue. It should be emphasized that plasmalemmasomes are found at the same frequency in healthy and CEVd-infected *Gynura* [34, 35]. These observations contradict earlier findings that plasmalemmasomes are only present in tissue infected with CEVd and PSTVd [36], and the conclusion that they are primary cytopathic effects of viroid infection.

The results shown in Fig. 8 are supported by examination of tomato leaves where plasmalemmasomes occur regularly in PSTVd-infected, healthy, and artificially stunted uninfected plants [12]. There are, however, clear differences in that the healthy plants show rather fewer and less prominent plasmalemmasomes. In infected or artificially stunted plants these structures appear morphologically and numerically similar (Fig. 9A, B).

One should note that some authors believe that plasmalemmasomes are fixation artifacts, whereas other authors have found plasmalemmasomes under standardized preparation conditions (cf. Discussion in [34]). This controversy may perhaps be clarified when the new cryotechniques for sample preparation are

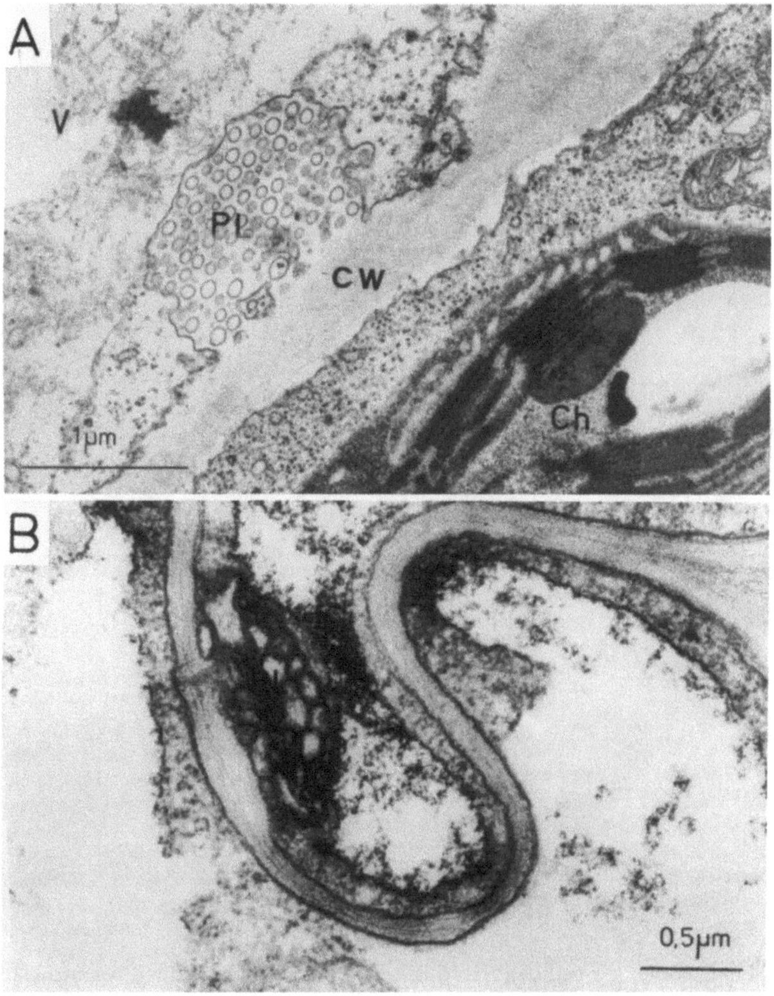

Fig. 8. Cytopathic effects of viroid infection. *CW* cell wall; *Ch* chloroplast; *V* vacuole. **A** Vesicular plasmalemmasome (*PL*) in tissue of healthy *Gynura aurantiaca*. **B** Vesicular PL in CEVd-infected *G. aurantiaca*. Note the irregular internal structures and association with malformed swollen cell walls. [2]

applied. Since plasmalemmasome malformation always occurred in connection with cell wall distortions, one may argue that the primary cytopathic alteration is in the cell wall, and this leads, perhaps amplified by preparation artifacts, to structural irregularities of the plasmalemmasomes.

In the chlorotic spots characteristic leaf tissue of "Mistletoe" chrysanthemum infected with CSVd (chrysanthemum stunt viroid), the number of chloroplasts was found to be reduced to about one third if compared with the adjacent green tissue or tissue from healthy plants. Electron microscopy showed that the thylakoid

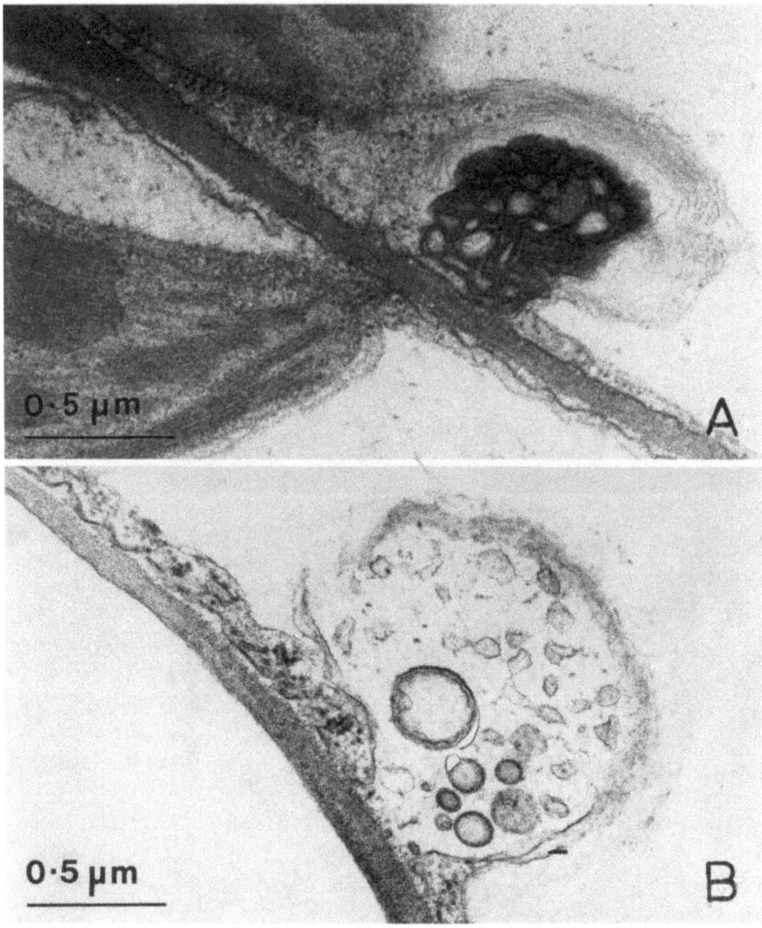

Fig. 9. A Plasmalemmasome in a cell of PSTVd-infected and severely stunted tomato leaf. **B** Plasmalemmasome in a cell of an uninfected artificially stunted tomato plant leaf

membrane system of these chloroplasts is irregular and filled with osmiophilic inclusions. In addition, parenchyma cells contain extreme accumulation of microfilaments which are rarely found in healthy tissue [37]. The cytopathic effects observed are not always "viroid-specific". The literature [37] describes similar structural changes in virus-infected and in senescent plants.

5 Prospects

When summarizing the state of knowledge of viroids one has to conclude that there is a great deal of information on viroid structure and location, and a fair understanding of viroid replication; several details of their cytopathology and symptomatology have also emerged. However, the link between the molecular structure of viroids and their cytopathic effects in plants is still missing. Further research should be directed towards understanding how viroids cause disease, with the aim of preventing the losses caused by these pathogens in many parts of the globe.

References

1. Diener TO (1979) Viroid and Viroid diseases. Wiley, New York
2. Sänger HL (1982) Biology, structure, functions and possible origin of viroids. In: Parthier B, Boulter D (eds) Encyclopedia of plant physiology, New Series, Vol 14B. Springer, Berlin Heidelberg New York, pp 368–454
3. Riesner D, Gross HJ (1983) Viroids, Annu Rev Biochem 54: 531–564
4. Diener TO (ed) (1987) The Viroids. Plenum, New York
5. Riesner D, Steger G (1980) In: Saenger W (ed) Viroids and viroid-like RNAs. Landolt-Börnstein, New series in biophysics-nucleic acids, Vol VII/1. Springer, Berlin Heidelberg New York Tokyo (in press)
6. Sogo JM, Koller T, Diener TO (1973) Potato spindle tuber viorid. X. Visualization and size determination by electron microscopy, Virology 55: 70–80
7. Sänger HL, Klotz G, Riesner D, Gross HJ, Kleinschmidt AK (1976) Viroids are single stranded covalently closed circular RNA molecules existing as highly base-paired rod-like structures. Proc Natl Acad Sci USA 73: 3852–3856
8. Riesner D, Henco K, Rokohl U, Klotz G, Kleinschmidt AK, Domdey H, Jank P, Gross HJ, Sänger HL (1979) Structure and structure formation of viroids. J Mol Biol 133: 85–115
09. McClements WL, Kaesberg P (1977) Size and secondary structure of potato spindle tuber viroid. Virology 76: 477–484
10. Randles JW, Hatta T (1979) Circularity of the ribonucleic acids associated with cadang-cadang disease. Virology 96: 47–53
11. Palukaitis P, Symons RH (1980) Purification and characterization of the circular form of chrysanthemum stunt viroid. J Gen Virol 46: 477–489
12. Gruner R, Santore UJ (1990) Comparative observations of plasmalemmasomes in PSTVd-infected, healthy and artificially stunted healthy tomato plants Phytopathol. (in press)
13. Ohno T, Akiya J, Higuchi M, Okada Y, Yoshikawa N, Takahashi T, Hashimoto J (1982). Purification and characterization of hop stunt viroid. Virology 118: 54–63
14. Chen W, Tien P, Zhu YX (1983) Viroid-like RNAs associated with burdock stunt disease. J Gen Virol 64: 409–415
15. Riesner D (1987) Structure formation. In: Diener TO (ed) The viroids Plenum, New York, pp 63–98
16. Steger G, Hoffmann H, Förtsch J, Gross HJ, Randles JW, Sänger HL, Riesner D (1984) Conformational transitions in viroids and virusoids: Comparison of results from energy minimization algorithm and from experimental data. J Biomol Struct Dyn 2: 548–571
17. Gross HJ, Riesner D (1980) Viroids: A class of subviral pathogens. Angew Chem Int Ed Engl 19: 231–243
18. Hecker R, Wang Z, Steger G, Riesner D (1988) Analysis of RNA structures by temperature-gradient gel electrophoresis: replication and processing. Gene 72: 59–74
19. Loss P (1990) Thesis. Heinrich-Heine-Universität Düsseldorf
20. Mühlbach HP, Sänger HL (1979) Viroid replication is inhibited by α-amanitin. Nature 278: 185–187

21. Rackwitz, HR, Rohde W, Sänger HL (1981) DNA-dependent RNA polymerase II of plant-origin transcribes viroid RNA into full-length copies. Nature 291: 297–301
22. Goodman ThC, Nagel L, Rappold W, Klotz G, Riesner D (1984) Viroid replication: Equilibrium association constants and comparative activity measurements for the viroid-polymerase interaction. Nucleic Acids Res 12: 6231–6246
23. Diener TO (1971) Potato spindle tuber virus: A plant virus with properties of a free nucleic acid. III. Subcellular location of PSTV-RNA and the question of whether virions exist in extracts or in situ. Virology 43: 75–89
24. Sänger HL (1972) An infectious and replicating RNA of low molecular weight: the agent of the exocortis disease of citrus. Adv Biosci 8: 103–116
25. Harders J, Lukacs N, Robert-Nicoud M, Jovin TM, Riesner D (1989) Imaging of viroids in nuclei from tomato leaf tissue by in situ hybridization and confocal laser scanning microscopy. EMBO J 8: 3941–3949
26. Frasch M (1985) Thesis, Eberhard-Karls-Universität, Tübingen, FRG Charakterisierung chromatinassoziierter Kernproteine von *Drosophila melanogaster* mit Hilfe monoklonaler Antikörper
27. Schumacher J, Sänger HL, Riesner D (1983) Subcellular localization of viroids in highly purified nuclei from tomato leaf tissue. EMBO J 2: 1549–1555
28. Riesner D (1987) Subcellular location and in situ association with cellular components. In: Diener TO (ed) The viroids. Plenum, New York, pp 99–116
29. Jovin TM, Arndt-Jovin DJ (1989) Luminescence digital imaging microscopy. Annu Rev Biophys Chem 18: 271–308
30. Binfrt M, Tourmente S, Roth J, Renaud M, Gehring WJ (1986) In situ hybridization at the electron microscope level: Localization of transcripts on ultrathin sections of Lowicryl K4M-embedded tissue using biotinylated probes and protein A-gold complexes. J Cell Biol 102: 1646–1653
31. Haas B, Klanner A, Ramm K, Sänger HL (1988) The 7SRNA from tomato leaf tissue resembles a signal recognition particle RNA and exhibits a remarkable sequence complementary to viroids. EMBO J 7: 4063–4074
32. Jacab G, Kiss T, Solymosy F (1986) Viroid pathogenicity and pre-rRNA processing: A model amendable to experimental testing. Biochem Biophys Acta 868: 190–197
33. Hardners J (1989) Thesis. Heinrich-Heine-Universität Düsseldorf. Sub-zelluläre Lokalisation der Viroide mit biochemischen und biophysikalischen Methoden
34. Wahn K, Rosenberg de Gomez F, Sänger HL (1980) Cytopathologie von viroid-infiziertem Pflanzengewebe. I Veränderungen des Plasmalemmas und der Zellwand bei *Gynura aurantiaca* DC nach Infektion mit dem Viroid der Citrus Exocortis Krankheit (CEV). Phytopathol Z 98: 1–18
35. Wahn K, Rosenberg de Gomez F, Sänger HL (1980) Cytopathic changes in leaf tissue of *Gynura aurantiaca* infected with the viroid of the exocortis disease of citrus. J Virol 49: 355–365
36. Semancik JS, Vanderwoude WJ (1976) Exocortis viroid: Cytopathic effects at the plasma membrane in association with pathogenic RNA. Virology 69: 719–726
37. Rosenberg de Gomez F, Wahn K, Sänger HL (1985) Cytopathologie von viroid-infiziertem Pflanzengewebe. II. Licht- und elektronenmikroskopische Untersuchungen an Blattgewebe von *Chrysanthemum morifolium* der Sorte 'Mistletoe' nach Infektion mit dem Viroid der Stauchekrankheit der Chrysantheme (CSV). Phytopathol Z 114: 41–68

Chapter 13

J. LHERMINIER[1], E. BOUDON-PADIEU[1], R. MEIGNOZ[1], A. CAUDWELL[1] and R.G. MILNE[2]

Immunological Detection and Localization of Mycoplasma-Like Organisms (MLOs) in Plants and Insects by Light and Electron Microscopy

1 Introduction

Yellows diseases of plants were reported long ago on many plant species [37, 38]. Kunkel, in the 1920's [23], suspected a common and distinctive class of the causal agents. These yellows diseases had many features recognized as those of viruses, such as viruslike symptoms, *leafhopper vectors*, host specificity, and filtrability of the agent through bacterial filters. Nevertheless, there were symptoms such as greening or *phyllody* of the inflorescence, which were very particular. Characterization of the pathogen was impossible. Yet extracts were infectious when injected into the abdomen of healthy leafhopper vectors, although such infectivity was highly labile and temperature-sensitive.

Electron microscopy afforded a breakthrough in the etiology of these diseases when in 1967 Doi et al. [15] showed by electron microscopy studies of thin sections that phloem of infected plants contained bodies similar in morphology to animal and human mycoplasmas. These so-called mycoplasma-like organisms or MLOs have since been found associated with more than a hundred yellows, phyllody, or witches' broom diseases on many plant species all over the world. They have also been found in the salivary glands of infective insect vectors. An exhaustive review of these diseases and their respective hosts was given recently [30]. MLOs are wall-less prokaryotes, now classified in the class Mollicutes, without vacuoles or internal membranes, and are heterogeneous in electron density, shape, and size (50–1000 nm in diameter) [30].

However, although many lines of evidence may implicate MLOs as the cause of a given disease, Koch's postulates have never been fulfilled. MLOs occur widely in "symptomless" plants from the field. They are difficult to find in some woody plants affected by MLO-like yellows diseases. If not seen in numbers, MLOs are uncertainly identified among other membrane-bound vesicles. They are either

[1] INRA, Station de Recherches sur les Mycoplasmes et les Arbovirus des Plantes, F-21034 Dijon, Cedex, France
[2] Istituto di Fitovirologia Applicata, CNR, Strada delle Cacce 73, I-10135 Torino, Italy

scattered or observed as abnormal or degenerated profiles, that are difficult to identify. Despite many attempts, plant-infecting mycoplasmas (excepting spiroplasmas) have never been cultivated in vitro and have never been purified, so they could not be returned to healthy host plants to reproduce the original disease. One important problem is that all MLOs look alike in ultrathin sections in the electron microscope, whereas in negative stain in dip preparations they are not clearly distinguishable from host elements. Many MLOs cause similar symptoms in herbaceous hosts such as periwinkle (*Catharanthus*), so that MLO taxonomy and diagnostics have been in disarray.

With the light microscope, MLOs have mainly been detected using histochemical methods [7] but it has often been necessary to confirm the results by electron microscopy. Unfortunately, this can at present be done only on similar tissues but not on the same or contiguous samples. So, MLOs have been detected in situ in plants and insects mainly by thin sectioning and electron microscopy. However, the difficulty has frequently been pointed out of recognition of MLOs because of the presence of small pleomorphic membrane-limited vesicles in tissues, and the variability of shape, size, and electron density of true MLOs [12, 14, 16, 17, 20, 28, 31, 36].

We wish to present, in two parts, a selection of recent findings: first the results obtained in vitro by immunosorbent electron microscopy, and then those on in situ immunolabeling of thick or thin sections. Other aspects of MLO diagnostics have been reviewed earlier [18, 19, 29].

2 In Vitro Methods

Immunosorbent electron microscopy (ISEM) was the first serological technique to be successfully applied to MLOs although production of specific antibodies was hindered by the difficulty of obtaining purified antigen preparations. In 1976, Derrick and Brlansky [13] reported the detection by ISEM of corn stunt spiroplasma and in 1982, Caudwell and co-workers succeeded in serological trapping of grapevine Flavescence Dorée (FD)-MLO [3, 4, 5]. Similar work was done with aster yellows and peach eastern X MLOs [19, 27, 29, 32, 33, 34].

Caudwell and co-workers raised antibodies in rabbits immunized with enriched FD-MLO preparations either from infected *Vicia faba* L. plants or from infected *Euscelidius variegatus* Kbm leafhoppers [2]. MLOs were trapped from infected leafhopper extracts using crude antiserum against FD-infected *V. faba* [4, 5] and from infected *V. faba* extracts using antiserum against FD-infected *E. variegatus* [3, 5]. Healthy controls or use of pre-immune serum showed trapping of some undefined structures, but true positive reactions included the presence of a great number of distinctive *MLO* profiles.

Trapping by ISEM resulted in an increasing amount of MLOs on the grid, while much interfering cell debris was washed out. However, it had to be demonstrated that in all cases MLOs were retained on the grid by specific absorption due to the reactivity of antisera. This was done by surface decorating with antibodies against FD-MLO labeled with gold-conjugate (Fig. 1), or by cross-reacting experiments with antisera and MLO preparations from diverse origins [33, 34].

The ISEM technique is more convenient than other serological methods when antibody preparations lack specificity. Sinha and Chiykowski [34] therefore applied the technique to peach eastern X in celery, and antibody specificity was improved by better purification of MLO, which reduced the level of nonspecific reactions. MLOs were detected by ISEM in the leafhopper vectors of peach eastern X disease and Aster yellows, and appeared to be related to the infectivity of the hoppers [35].

Milne R.G. and Lenzi R. (unpubl.) have used ISEM in attempts to trap primula yellows MLOs from sap of infected *Catharanthus* plants, using the monoclonal antibody of Clark et al. [8]. In addition, they tested tomato big bud MLOs using an antiserum raised by Clark and colleagues [8] against partially purified infected plant material and later absorbed with healthy plant antigens. In both these cases, MLO bodies could not clearly be distinguished from vesicles of host origin, and gold-labeled decoration of the trapped MLOs was not very successful. Best results were obtained by placing 10 μl of MLO preparation in 100 μl of distilled water, causing osmotic bursting, and thus multiplying the number of MLO-specific "units" in the preparation. They could be more easily trapped and decorated with gold (see Milne, Chap. 7, this Vol, Fig. 17), although they no longer looked like MLOs and thus lost their distinctive morphology. The "chains of pearls" were labeled (Fig. 2).

Microscopy has been combined with serology in a different context to demonstrate the purification of *Aster yellows* MLO by affinity chromatography [22].

3 In Situ Methods

In situ indirect immunofluorescence staining has been used to provide a rapid method with thick hand-sections from infected plants. Monoclonal antibodies have been successful with Aster yellows MLO; comparing monoclonal and polyclonal antibodies Lin and Chen showed that background fluorescence occurring on cell walls and membranes with the polyclonal was absent with the monoclonal [26]. MLOs of Eastern Aster yellows, Western Aster yellows, potato witches' broom and clover proliferation have been detected in the same way in infected *Catharanthus*, using polyclonal antibodies [11], although the authors noted autofluorescence in the xylem. Similar staining was successful with a monoclonal raised against the MLO of maize bushy stunt [6]. The in situ immunofluorescence technique, with

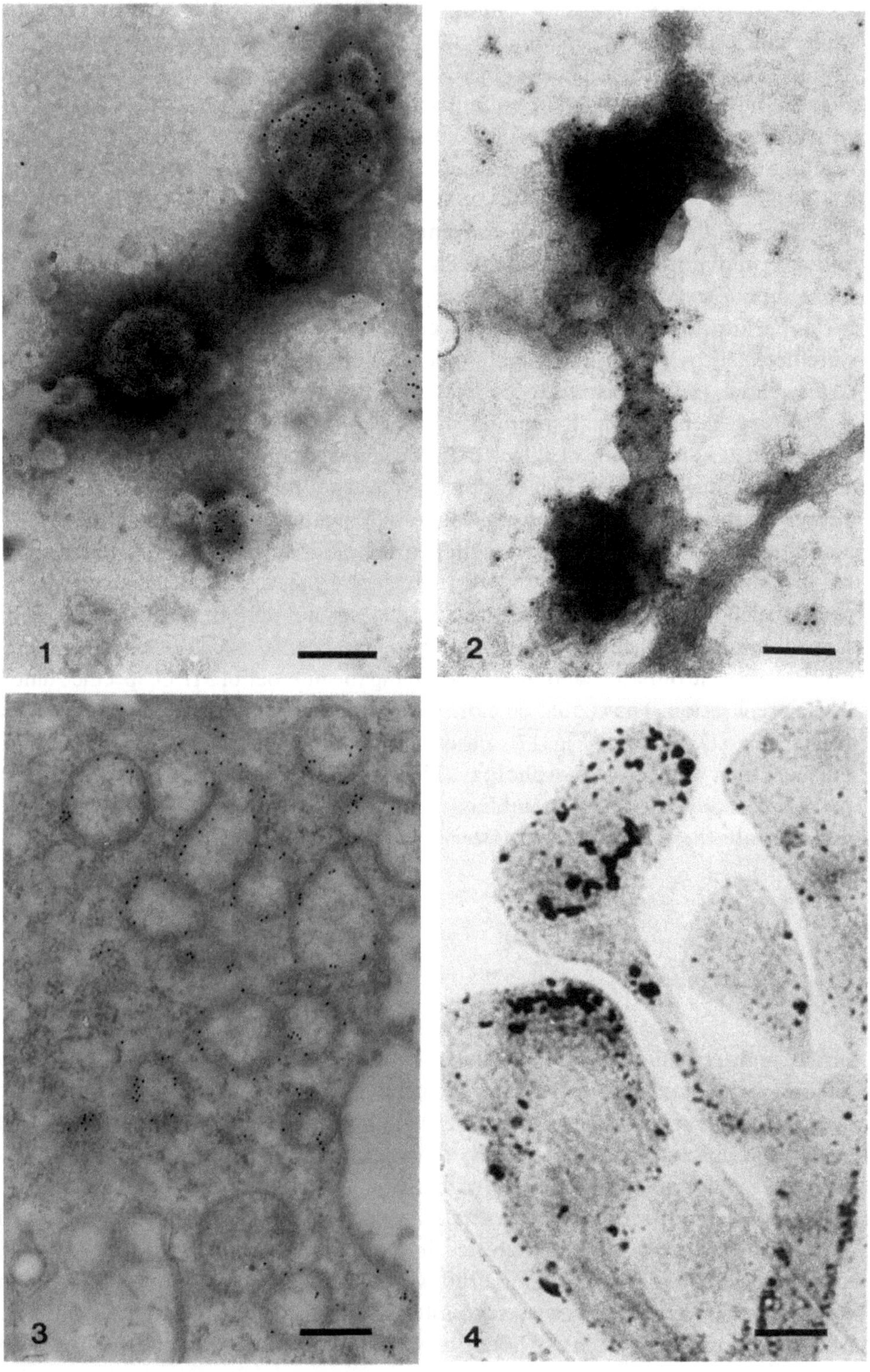
1
2
3
4

polyclonal antibodies, was also used to detect FD-MLO in excised salivary glands of the vector *E. variegatus* [24]. More recently, application of the method to tomato stolbur and faba-bean phyllody has been reported [9].

A post-embedding immunolabeling technique was used for the first time to identify the FD-MLO in salivary glands of the leafhopper vector [25]; gold probes and use of acrylic resins (LK4M or LR White) or ultrathin cryosections allowed accurate and highly specific detection of the MLO. Labeling was mainly associated with the membrane (Fig. 3) and allowed identification of electron-dense forms as true MLOs.

Post-embedding gold immunolabeling was also studied by light microscopy [25]. Historesin-embedded salivary glands of FD infected *E. variegatus* were treated with rabbit polyclonal anti-FD-MLO antibodies, then with gold-conjugated anti-rabbit IgG, and the gold signal was intensified with silver [10, 21] (Fig. 4). Specificity of the silver stain was checked on semi-thin serial sections from LK4M or LR White samples previously used for gold immunolabel electron microscopy. The results suggested that only certain cells of the posterior lobes of the salivary glands were infected. The method is highly specific, simple, short, and allowed positive identification of FD-MLOs with the light microscope.

Classical *post-embedding labeling* is convenient but aldehyde fixation followed by embedding in acrylic resins, gives relatively poor preservation of fine detail and especially of membranes. Post-embedding labeling of osmium-fixed epoxy-embedded material is sometimes possible [1] but was found unsuitable for FD-MLOs [25]. Post-embedding techniques can of course only be used to label antigens that lie on the surface of the section, and not in the bulk of material within the section, that contributes most of the electron-density and information to the image.

Milne and Lenzi (unpubl.) attempted *pre-embedding labeling* of MLOs, using primula yellows and tomato big bud as subjects, and the monoclonal and polyclonal antibodies of Clark mentioned above. The method was to fix the material in

Fig. 1. Detection by ISEM of the Flavescence Dorée MLO in infective *Euscelidius variegatus* extracts. Antigens were trapped with polyclonal rabbit antibodies against FD-infected broadbeans. Surface decoration was then done with the same antibodies followed by protein A-gold, and the preparation was stained with ammonium molybdate. Bar = 0.30 μm. (INRA, DIJON)

Fig. 2. Detection by ISEM of primula yellows MLO from infected *Catharanthus* extracts. Partially purified and osmotically disrupted MLO fragments were trapped with monoclonal antibodies followed by 5 nm gold-conjugated goat antimouse antibodies, and staining with 0.2% aqueous uranyl acetate. Bar = 0.10 μm. (CNR, TORINO)

Fig. 3. Post-embedding indirect gold immunolabeling of FD-MLO in *E. variegatus* salivary glands fixed with 4% paraformaldehyde-0.1% glutaraldehyde followed by LR White embedding at −15 °C. Bar = 0.30 μm. (IRNA, DIJON)

Fig. 4. Same material as in Fig. 3; embedding in LKB-historesin and immunogold labeling revealed by the silver enhancement technique. Bar = 25 μm. (INRA, DIJON)

aldehyde, rinse in buffer, slice finely so as to expose many cut ends of the phloem, apply the label (first antibody, second gold-labeled antibody), fix in osmium, and embed in Epon. The label is thus embedded within the material. Excellent labeling of MLO membranes was achieved, confirming that the target antigens were exposed on the MLO surface (see Milne, Chap. 7, this Vol, Fig. 16). As noted above, small dense forms were labeled, as were the "chains of pearls" connecting the more obvious MLO bodies. Background label was absent, not only with the monoclonal but also the polyclonal antibodies. Absence of background with the polyclonal may be connected to the fact that only unbroken surfaces were exposed.

Clearly, the pre-embedding approach is only adapted to experimental situations where MLOs are present in huge amounts; it consumes greater volumes of reagents, and is cumbersome in terms of experimental tactics. However, it may well be useful for diagnostic or taxonomic purposes, especially in cunjunction with, and as a check on post-embedding methods.

References

1. Bendayan M, Zollinger M (1983) Ultrastructural localization of antigenic sites on osmium-fixed tissues applying the protein A-gold technique. J Histochem Cytochem 31: 101–109
2. Boudon-Padieu E, Larrue J, Caudwell A (1989) ELISA and Dot-Blot detection of Flavescence Dorée MLO in individual leafhopper vectors during latency and inoculative state. Curr Microbiol 19: 357–364
3. Caudwell A, Meignoz R, Kuszala C, Larrue J, Fleury A, Boudon E (1982) Purification sérologique et observation ultramicroscopique de l'agent pathogène (MLO) de la flavescence dorée de la vigne dans les extraits liquides de fèves malades. C R Soc Biol 176: 723–729
4. Caudwell A, Meignoz R, Kuszala C, Schneider C, Larrue J, Fleury A, Boudon E (1982) Purification immunologique et observation ultramicroscopique en milieu liquide de l'agent pathogène (MLO) d'une jaunisse végétale, la flavescence dorée de la vigne. CR Acad Agric Fr 68: 407–415
5. Caudwell A, Meignoz R, Kuszala C, Schneider C, Larrue J, Fleury A, Boudon E (1983) Serological purification and visualization in the electron microscope of the grapevine Flavescence Dorée pathogen (MLO) in diseased plants and infectious vectors extracts. Yale J Biol Med 56: 936–937
6. Chen TA, Jiang XF (1988) Monoclonal antibodies against the maize bushy stunt agent. Can J Microbiol 34: 6–11
7. Chen TA, Lei JD, Lin CP (1989) Detection and identification of plant and insect mollicutes. In: Whitcomb RF, Tully JG (eds) The mycoplasmas, Vol 5. Academic Press, San Diego pp 393–424
8. Clark MF, Morton A, Buss SL (1989) Preparation of mycoplasma immunogens from plants and a comparison of polyclonal and monoclonal antibodies made against primula yellows MLO-associated antigens. Ann Appl Biol 114: 111–124
9. Cousin MT, Dafalla G, Demazeau E, Theveu E, Grosclaude J (1989) In situ detection of MLOs for Solanaceae stolbur and *Faba* bean phyllody by indirect immunofluorescence. J Phytopathol 124: 71–79
10. Danscher G, Norgaard RJO (1983) Light microscopic visualization of colloidal gold on resin-embedded tissue. J Histochem Cytochem 31: 1394–1398

11. Da Rocha A, Ohki ST, Hiruki C (1986) Detection of mycoplasmalike organisms in situ by indirect immunofluorescence microscopy. Phytopathology 76: 864–868
12. Davis RE, Whitcomb RF (1971) Mycoplasmas, Rickettsiae and Chlamidiae: possible relation to yellows diseases and other diseases of plants and insects. Annu Rev Phytopathol 9: 119–154
13. Derrick KS, Brlansky RH (1976) Assay for viruses and mycoplasmas using serologically specific electron microscopy. Phytopathology 66: 815–820
14. De Zoeten GA, Gaard G, Diez FB (1972) Nuclear vesiculation associated with pea enation mosaic virus-infected plant tissue. Virology 44: 638–647
15. Doi Y, Teranaka M, Yora K, Asuyama H (1967) Mycoplasma or PLT group-like microorganisms found in the phloem elements of plants infected with mulberry dwarf, potato witches'broom, aster yellows, or Paulownia witches'broom. Ann Phytopathol Soc Jpn 3: 259–266
16. Giannotti J, Vago C, Devauchelle G, Marchoux G (1968) Recherches sur les micro-organismes de type mycoplasme dans les cicadelles vectrices et dans les végétaux atteints de jaunisses. Entomol Exp Appl 11: 470–474
17. Hampton RO, Florance ER, Whitcomb RF, Seidler RJ (1976) Evidence suggesting nonassociation of mycoplasma with Pea disease. Phytopathology 66: 1163–1168
18. Hiruki C (ed) (1988a) Tree mycoplasmas and mycoplasma diseases. University of Alberta Press, Edmonton, pp 245
19. Hiruki C (1988b) Rapid and specific detection methods for plant mycoplasmas. In: Maramorosch K, Raychauhuri SP, Mycoplasma diseases of crops. Springer, Berlin Heidelberg New York Tokyo, pp 77–101
20. Hirumi H, Maramorosch K (1969) Mycoplasma-like bodies in the salivary glands of insect vectors carrying the aster yellows agent. J Virol 3: 82–84
21. Holgate CS, Jackson P, Cowen PN, Bird CC (1983) Immunogoldsilver staining: a new method of immunostaining with enhanced sensitivity. J Histochem Cytochem 31: 938–944
22. Jiang YP, Lei JD, Chen TA (1988) Purification of Aster yellows agent from diseased lettuce using affinity chromatography. Phytopathology 78: 828–831
23. Kunkel LO (1926) Studies on aster yellows. Am J Bot 13: 646–705
24. Lherminier J, Terwisscha van Scheltinga T, Boudon-Padieu E, Caudwell A (1989) Rapid immunofluorescent detection of the grapevine Flavescence Dorée mycoplasmalike organism in the salivary glands of the leafhopper *Euscelidius variegatus* Kbm. J Phytopathol 125: 353–360
25. Lherminier J, Prensier G, Boudon-Padieu E, Caudwell A (1990) Immunolabeling of Grapevine Flavescence Dorée MLO in salivary glands of *Euscelidius variegatus*: a light and electron microscopy study. J Histochem Cytochem 38: 79–89
26. Lin CP, Chen TA (1986) Comparison of monoclonal antibodies and polyclonal antibodies in detection of the aster yellows mycoplasmalike organism. Phytopathology 76: 45–50
27. Maramorosch K, Raychaudhuri SP (eds) (1988) Mycoplasma diseases of crops. Springer Berlin Heidelberg New York Tokyo, pp 456
28. Maramorosch K, Shikita E, Granados RR (1968) Structures resembling mycoplasma in diseased plants and in insect vectors. Trans NY Acad Sci II 30: 841–855
29. Markham PG (1988) Detection of mycoplasmas and spiroplasmas in insects. In: Hiruki (ed) Tree mycoplasmas and mycoplasma diseases. University of Alberta Press, Edmonton, pp 157–177
30. McCoy RE, Caudwell A, Chang CJ, Chen TA, Chiykowski LN, Cousin MT, Dale JL, de Leeuw GTN, Golino DA, Hackett KJ, Kirkpatrick BC, Marwitz R, Petzold H, Sinha RC, Sugiura M, Whitcomb RF, Yang IL, Zhu BM, Seemüller E (1989) Plant diseases associated with mycoplasma-like organisms. In: Whitcomb RF, Tully JG (eds) The mycoplasmas, Vol 5. Academic Press, San Diego, pp 545–640
31. Raine J, Forbes AR, Skelton FE (1976) Mycoplasma-like bodies, Rickettsia-like bodies and salivary bodies in the salivary glands and saliva of the leafhopper *Macrosteles fascifrons* (Homoptera: Cicadellidae). Can Entomol 108: 1009–1019

32. Sinha RC (1988) Serological detection of mycoplasmalike organisms from plants affected with yellows diseases. In: Hiruki C (ed) Tree mycoplasmas and mycoplasma diseases. University of Alberta Press, Edmonton, pp 143–156
33. Sinha RC, Benhamou N (1983) Detection of mycoplasmalike organism antigens from aster yellows-diseased plants by two serological procedures. Phytopathology 73: 1199–1202
34. Sinha RC, Chiykowski LN (1984) Purification and serological detection of mycoplasma-like organisms from plants affected by peach eastern X disease. Can J Plant Pathol 6: 200–205
35. Sinha RC, Chiykowski LN (1986) Detection of mycoplasma like organisms in leafhopper vectors of aster yellows and peach X disease by immunosorbent electron microscopy. Can J Plant Pathol 8: 387–393
36. Sinha RC, Paliwal YC (1969) Association, development and growth cycle of mycoplasma-like organisms in plants affected with clover phyllody. Virology 39: 759–767
37. Subba Rao M, Muniyappa V (1988) Epidemiology of sandal spike disease. In: Hiruki C (ed) Tree mycoplasmas and mycoplasma diseases. University of Alberta Press, Edmonton, pp 57–68
38. Wang MQ, Maramorosch K (1988) Earliest historical recond of a tree mycoplasma disease: beneficial effect of mycoplasmalike organisms on peonies. In: Maramorosch K, Raychaudhuri SP (eds) Mycoplasma diseases of crops. Springer, Berlin Heidelberg New York Tokyo, pp 349–356

Chapter 14 Ian Brown and John Mansfield

Interactions Between Pseudomonas and *Phaseolus vulgaris*

1 Introduction

The value of electron microscopy to studies of the processes of resistance to fungi has, in part, been due to the existence of clearly defined stages in fungal development characteristic of successful parasitism, e.g. the formation of penetration pegs and haustoria [2, 8, 16, 18, 23]. In contrast to the successes achieved with fungi, attempts to use microscopy to develop an understanding of specificity in the interactions between plants and phytopathogenic bacteria, most species of which are confined to the intercellular spaces in resistant and susceptible plants, have produced less conclusive results [1, 4, 7, 9, 11, 13, 25, 27–29]. We have therefore developed histochemical and quantitative approaches to study bacterial growth and the plant's response in interactions between pseudomonads and mesophyll cells of the French bean, *Phaseolus vulgaris* [3, 17]. The plant/bacterium interactions which have been examined are summarized in Table 1. Different types of interaction were compared, *viz* the restriction of a saprophytic bacterium, non-host resistance and race specific resistance both expressed by the hypersensitive reaction (HR), and the compatible interaction between the halo-blight bacterium *Pseudomonas syringae* pv. *phaseolicola* and susceptible bean cultivars.

In order to visualize polysaccharides of bacterial and plant origin, ruthenium red and the periodic acid, thiocarbohydrazide, silver proteinate (PATAg) staining procedure were used. Ruthenium red strongly stains the highly branched acidic polymers present in many forms of extracellular polysaccharide (EPS) whereas the PATAg stain is more specific, reacting predominantly with 1,4 linked molecules notably cellulose and pectic materials [24]. More recently, a specific defence response, the accumulation of hydroxyproline-rich glycoproteins (HRGPs) has also been analysed using post-embedding immunogold labelling with antibodies to melon $HRGP_{2b}$ [20, 22].

Some of the images obtained from our studies of leaves of the bean cv. Tendergreen are illustrated in Figs. 1–4.

Department of Biochemistry and Biological Sciences, Wye College (University of London), Ashford, Kent, TN25 5AH, UK

Table 1. Summary of interactions examined

Bacterium	Characteristic	Reaction[a] observed in leaves of bean cultivars	
		Tendergreen	Red Mexican
P. fluorescens	Saprophyte	NS	NS
P.s. pv. *coronafaciens*	Oat halo-blight pathogen	HR	HR
P.s. pv. *phaseolicola*	Bean halo-blight pathogen		
Race 1		S	HR
Race 3		HR	S
Race 3Ml[b]		S	S
Race 1 (p3A1000)[c]		HR	HR
Race 1 (p3A001)[d]		S	HR

[a] Symptom appearance: NS, no symptoms; HR, hypersensitive reaction, collapse and browning of the inoculation site; S, susceptible response, delayed collapse of infiltrated tissue and the formation of a yellow halo around the lesion.
[b] *P.s.* pv. *phaseolicola* race 3Ml is a nitrosoguanidine-derived mutant of race 3 [10].
[c] *P.s.* pv. *phaseolicola* race 1 (p3A1000) is a transconjugant harbouring the gene for avirulence from race 3 in the genomic clone p3A1000 [12].
[d] *P.s.* pv. *phaseolicola* race 1 (p3A001) is a transconjugant containing cloned DNA from race 3 which does not affect the virulence of race 1.

2 The Saprophyte

Pseudomonas fluorescens fails to multiply or produce significant amounts of EPS in bean. Following inoculation by injection, bacterial cells seem tightly bound at junctions between mesophyll cells (Fig. 1a). Amorphous deposits, which are PATAg-positive (Fig. 1b), accumulate around bacteria; immunocytochemistry has demonstrated the presence of HRGPs within the encapsulating material. At some sites, HRGPs and PATAg staining material also accumulate within mesophyll cells adjacent to the bacteria, but such deposits are rarely associated with the formation of distinct paramural papillae. The accumulation of HRGPs in material encapsulating cells of *P.s. fluorescens* (Fig. 1c) is consistent with the involvement of these

Fig. 1. Cells of *P. fluorescens* at mesophyll cell junctions in leaves of cv. Tendergreen. **a** Examined 24 h after inoculation. The shape and position of the bacteria (*B*) suggest a close association with the plant cell wall. The tissue was stained with ruthenium red; no fibrillar material is present but small amorphous deposits (*arrows*) are associated with the bacteria. Bar = 0.5 μm. **b** Examined 12 h after inoculation. The bacterium (*B*) is surrounded by PATAg-positive material (*arrows*) which appears to be continuous with the plant cell wall. A Golgi apparatus (*asterisk*) containing PATAg-positive material is present in an adjacent plant cell. Bar = 0.5 μm. **c** Examined 10 h after inoculation. Immunogold labelling shows that HRGP is present in the cell wall adjacent to bacteria (*B*) and in the encapsulating material (*arrows*) in the intercellular space. Bar = 0.5 μm

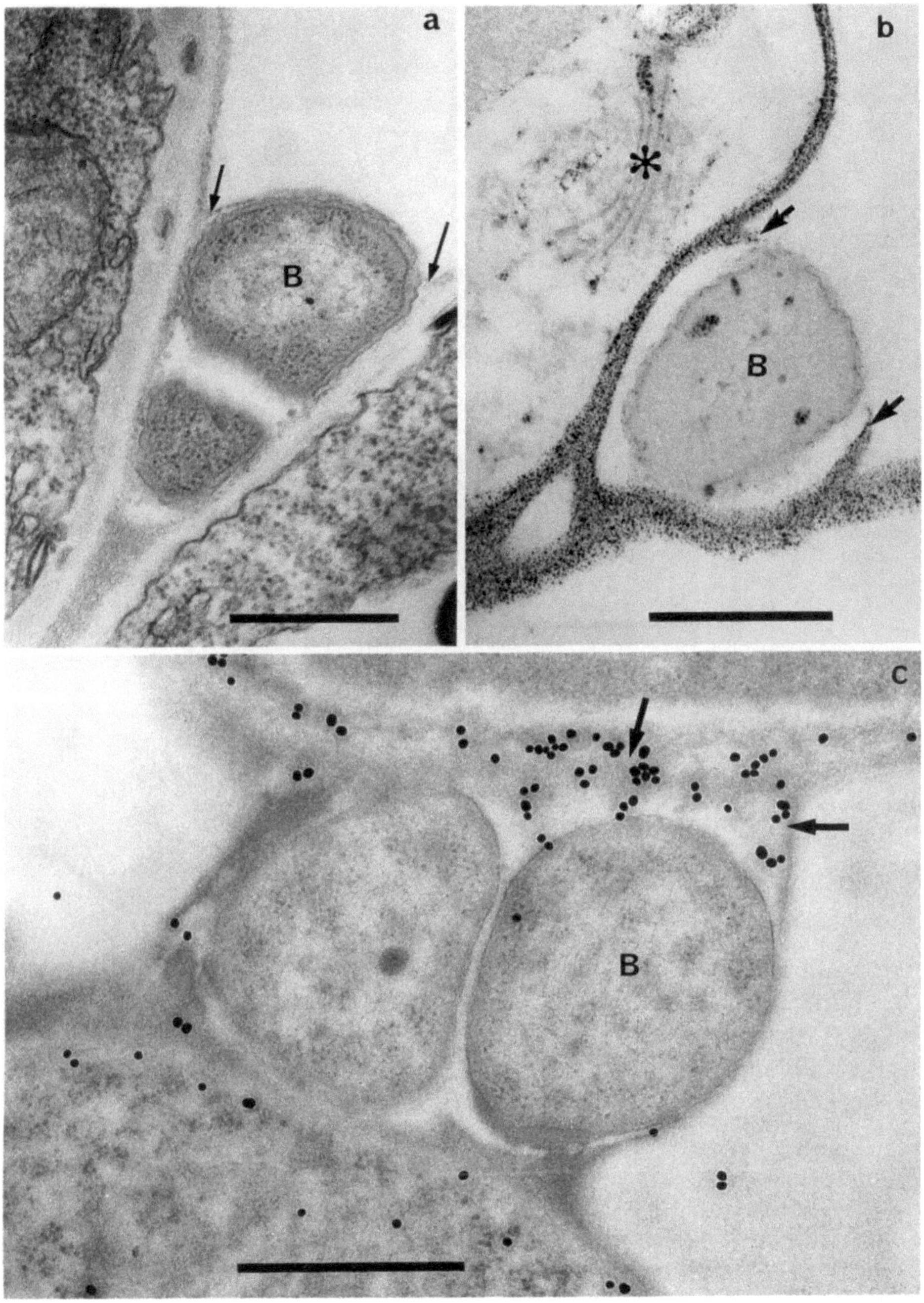
a
B
b
B
c
B

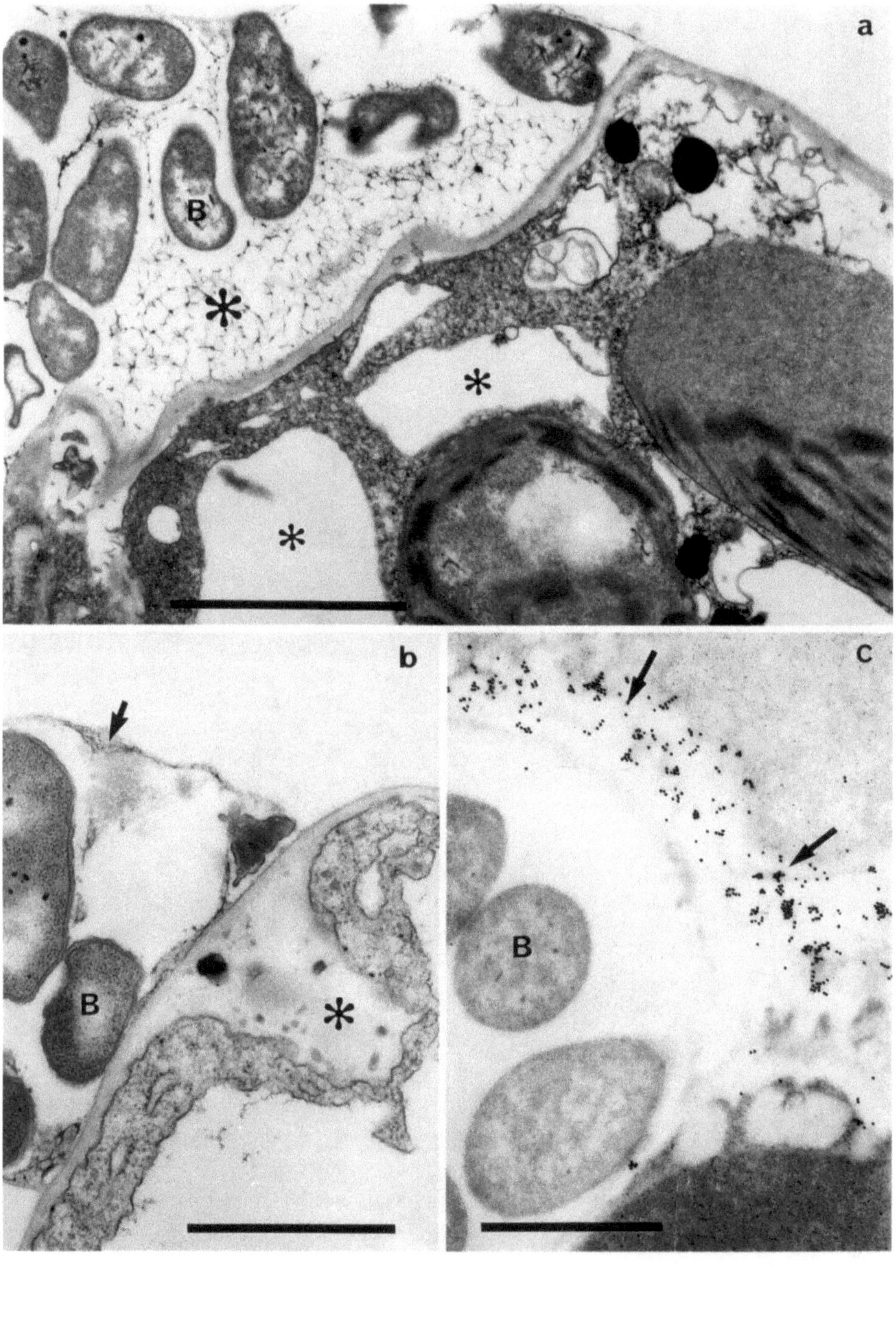
a
B
*
*
*
b
B
*
c
B

glycoproteins in agglutination and immobilization of saprophytic bacteria on plant cell walls. However, it is possible that bacterial growth is restricted before encapsulation.

3 The Hypersensitive Reaction

The HR has been studied as an expression of non-host resistance to the oat halo-blight pathogen *P.s.* pv. *coronafaciens* and also varietal resistance to avirulent races of *P.s.* pv. *phaseolicola* including transconjugants containing a cloned gene for avirulence [12]. During these incompatible interactions bacteria multiply following inoculation and produce fibrillar EPS, which stains with ruthenium red. Colony growth is restricted following plant cell collapse and the accumulation of phytoalexins during the first and second days after inoculation [10, 17].

The sequence of responses observed in cells undergoing the HR are very similar irrespective of the pathovar or race inoculated (Fig. 2a). The first response is the apparent convolution of the plasma membrane adjacent to bacteria. At sites of convolution electron-translucent papillae are formed. Papillae contain PATAg staining material, some of which may be HRGP as indicated by immunogold labelling (Fig. 2b, c). Papillae disperse during cell collapse (Fig. 2a) which follows irreversible membrane damage, as indicated by the failure of responding cells to plasmolyse during fixation in the presence of 0.9 M KNO_3 [3].

Immunocytochemical studies have shown that the major accumulation of HRGPs occurs in the walls of live cells adjacent to those undergoing the HR and that this is often associated with the formation of paramural deposits. Striking deposition of HRGPs is illustrated in Fig. 3. Accumulation of HRGPs in papillae and plant cell walls is unlikely to contribute to restricting growth of bacteria within the intercellular spaces. The accumulation of diffusible inhibitors such as phytoalexins, the synthesis of which is co-ordinately induced with that of HRGP in bean during the HR [10, 15, 26], is probably of more significance in resistance to pathovars of *P. syringae*. Some amorphous PATAg-positive material accumulates in the

◄

Fig. 2. a Cells of *P.s.* pv. *phaseolicola* race 1 (p3A1000) between mesophyll cells of cv. Tendergreen 12 h after inoculation, tissue was stained with ruthenium red. Bacteria (*B*) are surrounded by dark-staining fibrillar EPS (*large asterisk*). The mesophyll cell is in the early stages of hypersensitive collapse and many secondary vacuoles (*small asterisks*) are present in the dark-staining cytoplasm. Bar = 2 μm. **b** Cells of *P.s.* pv. *phaseolicola* race 1 (p3A001) adjacent to a mesophyll cell of cv. Tendergreen 12 h after inoculation. The tissue was stained with ruthenium red which reveals fibrillar EPS (*arrow*) surrounding bacteria. A large paramural papilla (*asterisk*) is present in the mesophyll cell adjacent to the bacterial colony. Bar = 1 μm. **c** Cells of *P.s.* pv. *phaseolicola* race 3 at a mesophyll cell junction in leaf tissue of cv. Tendergreen 24 h after inoculation. Immunogold labelling shows that HRGP is present in paramural papillae (*arrows*). Bar = 1 μm

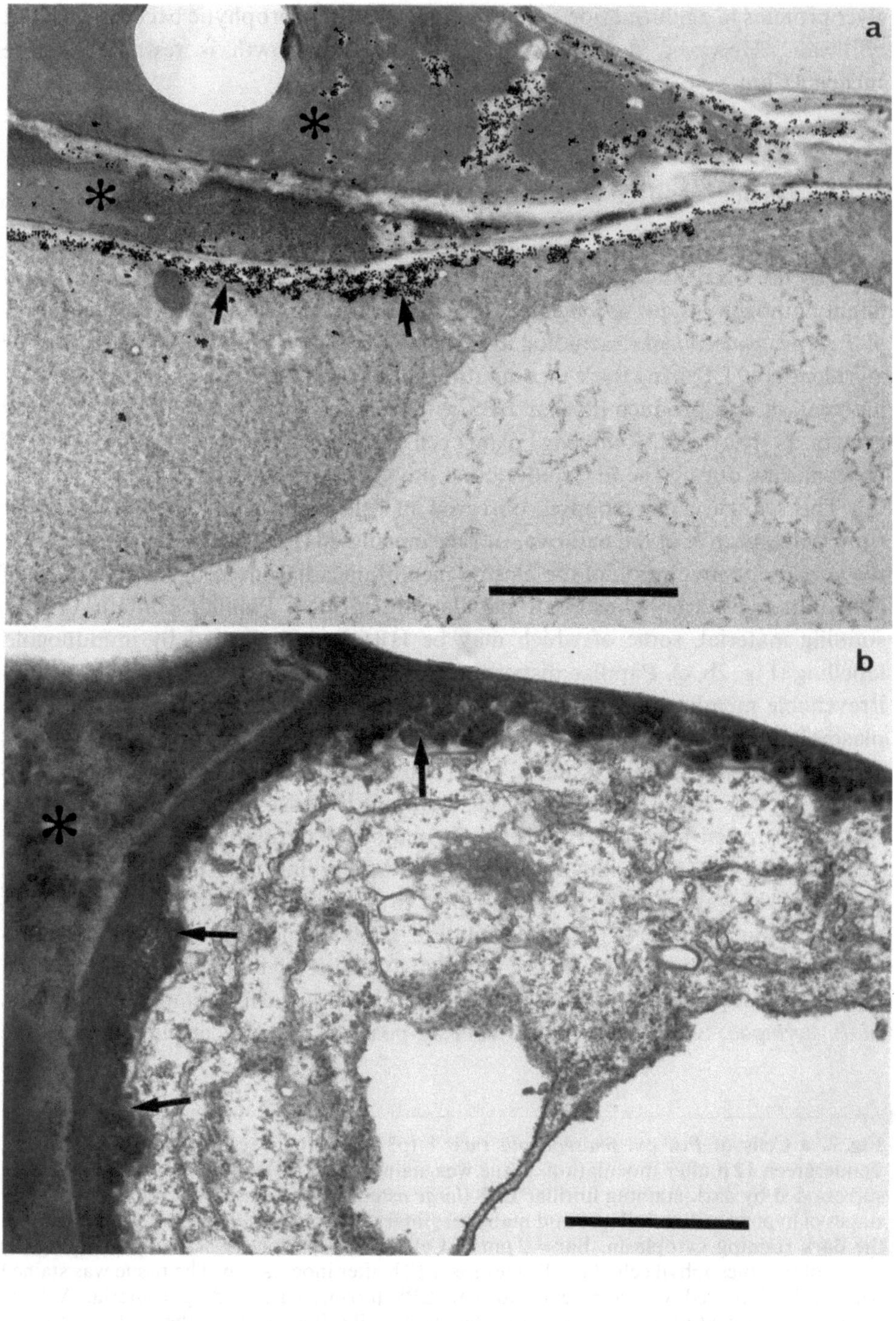
a
b

intercellular space on the cell wall and, to a lesser extent, within colonies of avirulent isolates of *P. syringae* (Fig. 4b). The amorphous deposits do not correspond to sites of gold labelling using antibodies to $HRGP_{2b}$.

4 The Compatible Interaction

Apparent convolution of the plasma membrane and papilla deposition also occur in cells adjacent to colonies of virulent bacteria. These non-specific responses are not followed by rapid cytoplasmic collapse as observed in the HR and, perhaps in consequence, papillae often appear more pronounced during the compatible interaction (Fig. 2b). In leaves, mesophyll cells collapse during the second day after inoculation but in pods, mesocarp cells within the large water-soaked lesions that develop do not undergo cytoplasmic disruption until 4 or 5 days after exposure to virulent bacteria [3, 10]. Changes in ultrastructure similar to those occurring during senescence are observed [10]. Thus, during compatible interactions, *P.s.* pv. *phaseolicola* may be considered to be growing biotrophically within susceptible tissues. HRGPs accumulate in papillae but the labelling in plant cell walls is much less pronounced during compatible than incompatible interactions [22]. Similarly, PATAg staining material, although present in papillae (Fig. 4a), is rarely observed in the intercellular spaces.

5 Quantitative Analyses Using Transconjugants

The transconjugants of race 1 used for this work are effectively "isogenic lines" differing only in the presence or absence of the cloned gene for avirulence from race 3 (Table 1). Comparisons can therefore be made using race 1 (p3A1000) and race 1 (p3A001) without the influence of differing genetic backgrounds such as occur when races of diverse origin are used. Following inoculation of cv. Tendergreen, data were collected using the methods described in Brown and Mansfield [3]. No differences were found between the avirulent and virulent transconjugants in terms of the attachment of bacteria to plant cell walls, the numbers of bacteria within colonies or production of fibrillar EPS during the first 12 h after inoculation.

Plant cells adjacent to colonies of virulent or avirulent transconjugants of race 1 underwent the same sequence of responses as reported for wild type strains race 1

◀

Fig. 3. The interface between collapsed and live mesophyll cells at the edge of a hypersensitive lesion in cv. Tendergreen 2 days (**a**) and 4 days (**b**) after inoculation with *P.s.* pv. *phaseolicola* race 3. **a** Immunogold labelling reveals heavy deposition of HRGP (*arrows*) in a healthy cell adjacent to collapsed cells (*asterisks*). Bar = 2 μm. **b** In conventionally fixed and stained tissue, electron dense paramural deposits (*arrows*) are also observed in a healthy cell adjacent to a collapsed cell (*asterisk*). Bar = 1 μm

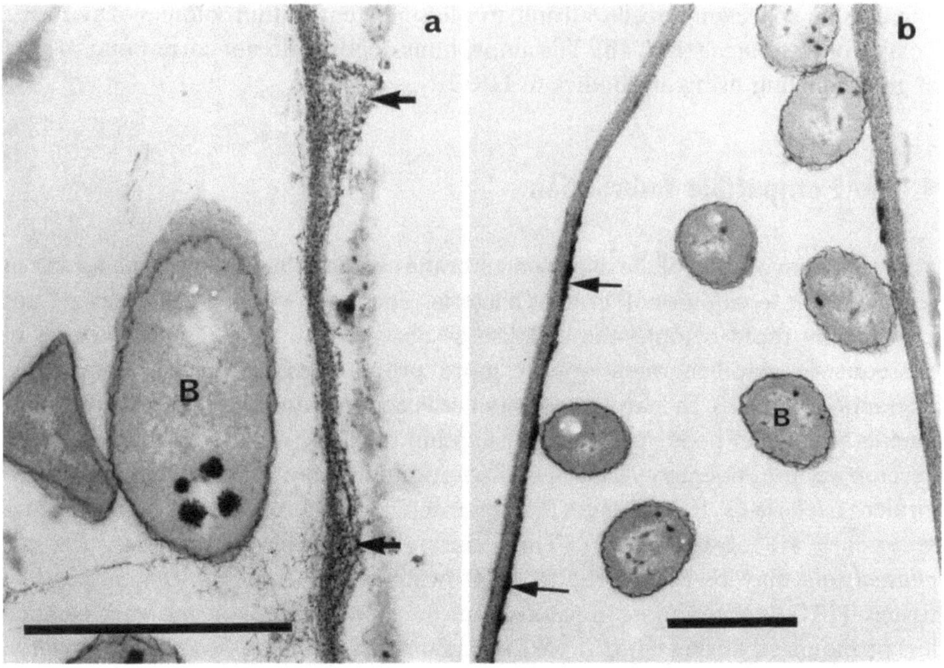

Fig. 4. Sections stained by the PATAg procedure. **a** Cells of *P.s.* pv. *phaseolicola* race 1 (p3A001) adjacent to a mesophyll cell of cv. Tendergreen 6 h after inoculation. Paramural papillae (*arrows*) contain PATAg-stained material. Bar = 1 μm. **b** Cells of *P.s.* pv. *phaseolicola* race 1 (p3A1000) between mesophyll cells of cv. Tendergreen 12 h after inoculation. Amorphous, PATAg-stained material is present on the outside of the plant cell walls (*arrows*) in the incompatible interaction. Bar = 1 μm

and race 3 respectively [3]. Figure 5 shows that the first sign of an obvious differential response and onset of the HR was the rupture of the tonoplast in cells adjacent to race 1 (p3A1000). The distribution of PATAg staining deposits was also different in the compatible and incompatible interactions; in particular, the quantitative assessment demonstrated that during the HR, PATAg staining material accumulated on the outside of the plant cell wall (Figs. 4 and 6).

6 Concluding Remarks

The main aim of our experiments was to use histochemical and quantitative analyses to reveal a factor clearly associated with induction of the HR. For example, we hoped to find localized responses specific to avirulent, HR-inducing bacteria and perhaps indicative of the activity of a cytotoxic metabolite. Such a dramatic result was not obtained, instead (although reactions to the saprophyte and pathogens were

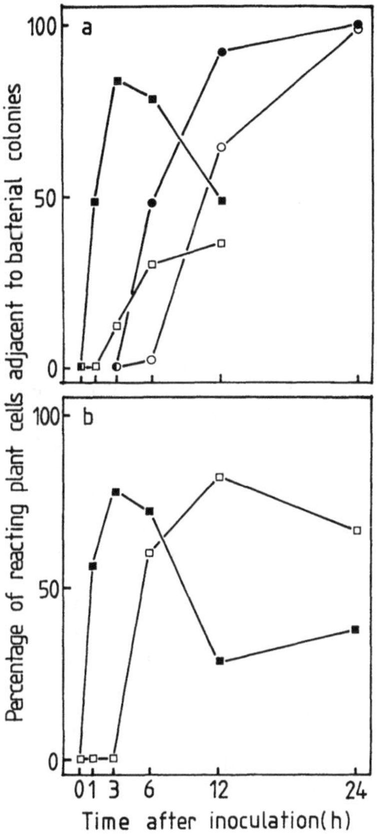

Fig. 5. Percentage of mesophyll cells of cv. Tendergreen adjacent to colonies of **a** *P.s.* pv. *phaseolicola* race 1 (p3A1000), incompatible interaction; **b** race 1 (p3A001), compatible interaction; showing convolution of the plasma membrane (*solid squares*), deposition of paramural deposits (*open squares*), rupture of the tonoplast (*solid circles*) and cytoplasmic collapse (*open circles*). At least 30 colonies were examined for the calculation of each percentage

distinct) we found that many plant responses were non-specific, e.g. apparent membrane convolution and papilla deposition. It was also clear that irreversible damage to the plasma membrane occurred several hours after the initial localized convolution. It is therefore likely that irreversible membrane damage is caused by an active response by the plant rather than by the direct action of a bacterial toxin [3, 14, 30, 31].

The significance of the accumulation of PATAg-positive material outside plant cells during the HR merits more detailed examination. We have suggested that the PATAg-positive staining detected in the intercellular space may be associated with the release of oligosaccharides from the plant cell wall by avirulent bacteria [3]. Such oligosaccharides may have a role as elicitors of the HR, as proposed by Albersheim and colleagues [5, 6]. However, before our suggestion can be substantiated, more specific histochemical and immunocytochemical analyses of polysaccharides are needed to define the structure of the bacterium/plant interface in more detail [19, 21, 32].

A priority for future analysis of the HR is to examine the function of the product of the avirulence gene cloned from race 3 [12]. Ultrastructural immuno-

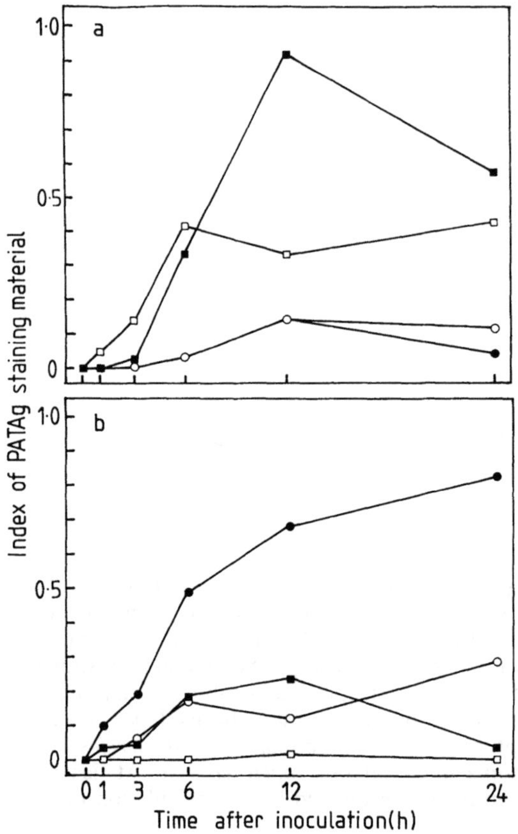

Fig. 6. Accumulation of PATAg staining material in cells of cv. Tendergreen adjacent to colonies of *P.s.* pv. *phaseolicola* race 1 (p3A001) (*squares*, compatible) and race 1 (p3A1000) (*circles*, incompatible). Density of staining at each site was scored on a 0–3 scale. At least 30 colonies were examined for the calculation of each index. **a** PATAg staining material inside the cell wall associated with papillae (*solid symbols*) or occurring as a layer inside the plant cell wall adjacent to colonies (*open symbols*). **b** PATAg staining material on the outside of the plant cell wall (*solid symbols*) and associated with bacterial colonies (*open symbols*)

cytochemistry using antibodies to the protein product of the avirulence gene will be an essential part of this work, allowing the localization and timing of appearance of the product to be investigated. It is hoped that integration of ultrastructural and molecular studies will allow the biochemical regulation of the HR to be elucidated.

Acknowledgments. We wish to acknowledge support from the Agricultural and Food Research Council, and Science and Engineering Research Council.

References

1. Al-Issa AN, Sigee DC (1982) The hypersensitive reaction in tobacco leaf tissue infiltrated with *Pseudomonas pisi*. 1. Active growth and division in bacteria entrapped at the surface of mesophyll cells. Phytopathol Z 104: 104–114
2. Bracker CE, Littlefield LJ (1973) Structural concepts of host-pathogen interfaces. In: Byrde RJW, Cutting CV (eds) Fungal pathogenicity and the plant's response. Academic Press, London, pp 159–317

3. Brown IR, Mansfield JW (1988) An ultrastructural study, including cytochemistry and quantitative analyses, of the interactions between pseudomonads and leaves of *Phaseolus vulgaris* L. Physiol Mol Plant Pathol 33: 351–376
4. Cason ET, Richardson PE, Essenberg MK, Brinkerhoff LA, Johnson WM, Venere RJ (1978) Ultrastructural cell wall alterations in immune cotton leaves inoculated with *Xanthomonas malvacearum*. Phytopathology 68: 1015–1021
5. Davis KR, Darvill AG, Albersheim P (1986) Host-pathogen interactions. XXX. Characterization of elicitors of phytolexin accumulation in soybean released from soybean cell walls by endopolygalacturonic acid lyase. Z Naturforsch 41c: 39–48
6. Doares SH, Bucheli P, Albersheim P, Darvill AG (1989) Host-pathogen interactions XXXIV. A heat-labile activity secreted by a fungal phytopathogen releases fragments of plant cell walls that kill plant cells. Mol Plant-Microbe Interact 2: 346–353
7. Ebrahim-Nesbat F, Slusarenko AJ (1983) Ultrastructure of the interaction of cells of *Pseudomonas phaseolicola* with cell walls of a resistant and susceptible bean cultivar. Phytopathol Z 108: 148–159
8. Fernandez MR, Heath MC (1986) Cytological responses induced by five phytopathogenic fungi in a non-host plant *Phaseolus vulgaris*. Can J Bot 64: 754–764
9. Goodman RN, Huang PY, White JA (1976) Ultrastructural evidence of immobilization of an incompatible bacterium *Pseudomonas pisi* in tobacco leaf tissue. Phytopathology 66: 754–764
10. Harper S, Zewdie N, Brown IR, Mansfield JW (1987) Histological, physiological and genetical studies of the response of leaves and pods of *Phaseolus vulgaris* to three races of *Pseudomonas syringae* pv. *phaseolicola* and to *Pseudomonas syringae* pv. *coronafaciens*. Physiol Mol Plant Pathol 31: 153–172
11. Hildebrand DC, Alosi MC, Schroth MN (1980) Physical entrapment of pseudomonads in bean leaves by films formed at air-water interfaces. Phytopathology 70: 98–109
12. Hitchin FE, Jenner CE, Harper S, Mansfield JW, Barber CE, Danield MJ (1989) Determinant of cultivar-specific avirulence cloned from *Pseudomonas syringae* pv. *phaseolicola* race 3. Physiol Mol Plant Pathol 34: 309–322
13. Jones SB, Fett WF (1985) Fate of *Xanthomonas campestris* infiltrated into soybean leaves: an ultrastructural study. Phytopathology 65: 733–741
14. Keen NT, Tamaki S, Kobayashi M, Stayton D, Gerhold D, Shen H, Gold S, Lorang J, Thordal-Christensen H (1990) Characterization and function of avirulence genes from *Pseudomonas syringae* pv. *tomato*. In: Lugtenberg BJJ (ed) Signal molecules in plants and plant-microbe interactions. NATO ASI Series, Springer, Berlin Heidelberg, New York Tokyo, pp 183–188
15. Lamb CJ, Lawton MA, Dron M, Dixon RA (1989) Signals and transduction mechanisms for activation of plant defenses against microbial attack. Cell 56: 215–224
16. Lazarovits G, Higgins VJ (1976) Ultrastructure of susceptible, resistant and immune reactions of tomato to races of *Cladosporium fulvum*. Can J Bot 54: 235–249
17. Mansfield JW, Brown IR (1986) The biology of interactions between plants and bacteria. In: Bailey JA (ed) Biology and molecular biology of plant-pathogen interactions. Springer, Berlin Heidelberg New York Tokyo, pp 71–98
18. Mansfield JW, Richardson A (1981) The ultrastructure of interactions between *Botrytis* species and broad bean leaves. Physiol Plant Pathol 19: 41–48
19. Mario J, Foiry B (1987) Electron microscopic visualization of capsular material of *Pasteurella multocida* Types A and D labelled with polycationic ferritin. J Bacteriol 169: 3470–3472
20. Mazau D, Rumeau D, Esquerré-Tugayé M-T (1989) Two different families of hydroxyproline-rich glycoproteins in melon callus: biochemical and immunochemical studies. Plant Physiol 86: 540–546
21. Mendgen K, Lange M, Bretschneider K (1985) Quantitative estimation of the surface carbohydrates on the infection structures of rust fungi with enzymes and lectins. Arch Microbiol 140: 307–311

22. O'Connell RJ, Brown IR, Mansfield JW, Bailey JA, Mazau D, Rumeau D, Esquerré-Tugayé M-T (1990) Immunocytochemical localization of hydroxyproline-rich glycoproteins accumulating in melon and bean at sites of resistance to bacteria and fungi. Mol Plant-Microbe Interact 2: 33–40
23. O'Connell RC, Bailey JA, Richmond DW (1985) Cytology and physiology of infection of *Phaseolus vulgaris* by *Colletotrichum lindemuthianum*. Physiol Plant Pathol 27: 75–98
24. Roland J-C (1978) General preparation and staining of thin sections. In: Hall JL (ed) Electron microscopy and cytochemistry of plant cells. Elsevier/North-Holland, Amsterdam, pp 1–62
25. Sequeira L, Gaard G, De Zoeten GA (1977) Interaction of bacteria and host cell walls: its relation to mechanisms of induced resistance. Physiol Plant Pathol 10: 43–50
26. Showalter AM, Bell JN, Cramer CL, Bailey JA, Varner JE, Lamb CJ (1985) Accumulation of hydroxyproline-rich glycoprotein mRNAs in response to fungal elicitor and infection. Proc Natl Acad Sci USA 82: 6551–6555
27. Sigee DC (1984) Induction of leaf cell death by phytopathogenic bacteria. In: Davies I, Sigee DC (eds) Cell ageing and cell death. Cambridge University Press, Cambridge, pp 295–322
28. Sigee DC, Al-Issa AN (1982) The hypersensitive reaction in tobacco leaf tissue infiltrated with *Pseudomonas pisi*. 2. Changes in the population of viable, actively metabolic and total bacteria. Phytopathol Z 105: 71–86
29. Sing VO, Schroth MN (1977) Bacteria-plant cell surface interactions: active immobilization of saprophytic bacteria in leaves, Science 197: 759–761
30. Slusarenko AJ, Longland A (1986) Changes in gene activity during expression of the hypersensitive response in *Phaseolus vulgaris* cv. Red Mexican to an avirulent race 1 isolate of *Pseudomonas syringae* pv. *phaseolicola*. Physiol Mol Plant Pathol 29: 79–94
31. Slusarenko AJ, Longland A, Friend J (1986) Expression of plant genes in the hypersensitive reaction of French bean (*Phaseolus vulgaris*) to the plant pathogenic bacterium Pseudomonas syringae pv. *phaseolicola*. In: Lugtenberg B (ed) Recognition in microbe-plant symbiotic and pathogenic interactions. Springer, Berlin Heidelberg New York Tokyo, pp 367–376
32. Vian B (1986) Ultrastructural localisation of carbohydrates. Recent developments in cytochemistry and affinity methods. In: Bailey JA (ed) Biology and molecular biology of plant-pathogen interactions. Springer, Berlin Heidelberg New York Tokyo, pp 49–57

Chapter 15 Frank Gubler and Adrienne R. Hardham

The Fate of Peripheral Vesicles in Zoospores of *Phytophthora cinnamomi* During Infection of Plants

1 Introduction

Motile zoospores play a major role in the infection of plant roots by the dieback fungus, *Phytophthora cinnamomi* [22]. Under moist conditions, the fungus releases vast numbers of zoospores into the soil. The zoospores can be transported passively over considerable distances by underground water, thereby facilitating the spread of the pathogen. Once in the vicinity of a root, the zoospores exhibit positive chemotaxis to the root exudate and as a result rapidly accumulate along the zone of cell elongation just behind the root tip. Contact with root surface components triggers the encystment of the zoospores. Fucosyl residues on the root surface have been implicated in this recognition step [10] but so far no fucosyl-containing molecules which can trigger encystment have been isolated from the root surface. Encystment entails the rapid loss of motility and shedding of flagella, exocytosis of peripheral vesicles, change in cell shape and, after 5–10 min, formation of a microfibrillar cell wall. This is followed, about 20 min later, by germination and subsequent invasion of the root tissue.

Although the presence of numerous peripheral vesicles in zoospores of *P. cinnamomi* appears to be characteristic feature of Oomycete zoospores [2], very little is known about their roles in the infection process. Four different types of peripheral vesicles have been identified on the basis of cytological evidence in *P. cinnamomi* zoospores: peripheral cisternae, large peripheral vesicles, and two types of small peripheral vesicles [7]. Similar vesicles have been observed in other *Phytophthora* species. Proposed roles for these vesicles include: (1) storage sites for cyst coat components including an adhesive which binds the cyst to the root surface [1, 4–6, 15, 17, 19, 20]; (2) synthesis of cell wall components [3, 4, 9, 16]; (3) microbody function [3, 4, 11, 18]; (4) osmotic stabilization [2] and (5) protein storage [5]. However, in the absence of specific markers for the various types of vesicles it has proved difficult to obtain confirmation of these roles.

Plant Cell Biology Group, Research School of Biological Sciences, The Australian National University, Canberra, ACT 2601, Australia

We have recently raised monoclonal antibodies (mAbs) to components of *P. cinnamomi* zoospores and cysts in order to investigate the roles of these components in the infection process [8]. From this bank of mAbs we have identified a number of antibodies which binds to large peripheral vesicles, two types of small peripheral vesicles (dorsal and ventral vesicles), and peripheral cisternae ([5]; Hardham and Gubler unpubl.). In this chapter we describe experiments using these mAbs to determine the fate and possible roles of these vesicles.

2 Peripheral Vesicles in *Phytophthora cinnamomi* Zoospores

Zoospores are ovoid in shape with a tinsel and a whiplash flagellum projecting from a longitudinal groove on the ventral surface (Fig. 1). These cells lack a cell wall. A water expulsion vacuole, assumed to be important in osmoregulation, is situated close to the anterior end. The cytoplasm is characterized by a prominent nucleus, 8–11 stacks of Golgi cisternae, numerous mitochondria, microbodies, ribosomes, endoplasmic reticulum, finger print vesicles, and a variety of peripheral vesicles [7].

Four different peripheral vesicles have been identified in zoospores of *P. cinnamomi* on the basis of immunocytochemical staining and morphology ([5, 7]; Gubler and Hardham unpubl.). The distributions of these vesicles are shown in Fig. 2.

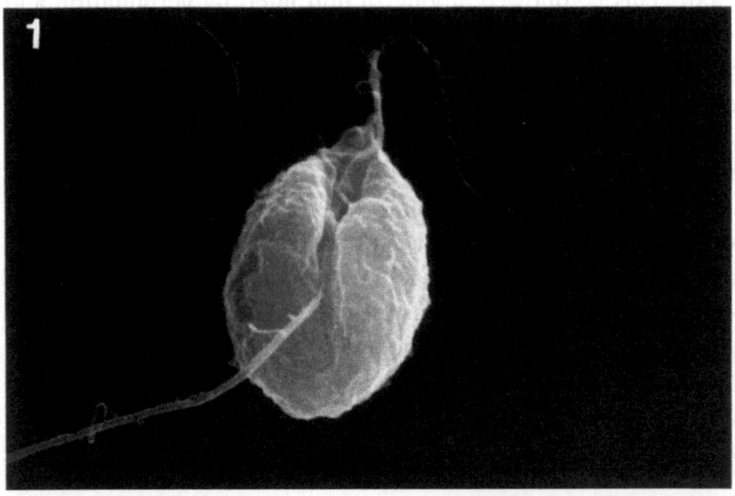

Fig. 1. Scanning electron microscope micrograph of a zoospore of *Phytophthora cinnamomi*. Two flagella emerge from the centre of a groove along the ventral surface. The cell was fixed with glutaraldehyde and freeze-dried prior to examination. × 4933

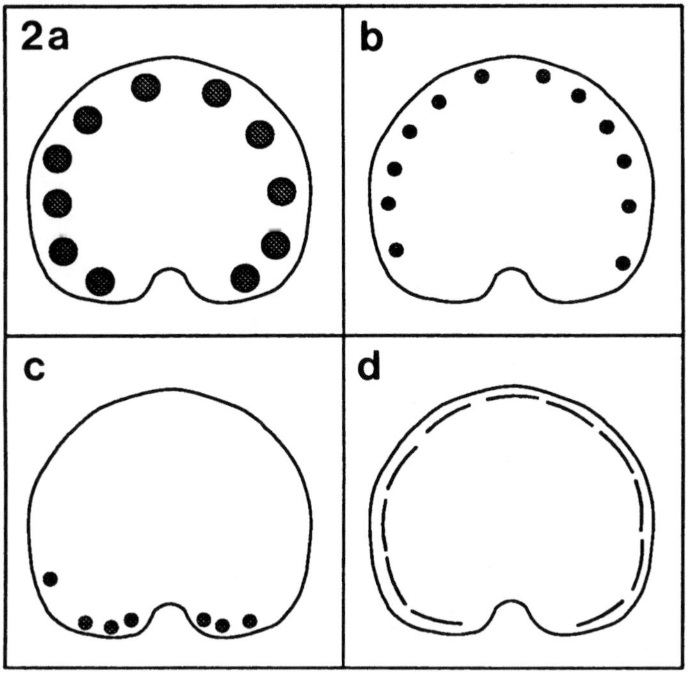

Fig. 2. Diagram of median transverse sections through a zoospore showing the distribution of the four types of peripheral vesicles. **a** Large peripheral vesicles; **b** dorsal vesicles; **c** ventral vesicles; **d** peripheral cisternae. The ventral surface with the groove is at the *bottom* of the profile

Large Peripheral Vesicles. mAb Lpv-1 labels large conspicuous vesicles which occur throughout the zoospore cortex except within the groove (Fig. 2a). These ellipsoidal vesicles appear to be in close contact with the plasma membrane, and in cells that are embedded at low temperature in Lowicryl K4M, the vesicles are characterized by moderately electron-dense contents (Fig. 3). Immunoblot analysis has shown that Lpv-1 recognizes three glycoproteins ($M_r > 300$ kDa) that contain mannosyl/glucosyl residues [5].

Dorsal Vesicles. mAb Cpa-2 labels small spherical vesicles which are common throughout the zoospore cortical cytoplasm except in the central regions of the ventral surface including the groove (Fig. 2b). This type of small peripheral vesicles predominates on the dorsal surface and will be referred to as dorsal vesicles. Double immunogold labeling experiments using Cpa-2- and Lpv-1-gold probes confirm that these antibodies recognize different peripheral vesicles (Fig. 4). In addition to immunological differences between these two types of vesicles, they also differ in their internal structure. In cells embedded in LR White resin, dorsal vesicles have a heterogeneous internal structure (Fig. 5). Immunoreactivity is restricted to electron-lucent regions which contain fine flocculent material, while electron-dense regions

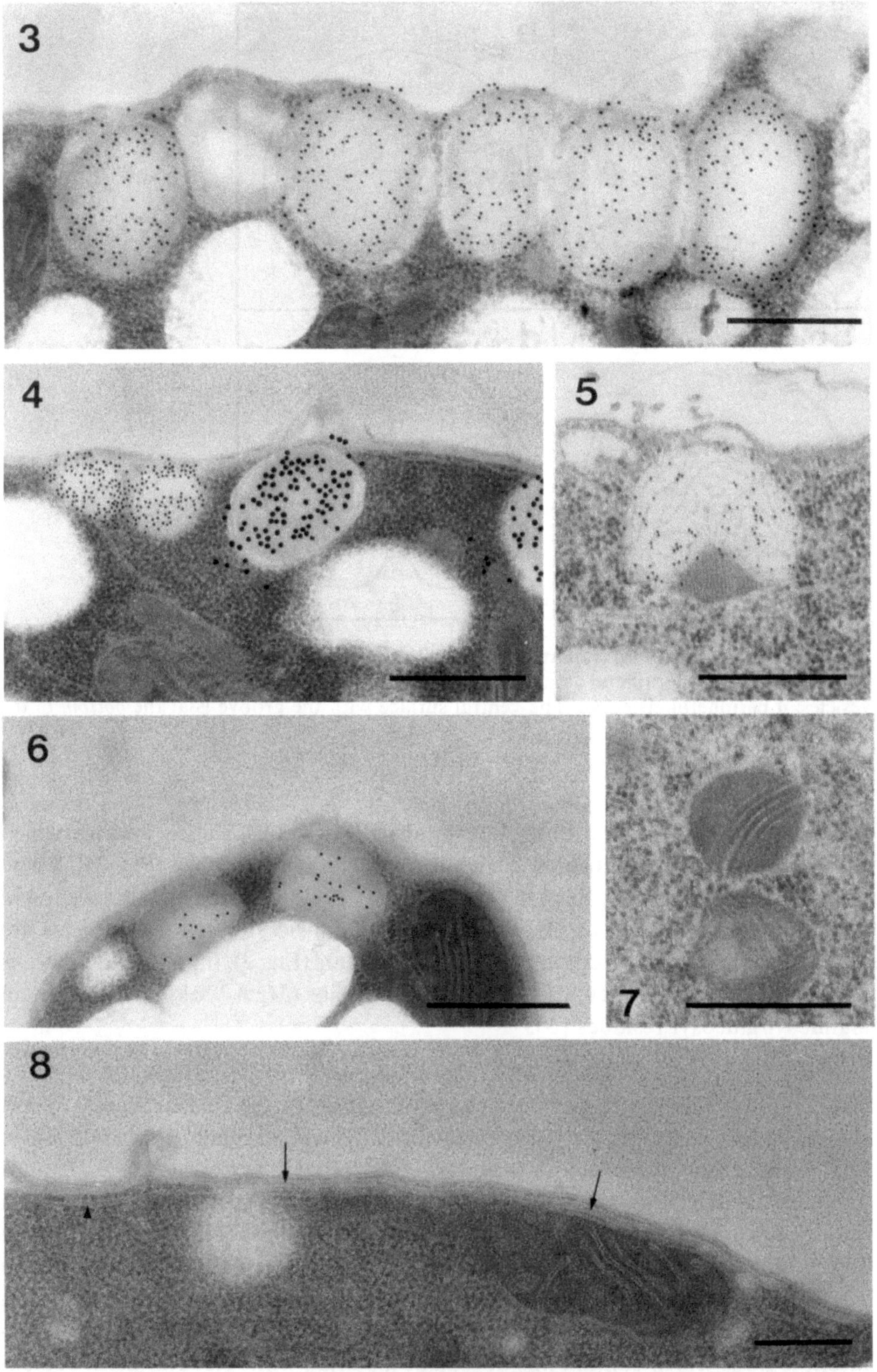
3
4
5
6
7
8

are poorly labeled. Fine laminations are often visible in the electron-dense material. Whether the electron-dense material is chemically identical to the flocculent material remains to be determined. It is possible that the immunoreactive material may be packaged differently in electron-dense regions so that the epitope is masked. Similar non-uniform Cpa-2 labeling patterns are observed over dorsal vesicles in cells embedded at low temperature in Lowicryl K4M but the ultrastructural details are obscure. Immunoblot analysis has shown that Cpa-2 labels 3 glycoproteins ($M_r > 300$ kDa) that contain N-acetyl-D-galactosaminosyl/galactosyl residues [5].

Ventral Vesicles. mAb Vsv-1 labels the second type of small peripheral vesicle which is localized at the ventral surface, especially along the edge of the groove (Fig. 2c). They are rarely observed along the dorsal surface, and like large and dorsal vesicles, are absent from the cell cortex within the groove. In cells embedded at low temperature in Lowicryl K4M, the contents of ventral vesicles are characterized by an outer region of moderately electron-dense material surrounding an electron-lucent band which contains the immunoreactive material (Fig. 6). The band appears to correspond to electron-dense laminations seen in conventionally embedded cells (Fig. 7). Like dorsal vesicles, ventral vesicles do not appear to be in contact with the plasma membrane. Immunoblot analysis has shown that Vsv-1 labels a protein with an apparent high molecular weight ($M_r > 200$ kDa; Hardham and Gubler unpubl.).

Peripheral Cisternae. These vesicles consist of long flat cisternae that lie directly below the plasma membrane (Fig. 8). They form a system of closely packed discs which extend throughout all regions of the zoospore periphery except the groove (Fig. 2d). This system appears to be interrupted only by other peripheral vesicles. Various investigators have noted that these cisternae are morphologically similar to the plasma membrane in Oomycete zoospores [4, 7]. Short strands of rough ER are commonly located adjacent to these cisternae. Cpw-1, a mAb to a cyst cell wall epitope, labels these vesicles and Golgi bodies very weakly (not shown).

◄

Figs. 3–8. Immunocytochemical and morphological characterization of peripheral vesicles in zoospores of *Phytophthora cinnamomi.* Cells were fixed in 1% glutaraldehyde and either embedded at low temperature in Lowicryl K4M (**3, 4, 6,** and **8**) or postfixed in osmium tetroxide and embedded in LR White resin (**5** and **7**). **3** Immunogold labeling of large peripheral vesicles with mAb Lpv-1, followed by sheep anti-mouse Ig-Au_{10}. Not all peripheral vesicles are labeled. **4** Double immunogold labeling of dorsal and large peripheral vesicles with mAb-gold probes, Cpa-2-Au_{10} and Lpv-1-Au_{18} respectively. **5** Immunogold labeling of dorsal vesicle with Cpa-2 followed by sheep anti-mouse Ig-Au_{10}. **6** Immunogold labeling of ventral vesicles with mAb Vsv-1, followed by sheep anti-mouse Ig-Au_{10}. **7** Ventral vesicles with electron-dense laminations. **8** Peripheral cisterna (*arrows*) underlying the plasma membrane of a zoospore. Short strands of ER (*arrowhead*) and mitochondria are often closely associated with these cisternae. Bar = 0.5 μm

3 Exocytosis of Peripheral Vesicles During Encystment

To determine the fate of the peripheral vesicles during encystment, zoospores were allowed to infect roots of *Eucalyptus seiberi* seedlings for periods up to 4 h prior to immunocytochemical analysis. In addition, encystment was triggered in vitro by addition of citrus pectin. On the basis of immunogold labeling of *P. cinnamomi* cysts with mAbs specific to components of peripheral vesicles (Cpa-2, Lpv-1, Vsv-1 and Cpw-1), three types of peripheral vesicles have been shown to undergo exocytosis during encystment ([5]; Hardham and Gubler, unpubl.).

The contents of the dorsal vesicles are secreted within 1–2 min of the induction of encystment. Glycoprotein components of these vesicles recognized by mAb Cpa-2 form irregular patches on the cyst surface (Fig. 9). By the start of cell wall synthesis (approximately 5 min after the onset of encystment), the patches coalesce to form a thin coat over most of the surface of the cyst. Close examination of young cysts on root surfaces shows that the distribution of these secreted glycoproteins is polarized (Fig. 10). The glycoproteins form almost a complete coat over the cells, but are notably absent from the surface which is in contact with the root surface. The glycoproteins remain on the cyst following germination and have been detected on the cyst surface up to 4 h after induction of encystment.

Immunogold labeling of cysts with mAb Vsv-1 shows that the ventral vesicles also undergo exocytosis during encystment. Within 1–2 min after induction of encystment, the Vsv-1-positive vesicles disappear from the periphery of the zoospore and the ventral surface becomes coated with material which reacts with Vsv-1. The polarized distribution of the secreted glycoprotein is also noticeable on the surface of cysts attached to roots. Immunogold labeling of 4-min cysts on *E. seiberi* roots with Vsv-1 followed by silver enhancement reveals that most of the glycoprotein is localized in the contact zone between the cyst and the root surface (Fig. 11). Double labeling of the cysts with Vsv-1 and Cpa-2 shows that these surface components,

Figs. 9–12. Immunolabeling of cysts with mAbs to components of the dorsal (Cpa-2) and ventral vesicles (Vsv-1). Cysts at various stages of encystment were fixed and embedded in Lowicryl K4M. **9** Immunogold labeling of 1-min cyst with Cpa-2, followed by sheep anti-mouse Ig-Au_{10}. The surface of the cyst is heavily labeled with gold particles. **10** Immunogold-silver staining of cysts on a *Eucalyptus seiberi* root with Cpa-2. The section has been incubated with Cpa-2, followed by goat anti-mouse IgG-Au_5 and then silver enhanced with IntenSE M (Janssen, Belgium). In many cysts the Cpa-2-reactive material forms a continuous coat over most of the cell surface but is notably absent from the side closest to the root surface. **11** and **12** Cysts on the surface of an *E. seiberi* root 4 min after the start of infection. The cysts have been double-labeled with Vsv-1 (**11**) and Cpa-2 (**12**). Note the presence of the root hairs (*h*) adjacent to the cysts. **11** Infected roots were fixed and incubated with Vsv-1 followed by immunogold-silver staining prior to embedding in Lowicryl K4M. Differential interference contrast optics. **12** Post-embedding immunofluorescence staining with Cpa-2. Control experiments using irrelevant mAbs showed there was no cross-reactivity between the antibodies used for pre-embedding labeling and those used for post-embedding labeling. Bars: **9** 0.5 μm; **10–12** 20 μm

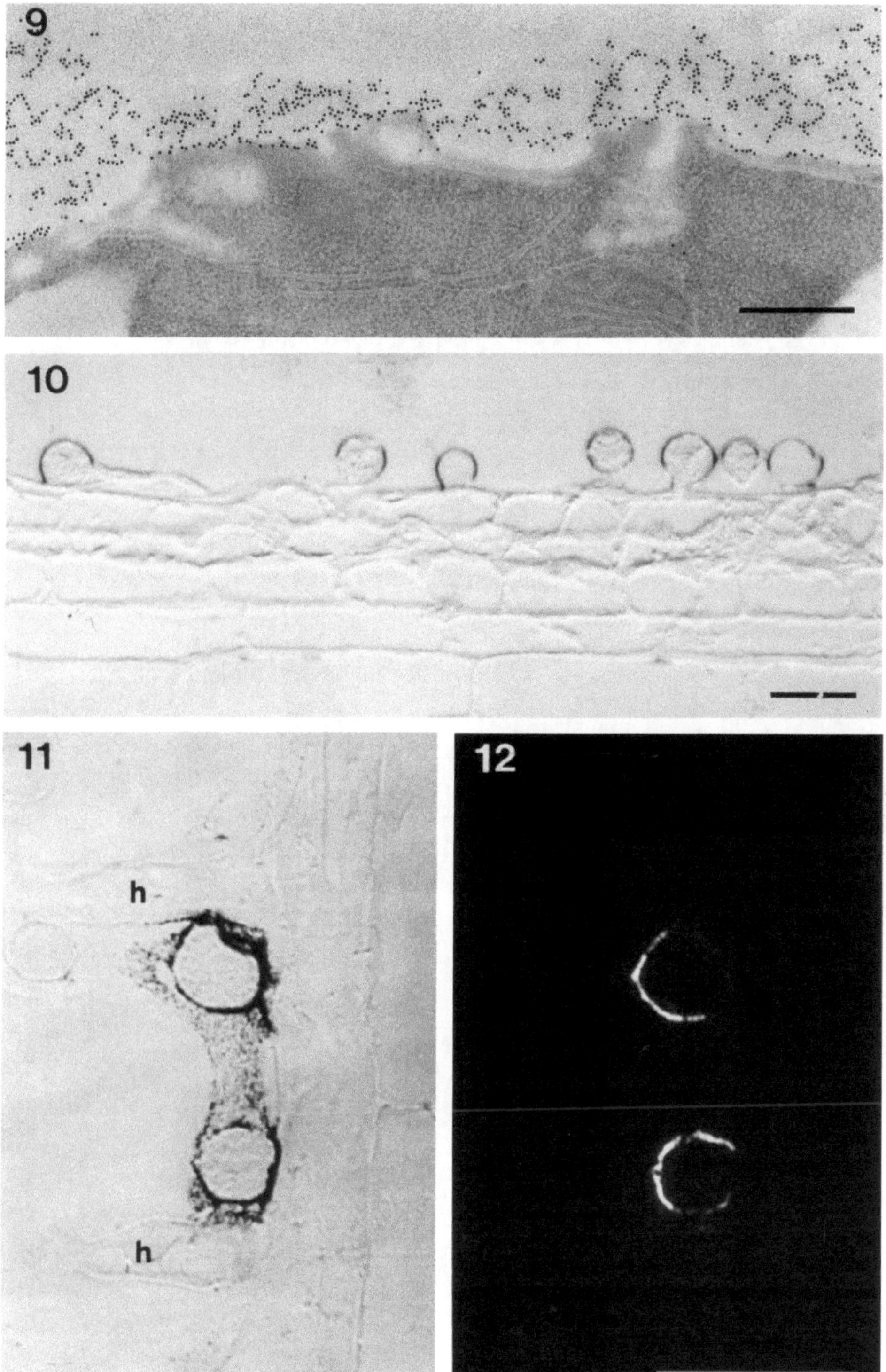
9
10
11
h
h
12

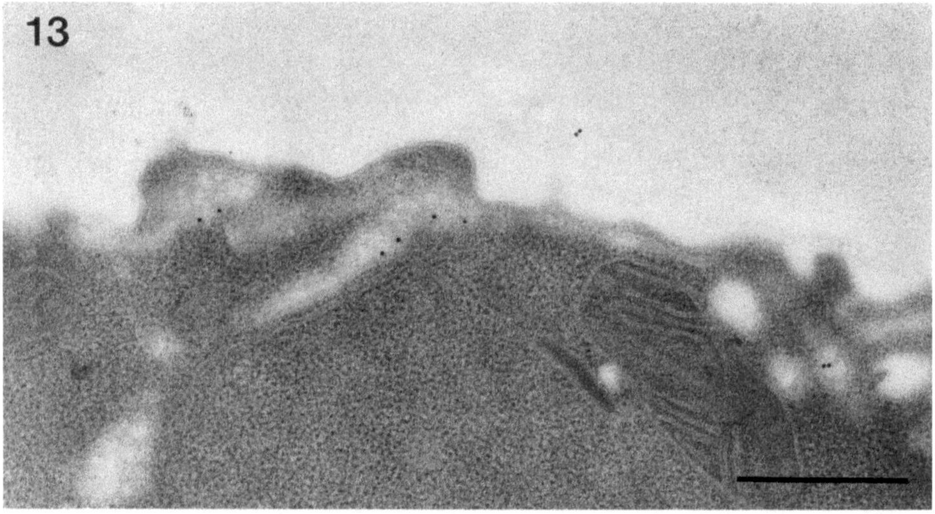

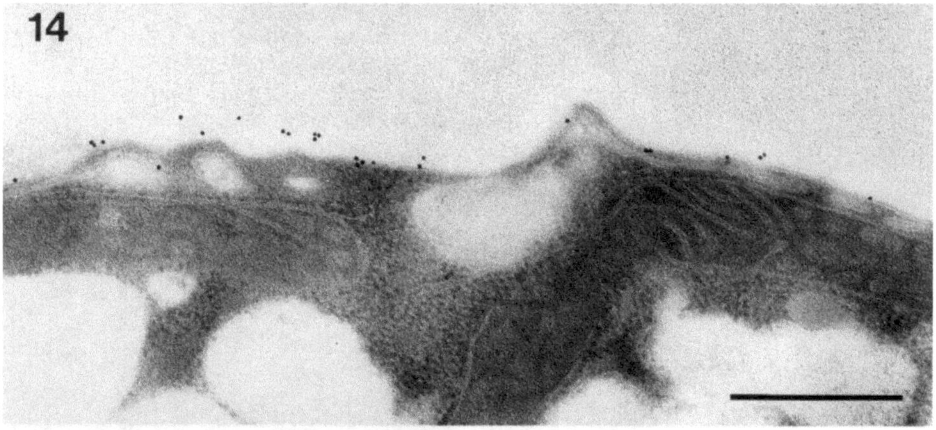

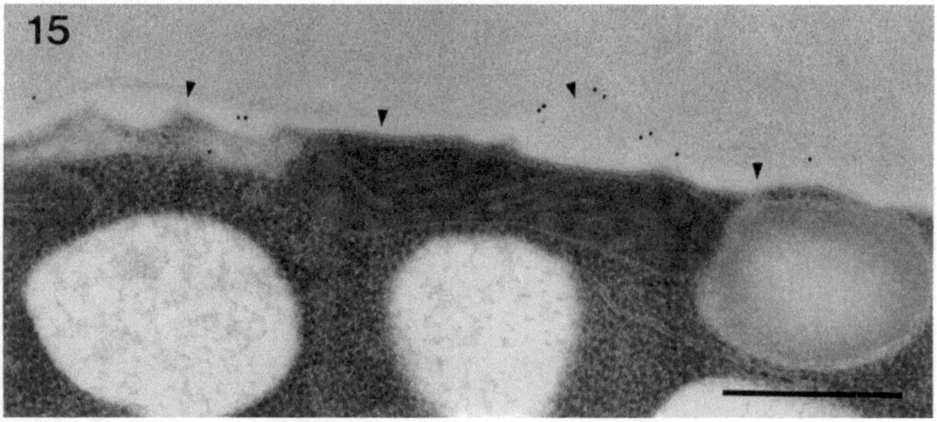

which originate in the ventral and dorsal vesicles respectively, form a thin heterogenous coat on the cyst surface (Figs. 11, 12).

The peripheral cisternae undergo marked changes during encystment. Within 1–2 min after the induction of encystment, the cisternae move away from the plasma membrane and fragment into small vesicles [4, 7]. mAb Cpw-1, which weakly labels the peripheral cisternae in zoospores, also labels these small vesicles as well as Golgi bodies in encysting cells (Figs. 13, 14). Between 2–5 min after the onset of encystment, many of these Cpw-1-positive vesicles undergo exocytosis, releasing their contents onto the surface of the cyst plasma membrane. By 10 min, the Cpw-1-positive material has become incorporated into the newly synthesized cell wall (Fig. 15). Double labeling experiments of walled cysts with Cpw-1 and Cpa-2 have confirmed that the secreted Cpw-1-positive material is present only in the cell wall and not in areas labeled by Cpa-2 (not shown).

Exocytosis of peripheral vesicles was also monitored using a quantitative immunofluorescence assay (Hardham and Gubler unpubl.). Cells at various stages of encystment were fixed and labeled with Cpa-2, Vsv-1, and Cpw-1, followed by sheep anti-mouse immunoglobulin-FITC. The intensity of fluorescent labeling was measured using a spectrofluorimeter. The results show that exocytosis of dorsal and ventral vesicles, along with the vesicles derived from the peripheral cisternae starts within 30s after induction of encystment with pectin. Secretion of components of the dorsal and ventral vesicles is finished after 2 min, while secretion of the cell wall component continues up to 5 min after exposure to pectin.

4 Fate of the Large Peripheral Vesicles

The fate of the large peripheral vesicles during encystment and cyst germination was monitored by immunogold labeling with mAb Lpv-1 ([5]; Gubler and Hardham, in press). As pointed out earlier, this antibody reacts with glycoproteins stored in these vesicles. No evidence for secretion of the contents of the large peripheral vesicles could be detected during encystment on the basis of immunogold staining and morphometric analysis. Double immunogold labeling experiments clearly showed that the large peripheral vesicles were retained within the cell while dorsal vesicles underwent exocytosis (Fig. 16). Morphometric analysis of immunolabeled sections showed that there was no change in the relative volume occupied by these vesicles during the first 10 min of encystment. During the first 5 min of encystment,

◀

Figs. 13–15. Cysts at various stages of encystment labeled with mAb Cpw-1. Thin sections of cysts embedded in Lowicryl K4M were incubated with Cpw-1, followed by sheep anti-mouse Ig-Au_{10}. **13** 1-min cyst. The gold particles are mostly associated with small vesicles in the cell cortex. **14** 2-min cyst. Cpw-1 labels the cell surface as well as a few vesicles. **15** Walled cyst. The gold particles are associated with the cell wall (*arrowheads*) Bars = 0.5 μm

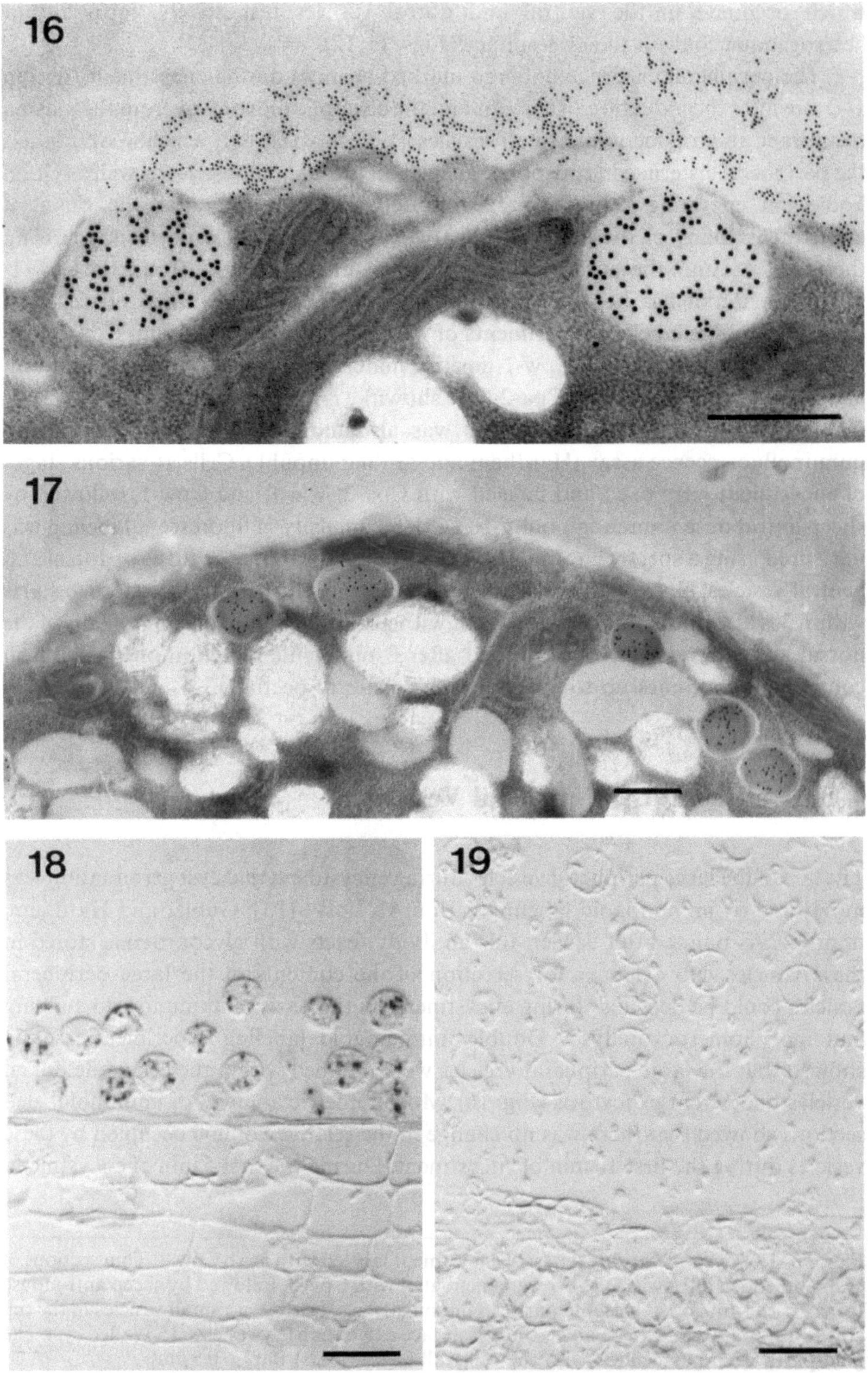
16
17
18
19

the large vesicles remained along the periphery of the cyst but subsequently migrated away from the cell periphery during the period of cell wall synthesis (Fig. 17).

Cyst germination had no detectable effect on the large peripheral vesicles (Gubler and Hardham submitted). Immunogold labeling of cysts on the root surface of *E. seiberi* with Lpv-1, followed by silver enhancement, showed that large peripheral vesicles were still present 30 min after germination (Fig. 18). However, 3.5 h after the onset of germination, the large peripheral vesicles were not detected in cysts or germ tubes (Fig. 19). There was no evidence of secretion of the contents of the large peripheral vesicles either on the root surface or in infected regions of the root. By this stage, germ tubes had penetrated into the root and caused considerable disruption in the cortex.

5 Roles of Peripheral Vesicles During Encystment

The use of monoclonal antibodies as cytological markers has greatly facilitated monitoring the fate of peripheral vesicles during encystment. Our results have shown that three types of peripheral vesicles undergo exocytosis and distribution of the secreted vesicle components on the cyst surface provides clues as to their possible functions.

In an earlier study, we proposed that the adhesive responsible for cyst attachment to the root surface was stored in dorsal vesicles [5]. We had shown that the contents of these vesicles were secreted during encystment and formed a thin coat on the cyst surface. The timecourse of secretion of these components was similar to that of the acquisition of cyst adhesiveness [6]. However, a close examination of the distribution of the secreted components on cysts now reveals that they are unlikely to be the adhesive responsible for cyst attachment. The vesicle components are rarely detected in the contact zone between the cyst and the root surface. Instead, these components form a thin coat over the parts of the cell facing away from the

◀

Figs. 16–19. Immunogold labeling of large peripheral vesicles in cysts with mAb Lpv-1. Cells embedded in Lowicryl K4M were used for immunolabeling. **16** Double immunogold labeling of 1-min cyst with Lpv-1-Au_{18} and Cpa-2-Au_{10}. The large peripheral vesicles labeled with 18 nm gold particles remain in the cell periphery. The contents of the dorsal vesicles, labeled with 10 nm gold particles, coat the surface of the cyst. **17** Immunogold labeling of 10-min cyst with Lpv-1 followed by sheep anti-mouse Ig-Au_{10}. The vesicles have migrated away from the cell periphery. **18** and **19** Immunogold-silver staining of large peripheral vesicles in *P. cinnamomi* infecting *E. seiberi* roots, 1 h (**18**) and 4 h (**19**) after the start of infection. Many large peripheral vesicles are still present in cysts 1 h after the start of infection (**18**), but by 4 h these had all disappeared (**19**). Semi-thin sections were incubated with Lpv-1 and goat anti-mouse Ig-Au_5, followed by silver enhancement with IntenSE M. Bars: **16** and **17** 0.5 μm; **18** and **19** 20 μm

root surface. The role of these cyst coat components remains obscure. It has been suggested that the cyst coat may act as a protective layer [4] or alternatively it may provide a suitable matrix for the assembly of the microfibrillar cell wall [1], but so far there has been no evidence to support these proposals.

The surface of the cyst closest to the root is coated with a protein component from the ventral vesicles. The timecourse of secretion of this protein correlates closely with the timecourse of acquisition of zoospore adhesiveness (Hardham and Gubler unpubl.). Together, these data strongly suggest that the ventral vesicles contain the adhesive secreted by the zoospores. These results are consistent with evidence from a recent study, which suggested that adhesive material is stored in vesicles on the ventral surface in secondary zoospores of *Saprolegnia ferax* [15].

The localization of a cell wall component in peripheral cisternae prior to its secretion supports earlier suggestions that the peripheral cisternae are involved in cell wall formation [3, 9]. The apparent increase in the density of Cpw-1 labeling of peripheral cisternae and Golgi bodies following induction of encystment, without any corresponding increase in labeling associated with ER, suggests that at least part of this cell wall component is synthesized in these organelles. However, further work is required to identify this cell wall component.

Results of quantitative immunofluorescence measurements show that exocytosis of the three types of peripheral vesicles starts almost simultaneously within the first minute of encystment. This indicates a common intracellular signal which triggers fusion of the dorsal and ventral vesicles with the plasma membrane, together with the vesiculation of the peripheral cisternae. There is indirect evidence that calcium may mediate this response. Zoospores of *P. palmivora* secrete up to 40% of their total calcium during the first 2 min of encystment [12]. Calcium efflux is a common feature of exocytosis in many secretory cells, and is preceded by a transient rise in cytosolic calcium, the presumed trigger of cortical granule exocytosis [13, 14, 21].

Contrary to previous proposals, our results clearly show that the large peripheral vesicles do not undergo exocytosis during encystment. It has generally been believed that in *Phytophthora* zoospores, adhesive material is stored in the large peripheral vesicles [19, 20]. This idea was based largely on cytological evidence which indicates that the vesicles disappear during encystment [4, 9, 17, 20]. However, the evidence from immunolabeling with the mAb specific to components of the large peripheral vesicles shows that these vesicles are still present within the cell 10 min after the induction of encystment and that their contents are absent from the cyst surface at all stages of encystment. The disappearance of the vesicles following germination is consistent with the proposal that these vesicles contain storage protein [5]. The mobilization of these reserves during the initial stages of invasion may help maintain active growth of the germ tube until sufficient nutrients can be obtained from the host plant.

Use of mAbs and immunocytochemistry constitutes a new approach to studies of the infection of plants by pathogenic organisms. The results obtained by labeling

P. cinnamomi cells with the 4 mAbs described in this article demonstrate the potential value of this approach. It has allowed us to identify unambiguously a variety of peripheral vesicles in *P. cinnamomi* zoospores and, for the first time, to follow the fate of these vesicles during the early stages of infection. The new information greatly augments our understanding of the roles of these vesicles in the infection process.

Acknowledgments. The authors thank Mrs. J. Duniec for excellent technical assistance. We also thank M.C. Webb for help with the video-microscope used to obtain Figs. 11 and 12, and Dr L.P. Lehnen for valuable comments on the manuscript. We acknowledge the support of a National Research Fellowship from the Australian Government. F.G. is currently supported by a Queen Elizabeth II Fellowship from the Australian Government.

References

1. Bartnicki-Garcia S, Hemmes DE (1976) Some aspects of the form and function of Oomycete spores. In: Weber DJ, Hess WM (eds) The fungal spore: form and structure. John Wiley, New York, pp 593
2. Beakes GW (1987) Oomycete phylogeny: ultrastructural perspectives. In: Rayner ADM, Brasier CM, Moore D (eds) Evolutionary biology of the fungi. Cambridge University Press, Cambridge, pp 405
3. Bimpong CE, Hickman CJ (1975) Ultrastructural and cytochemical studies of zoospores, cysts, and germinating cysts of *Phytophthora palmivora*. Can J Bot 53: 1310–1327
4. Grove SN, Bracker CE (1978) Protoplasmic changes during zoospore encystment and cyst germination in *Pythium aphanidermatum*. Exp Mycol 2: 51–98
5. Gubler F, Hardham AR (1988) Secretion of adhesive material during encystment of *Phytophthora cinnamomi* zoospores, characterized by immunogold labeling with monoclonal antibodies to components of peripheral vesicles. J Cell Sci 90: 225–235
6. Gubler F, Hardham AR, Duniec J (1989) Characterising adhesiveness of *Phytophthora cinnamomi* zoospores during encystment. Protoplasma 149: 24–30
7. Hardham AR (1987) Ultrastructure and serial section reconstruction of zoospores of the fungus *Phytophthora cinnamomi*. Exp Mycol 11: 297–306
8. Hardham AR, Suzaki E, Perikin JL (1986) Monoclonal antibodies to isolate-, species- and genus-specific components on the surface of zoospores and cysts of the fungus *Phytophthora cinnamomi*. Can J Bot 64: 311–321
9. Hemmes DE, Hohl HR (1971) Ultrastructural aspects of encystation and cyst-germination in *Phytophthora parasitica*. J Cell Sci 9: 175–191
10. Hinch JM, Clarke AE (1980) Adhesion of fungal zoospores to root surfaces is mediated by carbohydrate determinants of the root slime. Physiol Plant Pathol 16: 303–307
11. Ho HH, Zachariah K, Hickman CJ (1968) The ultrastructure of zoospores of *Phytophthora megasperma* var. *sojae*. Can J Bot 46: 37–41
12. Irving HR, Griffith JM, Grant BR (1984) Calcium efflux associated with encystment of *Phytophthora palmivora* zoospores. Cell Calcium 5: 487–500
13. Jaffe LF (1983) Sources of calcium in egg activation: A review and hypothesis. Dev Biol 99: 265–276
14. Kubota HY, Yoshimoto Y, Yoneda M, Hiramoto Y (1987) Free calcium wave upon activation in *Xenopus* eggs. Dev Biol 119: 129–136
15. Lehnen LP, Powell MJ (1989) The role of kinetosome-associated organelles in the attachment of encysting secondary zoospores of *Saprolegnia ferax* to substrates. Protoplasma 149: 163–174

16. Lunney CZ, Bland CE (1976) Ultrastructural observations of mature and encysting zoospores of *Pythium proliferum* de Bary. Protoplasma 90: 119–137
17. Pinto Da Silva P, Nogueira ML (1977) Membrane fusion during secretion. A hypothesis based on electron microscope observations of *Phytophthora palmivora* zoospores during encystment. J Cell Biol 73: 161–181
18. Powell MJ, Bracker CE (1986) Distribution of diaminobenzidine reaction products in zoospores of *Phytophthora palmivora*. Mycologia 78: 892–900
19. Sing VO, Bartnicki-Garcia S (1975a) Adhesion of *Phytophthora palmivora* zoospores: electron microscopy of cell attachment and cyst wall fibril formation. J Cell Sci 18: 123–132
20. Sing VO, Bartnicki-Garcia S (1975b) Adhesion of *Phytophthora palmivora* zoospores: detection and ultrastructural visualization of concanavalin A receptor sites appearing during encystment. J Cell Sci 19: 11–20
21. Swann K, Whitaker M (1986) The part played by inositol triphosphate and calcium in the propagation of the fertilization wave in sea urchin eggs. J Cell Biol 103: 2333–2342
22. Zentmyer GA (1980) *Phytophthora cinnamomi* and the diseases it causes. The American Phytopathological Society, St Paul, Minnesota

Chapter 16 Richard J. O'Connell and John A. Bailey

Hemibiotrophy in *Colletotrichum lindemuthianum*

1 Introduction

Plant pathogenic fungi have been classified into those obtaining their nutrients from living cells (biotrophs) and those obtaining nutrients from host cells that they have killed (necrotrophs) [20]. However, there is a large group of fungi, termed hemibiotrophs by Luttrell [22], which use both modes of nutrition at different stages of their development. After an initial biotrophic phase of varying duration, these fungi switch to necrotrophy. Such nutritional shifts are common among ascomycete and deuteromycete fungi that cause necrotic spotting on leaves, stems and fruits [22], and also occur in diseases caused by some members of the Albuginaceae and Peronosporaceae [8].

Infection by obligate biotrophs characteristically causes minimal damage to host tissues [20]. Likewise, in the initial stages of infection by hemibiotrophs, enzymic degradation of host cell walls is highly restricted and, although membrane permeability may increase, host protoplasts remain alive [8]. During their biotrophic phase, hemibiotrophs ramify through host tissues by means of subcuticular hyphae (e.g. *Rhynchosporium secalis* [7], *Venturia inaequalis* [9]), intercellular hyphae (e.g. *Cladosporium fulvum* [19], *Stemphylium botryosum* [32], *Leptosphaeria maculans* [14]) or intracellular hyphae (e.g. *Colletotrichum lindemuthianum* [30], *Gibberidea heliopsidis* [23], *Guignardia bidwellii* [22]). With the exception of some Oomycetes [6, 36], most hemibiotrophs do not produce specialised intracellular haustoria.

Depending on the pathogen species, the symptomless biotrophic phase may persist for a few days, e.g. *P. infestans* on potato and *R. secalis* on barley, or for several weeks, e.g. *V. inaequalis* on apple and *L. maculans* on oilseed rape [7, 14]. Subsequently, large numbers of host protoplasts die and a spreading lesion develops. Hyphae now ramify through the dead tissues both intra- and intercellularly. The switch to necrotrophic nutrition is often accompanied by tissue

Department of Agricultural Sciences, University of Bristol, AFRC Institute of Arable Crops Research, Long Ashton Research Station, Bristol BS18 9AF, UK

maceration and watersoaking [7], the resulting soft-rot symptoms resembling those caused by necrotrophic pathogens such as *Botrytis* or *Monilinia*.

A limited number of host cells may senesce and die in the final stages of infection by obligate biotrophs, but unlike hemibiotrophs, these fungi are unable to grow on dead tissues. Obligate biotrophs also differ from most hemibiotrophs in that they cannot be readily cultured on chemically defined media without loss of their pathogenicity or changes in their morphology [7].

Many features of hemibiotrophic nutrition are well illustrated by the ascomycete pathogen *Colletotrichum lindemuthianum* (perfect state *Glomerella cingulata*), the causal agent of anthracnose on French bean (*Phaseolus vulgaris*) seedlings. The present chapter reviews the cytology of this disease, including differentiation of fungal infection structures, changes in host ultrastructure induced by infection, and the nature of the fungal-plant interface. Similarities and contrasts with obligate biotrophic fungi are discussed.

2 Initial Penetration of Epidermal Cells

The unicellular conidium of *C. lindemuthianum* germinates in free water to produce a single germ-tube. Upon contact with a hard surface, the germ-tube tip swells and differentiates into a thick-walled, darkly melanised appressorium (4.0–5.0 μm), separated from the empty spore by a septum (Fig. 1a). Extracellular mucilage surrounding the appressorium (Fig. 2) is stained by periodic acid-silver methenamine (PA-AgMe), indicating the presence of carbohydrate containing vicinal

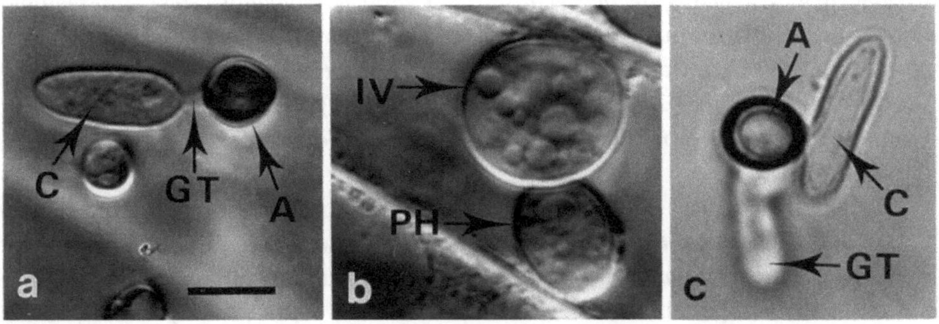

Fig. 1. Germination of conidia and appressoria (**a, b**) on hypocotyl tissue and (**c**) in vitro. **a** Conidium (*C*) on plant surface with short germ-tube (*GT*) subtending melanised appressorium (*A*), viewed with differential interference contrast microscopy. **b** Lower plane of focus of infection shown in **a**. Note large infection vesicle (*IV*) and primary hypha (*PH*) within epidermal cell. **c** Conidium (*C*) and appressorium (*A*) on surface of Formvar membrane supported on water agar (bright field microscopy). After penetration of membrane, narrow germ-tube (*GT*) has developed in water agar beneath. Bar = 5 μm (**a, b, c** same magnification)

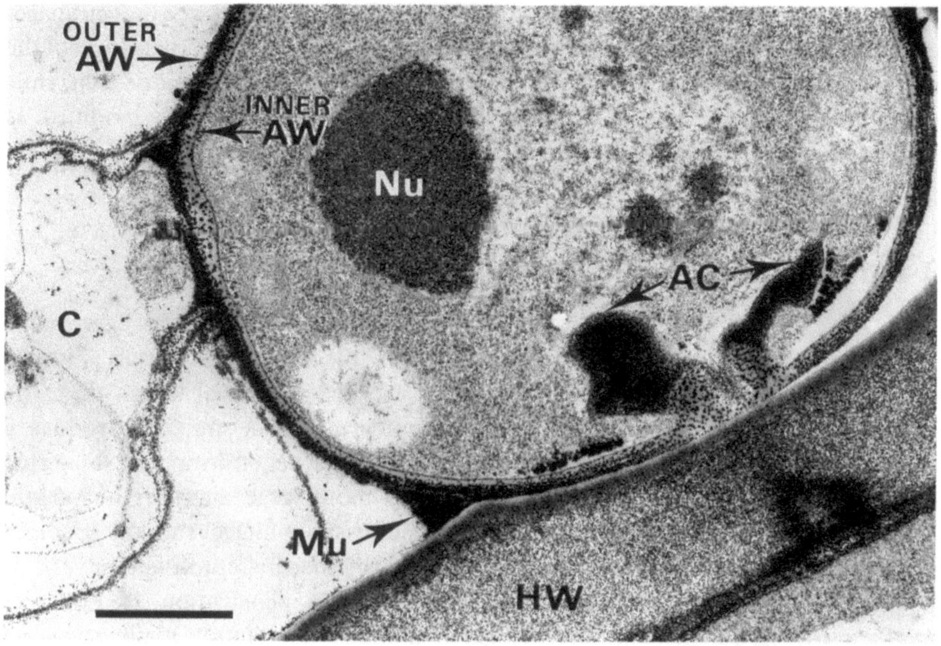

Fig. 2. Appressorium on surface of hypocotyl sectioned near penetration site, stained for carbohydrates with periodic acid-silver methenamine. Note that inner appressorial wall (*AW*) is stained less strongly than the outer *AW* and appressorial cone (*AC*), which surrounds the germ pore. Nucleolus ($N\mu$) and extracellular mucilage (*Mu*) are strongly stained. *C* conidium; *HW* host wall. Bar = 1 μm

hydroxyl groups [10]. This material may promote adherence of appressoria to surfaces. Mature appressoria have a two-layered wall (Fig. 2), the outer layer being more strongly stained by PA-AgMe than the inner, suggesting a difference in carbohydrate composition.

The germ-pore which develops in the ventral wall of the appressorium is surrounded by a funnel-shaped structure, termed an appressorial cone by Landes and Hoffmann [18]. This structure also stains strongly with PA-AgMe (Fig. 2). An infection peg emerges through the pore and penetrates bean epidermal cells directly. The wall of the infection peg is continuous with the cone rather than the appressorial wall [18, 27]. The infection peg is very narrow (approximately 100 nm) where it penetrates the cuticle, widening to 500 nm during penetration of the periclinal cell wall.

C. gloeosporioides, *C. lagenarium* and a few unrelated fungi that also penetrate epidermal cells directly have appressorial cones analogous to those of *C. lindemuthianum* [4, 9, 17, 41]. In these fungi the cone may function to focus the turgor pressure of the appressorial protoplast on a small area of host cuticle [39]. The ability of *C. lindemuthianum* infection pegs to penetrate thin (90–100 nm) Formvar plastic membranes (Fig. 1c, see also [26, 39]) suggests that mechanical force is an important factor in penetration. However, there is evidence that cutinases also have a role in cuticle penetration by *C. gloeosporioides* [11]. Penetration of the epidermal cell wall by *C. lindemuthianum* appears to involve localised enzymic degradation since the wall adjacent to the infection peg is chemically modified, as shown by its increased affinity for various cytochemical stains [29, 30].

3 Intracellular Biotrophic Phase

After penetration of the epidermal cell wall, the tip of the infection peg swells to form 9 large (6–9 μm diameter) globose infection vesicle inside the cell lumen (Fig. 1b). Many other fungi which directly penetrate epidermal cells similarly produce a swollen intracellular hypha or vesicle immediately after penetration, e.g. some rust fungi [13, 16] and Oomycetes [6, 24]. One or more large diameter (3.5–5 μm) primary hyphae (Figs. 1b, 3a, b) develop from the vesicle, which eventually becomes separated from both the infection peg and primary hyphae by septa. Neither vesicles nor primary hyphae are formed in vitro following penetration of Formvar membranes (Fig. 1c). Instead, narrow (2.5 μm diameter) filamentous hyphae develop beneath the membrane. Thus, in contrast to some rust fungi [12], the differentiation of typical intracellular infection structures by *C. lindemuthianum* appears to require contact with living host cells.

Primary hyphae colonise host cells entirely intracellularly, a mode of ramification seen in only a few other biotrophs [3, 13, 14] and hemibiotrophs [23]. Primary hyphae become greatly constricted during cell-cell penetration, although the pore formed in the host cell wall is considerably wider (0.9–1.8 μm diameter) than that produced by the infection peg (Fig. 3b). A perforate septum usually develops at the point of constriction (Fig. 4a). The microfibrils of the host cell wall show little physical deformation (Fig. 4a), and it appears that primary hyphae penetrate walls by means of localised enzymic dissolution [30].

The relationship between the intracellular primary mycelium and host cells is a complex and dynamic one. Host cells initially survive penetration: the thin parietal

Fig. 3. a Biotrophic and **b** necrotrophic primary hyphae in hypocotyl tissue. **a** Longitudinal section of epidermal cell (4 days after inoculation) containing infection vesicle (*IV*) and short primary hypha (*PH*). Thin layer of host cytoplasm (*HC*) is invaginated around fungus and has normal ultrastructure. **b** Longitudinal section of hypocotyl tissue 5 days after inoculation. Host wall is penetrated by narrow infection peg (*IP*) beneath appressorium (*A*). Epidermal cell (*1*) and cortical cell (*2*) contain infection vesicle (*IV*) and primary hyphae (*PH*). Plasma-membrane of both cells is ruptured; cytoplasm of cell 1 is completely disorganised; swollen chloroplasts (*Ch*) and nucleus remain in cell 2. Uninfected cells (*3* and *4*) have cytoplasm of normal ultrastructure. *PH* is constricted at site of cell wall penetration (*arrow*). *V* vacuole; *C* conidium. Bars = 5 μm

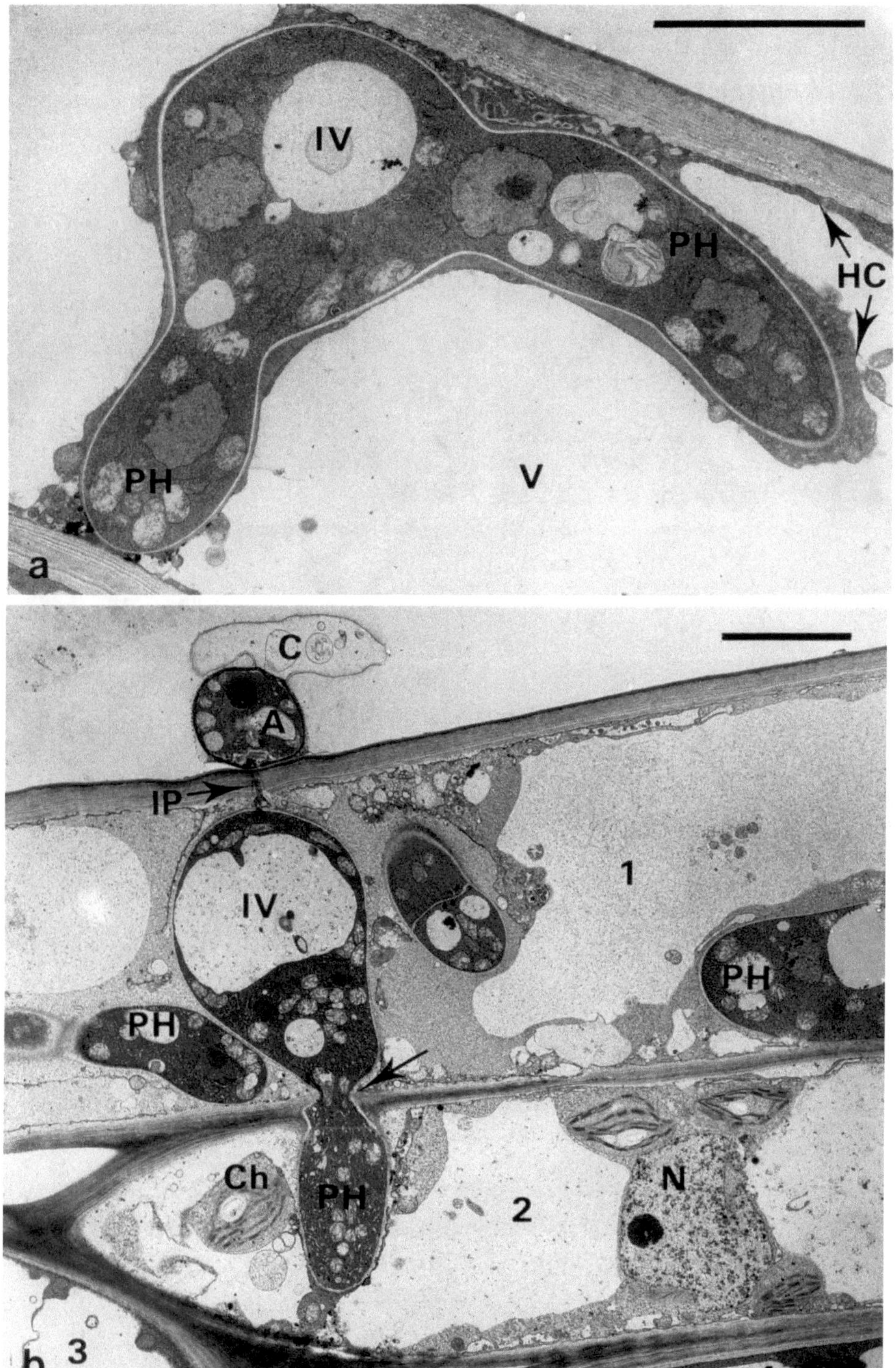
IV
PH
HC
PH
V
a
C
A
IP
1
IV
PH
PH
Ch
PH
2
N
b
3
4

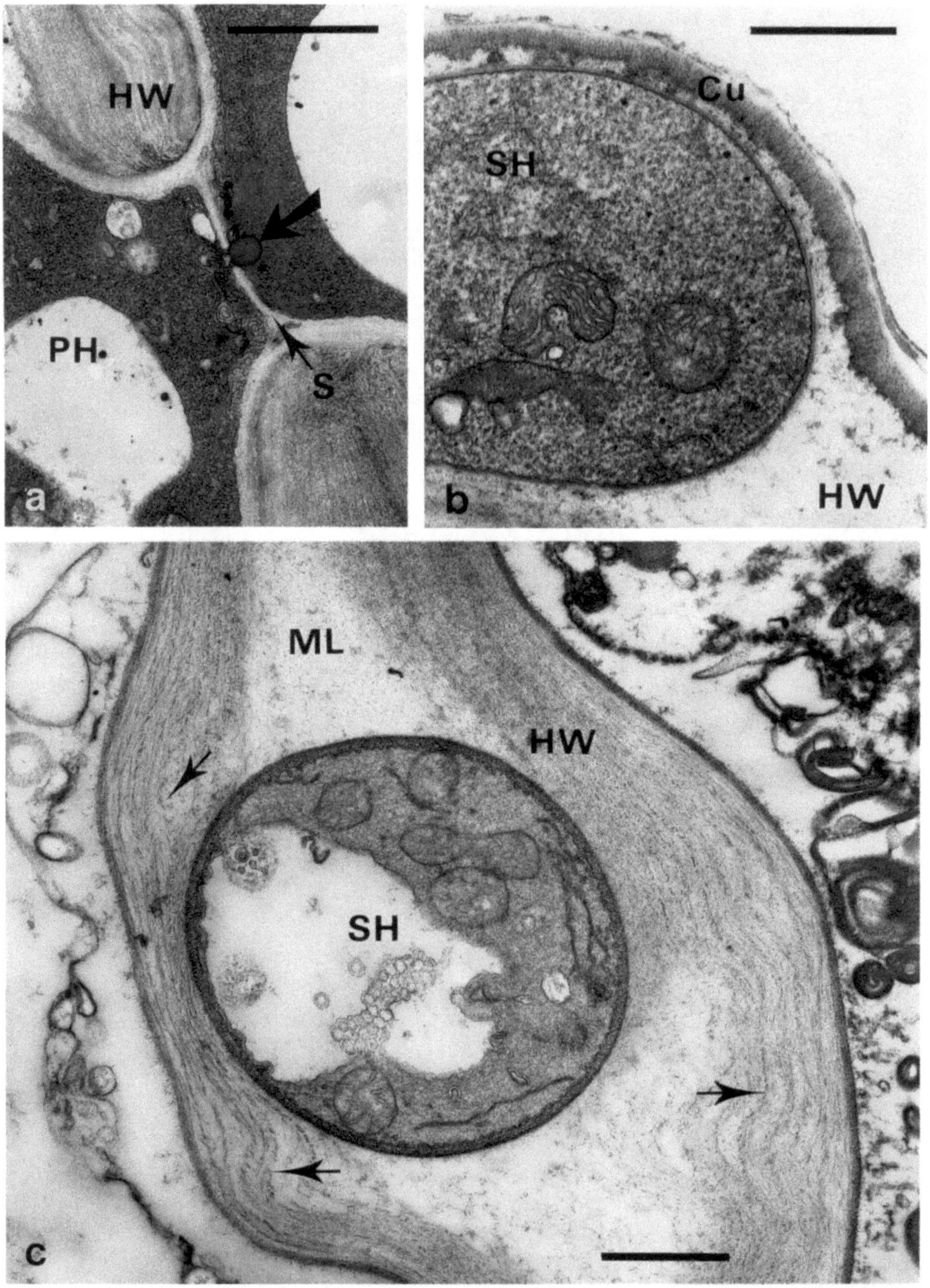
HW
PH
S
a
Cu
SH
HW
b
ML
HW
SH
c

layer of host cytoplasm becomes invaginated around the fungus and shows no detectable change in ultrastructure (Fig. 3a). However, this biotrophic relationship is transient, since infected cells lose the ability to plasmolyse normally and to exclude tannic acid (a permeability tracer) approximately 24 h after penetration. These changes in plasma membrane function are associated with an increase in the volume of cytoplasm and dilation of the endoplasmic reticulum, dictyosomes and chloroplast membranes [30]. Approximately 48 h after penetration, the plasma membrane and tonoplast lose their structural integrity, and by 60 h ribosomes have disappeared and most organelles are completely disorganised (Fig. 3b).

The sequence of a brief biotrophic phase, followed by loss of membrane function and gradual cytoplasmic disorganisation, is repeated as primary hyphae colonise other host cells. Consequently, young hyphae at the edge of the infection area may be feeding biotrophically while older hyphae at the centre are necrotrophic, obtaining nutrients from killed host cells. A similar phenomenon has been reported for other hemibiotrophs [7, 36]. Bean tissues show no visible symptoms at this stage, since cells killed by primary hyphae do not become brown.

4 The Host-Parasite Interface

The intracellular infection vesicles and primary hyphae of *C. lindemuthianum* remain external to the host plasma membrane, which becomes invaginated around the fungus. The hyphae develop within a carbohydrate-rich matrix, separating the fungal cell wall from the host plasma membrane [28]. Similar matrices are associated with intracellular hyphae of *C. gloeosporioides* [4] and many other biotrophic pathogens and symbionts [21, 40], see also Bonfante-Fasolo and Perotto, Chap. 20, and Harder and Chong, Chap. 16, this Vol. There is evidence that host products are deposited in the matrix by means of exocytosis, and immunogold cytochemistry indicates that fungal glycoproteins are also present [28, 31]. Hence, both host and fungus may contribute to forming the matrix around *C. lindemuthianum* hyphae. In contrast, micro-autoradiographic studies of

◀

Fig. 4. Penetration of host cell walls during **a** biotrophic phase and **b, c** necrotrophic phase. **a** Primary hypha (*PH*) penetrating host wall (*HW*) between two cortical cells (5 days after inoculation). Note constriction of *PH* and perforate septum (*S*) with associated Woronin body (*large arrow*). *HW* is not extensively degraded. **b** Secondary hypha (*SH*) within epidermal cell wall (*HW*), 8 days after inoculation. Note dissolution of *HW* (microfibrils not visible) and undamaged cuticle (*Cu*). **c** Secondary hypha (*SH*) within wall (*HW*) between two dead cortical cells (8 days after inoculation), stained for carbohydrates with periodic acid-thiocarbohydrazide-silver proteinate. Note swelling of *HW* and dissolution of middle lamella (*ML*). Solubilisation of other wall polymers has revealed the pattern of wall microfibrils (*arrows*). Bars = 1 μm. [Reprinted with permission of Keon JPR from *Fungal infection of plants*, ed by Pegg GF and Ayres PG (1987), Cambridge University Press]

Uromyces and *Erysiphe* suggest that the extrahaustorial matrix is solely a host product [25, 34].

The surface area of the host plasma membrane must increase to accommodate invagination by *C. lindemuthianum* hyphae, and new membrane material appears to be added by the fusion of cytoplasmic vesicles with the plasma membrane [28]. The invaginated region of the plasma membrane is indistinguishable from the wall-lining region: it appears normal when freeze-fractured, retains adenosine triphosphatase activity and stains normally with phosphotungstic acid [28]. This apparent absence of any structural, physiological or cytochemical modification contrasts with the high degree of specialisation induced in the invaginated host plasma membrane by haustoria of many obligate biotrophs [21, 40], see also Harder and Chong, Chap. 18, this Vol].

5 Necrotrophic Phase

Following establishment of a large intracellular and partially biotrophic primary mycelium, changes occur in the mode of parasitism and morphology of *C. lindemuthianum.* Narrow (1.5–2.5 μm diameter) secondary hyphae emerge as lateral branches from primary hyphae and ramify through host tissues both inter- and intracellularly [30]. Many hyphae grow within host cell walls (Fig. 4b, c). Development of the secondary mycelium is associated with rapid disruption of host protoplasts and dissolution of host cell walls in advance of infection [30]. Visible symptoms now appear as brown spreading lesions, within which host tissue is water-soaked and macerated. Degraded host cell walls are swollen and, after selective staining for carbohydrates [10], the pattern of wall microfibrils is clearly visible (Fig. 4c), probably because the matrix of pectic polysaccharides has been removed by fungal enzymes [33]. The host cuticle remains unaltered even after complete dissolution of underlying epidermal cell walls (Fig. 4b).

6 Basis of Nutritional Shifts

The processes which regulate the nutritional mode of *C. lindemuthianum* at different stages of its development are obscure at present, but with the data available it is possible to speculate. During the initial biotrophic phase of infection, *C. lindemuthianum* develops in contact with living host protoplasts, and cell wall dissolution is highly localised to sites of penetration. Being a facultative parasite capable of growth on simple media, the fungus may be able to obtain sufficient nutrients for its early development from the plant apoplast [15]. Subsequent changes in host cell permeability may also release nutrients into the apoplast.

However, in the absence of a specialised fungal-plant interface to control the efflux of solutes, the affected cells senesce and die. Thus biotrophy gives way to a benign form of necrotrophy, in which the fungus utilises nutrients released from dead infected cells, but without extensive digestion of their cell walls. It is therefore unlikely that biotrophy breaks down due to increased secretion of cell wall-degrading enzymes by the fungus, as suggested for other hemibiotrophs [7].

In contrast, however, cell wall-degrading enzymes probably play a crucial role in the later transition to destructive necrotrophy, by which the fungus fully utilises host tissues as a nutrient source through solubilisation of the structural polysaccharides of cell walls and disruption of large numbers of host protoplasts. The appearance of lesions coincides with an increase in endo-pectin lyase (PL) activity, and purified preparations of PL cause maceration and electrolyte leakage when applied to bean tissues [37, 38]. Being an inducible enzyme, PL is likely to be subject to catabolite repression [38]. Synthesis of PL may therefore be delayed until apoplast sugars are locally depleted by the primary mycelium. *C. lindemuthianum* also produces an endo-polygalacturonase in vitro, but as yet the enzyme has not been detected in tissue extracts, possibly due to the presence of an inhibitor [38].

Although *C. lindemuthianum* has the potential for highly destructive necrotrophic growth, it is unable to colonise bean tissues previously killed by mechanical wounding or freeze-thawing (Bailey, unpubl. data). An initial brief period of biotrophic growth appears to be essential for successful infection. Thus, the biotrophic phase may allow the pathogen to accumulate sufficient food reserves for subsequent large-scale synthesis and secretion of cell wall-degrading enzymes, which probably overcome active host defences by rapidly killing cells in advance of invasion [1].

7 Hemibiotrophy in Other *Colletotrichum* Species

Colletotrichum is a large genus causing diseases on a wide range of crop and weed species. At present it is unclear how widespread hemibiotrophy is among members of the genus, since most cytological studies have not assessed the viability of host cells during infection. However, plasmolysis experiments have shown that *C. pisi* infecting pea and *C. gloeosporioides* infecting cowpea are both hemibiotrophs [2]; A. Uronu and R.J. O'Connell unpubl. data. These fungi have very similar infection processes, which differ in several respects from that of *C. lindemuthianum*. Large-diameter primary hyphae develop within epidermal cells but no distinct infection vesicle is produced. Epidermal cells initially survive penetration but lose the ability to plasmolyse after approximately 24 h. The dead host cell becomes densely packed with primary hyphae, before narrow secondary hyphae ramify through the surrounding tissue, killing cells in advance of infection. In these two species biotrophy and the primary mycelium are confined to the initially infected epidermal cell.

Infection by a large number of other *Colletotrichum* species is characterised by an initial symptomless phase, but often this does not involve active biotrophic growth. Such infections are associated with appressoria, subcuticular hyphae or intracellular hyphae which lie dormant for long periods until the physiology of the host tissue changes, e.g. during fruit ripening or the senescence of vegetative organs [5, 35].

8 Conclusions

Hemibiotrophy is a strategy used by a large number of plant pathogenic fungi to invade host tissues. Ultrastructural studies have greatly increased our understanding of this fascinating group of pathogens, which are capable of both subtle growth inside living host cells and massive destruction of host tissues. Electron microscopy can be used to study the changing relationship between hemibiotrophs and host cells and to give insight into the regulation of biotrophy and necrotrophy. In the future we can anticipate increasing use of antibody and lectin probes to study the composition of fungal cell surfaces and the involvement of cell wall-degrading enzymes at different stages of infection.

Acknowledgments. The authors wish to thank Miss C. Nash for technical assistance. This research was funded by the Agricultural and Food Research Council.

References

1. Bailey JA, O'Connell RJ (1989) Plant cell death: a determinant of disease resistance and susceptibility. In: Graniti A, Durbin RD, Ballio A (eds) Phytotoxins and plant pathogenesis NATO ASI Series, Vol H27. Springer, Berlin Heidelberg New York Tokyo, pp 275–283
2. Bailey JA, Nash C, O'Connell RJ, Skipp RA (1990) Infection process and host specificity of a *Colletotrichum* species causing anthracnose disease of cowpea, *Vigna unguiculata*. Mycol Res (in press)
3. Bonde MR, Bromfield KR, Melching JS (1982) Morphological development of *Physopella zeae* on corn. Phytopathology 72: 1489–1491
4. Brown GE (1977) Ultrastructure of penetration of ethylene-degreened Robinson tangerines by *Colletotrichum gloeosporioides*. Phytopathology 67: 315–320
5. Cerkauskas RF (1988) Latent colonization by *Colletotrichum* spp.: epidemiological considerations and implications for mycoherbicides. Can J Plant Pathol 10: 297–310
6. Coffey MD, Wilson UE (1983) Histology and cytology of infection and disease caused by *Phytophthora*. In: Erwin DC, Bartnicki-Garcia S, Tsao PH (eds) *Phytophthora*—its biology, taxonomy, ecology and pathology. The American Phytopathological Society, St. Paul, pp 289–301

7. Cooke RC (1977) The biology of symbiotic fungi. John Wiley, London
8. Cooke RC, Whipps JM (1980) The evolution of modes of nutrition in fungi parasitic on terrestrial plants. Biol Rev 55: 341–362
9. Corlett M, Chong J (1977) Ultrastructure of the appressorium of *Spilocaea pomi*. Can J Bot 55: 5–7
10. Courtoy R, Simar LJ (1974) Importance of controls for the demonstration of carbohydrates in electron microscopy with the silver methenamine or the thiocarbohydrazide-silver proteinate methods. J Microsc 100: 199–211
11. Dickman MB, Patil SB, Kolattukudy PE (1982) Purification, characterization and role in infection of an extracellular cutinolytic enzyme from *Colletotrichum gloeosporioides* Penz. on *Carica papaya* L. Physiol Plant Pathol 20: 333–347
12. Freytag S, Bruscaglioni L, Gold RE, Mendgen K (1988) Basidiospores of rust fungi (*Uromyces* species) differentiate infection structures in vitro. Exp Mycol 12: 275–283
13. Gold RE, Mendgen K (1984) Cytology of basidiospore germination, penetration and early colonization of *Phaseolus vulgaris* by *Uromyces appendiculatus* var. *appendiculatus*. Can J Bot 62: 1989–2002
14. Hammond KE, Lewis BG (1987) The establishment of systemic infection in leaves of oilseed rape by *Leptosphaeria maculans*. Plant Pathol 36: 135–147
15. Hancock JG, Huisman OC (1981) Nutrient movement in host-pathogen systems. Annu Rev Phytopathol 19: 309–331
16. Hunt P (1968) Cuticular penetration by germinating uredospores. Trans Br Mycol Soc 51: 103–112
17. Koch E, Ebrahim-Nesbat F, Hoppe HH (1983) Light and electron microscopic studies on the development of soybean rust (*Phakopsora pachyrhizi* Syd.) in susceptible soybean leaves. Phytopathol Z 106: 302–320
18. Landes M, Hoffmann GM (1979) Zum Keimungs-und Infektionsverlauf bei *Colletotrichum lindemuthianum* auf *Phaseolus vulgaris*. Phytopathol Z 95: 259–273
19. Lazarovits G, Higgins VJ (1976) Histological comparison of *Cladosporium fulvum* race 1 on immune, resistant and susceptible tomato varieties. Can J Bot 54: 224–234
20. Lewis DH (1973) Concepts in fungal nutrition and the origin of biotrophy. Biol Rev 48: 261–278
21. Littlefield LJ, Bracker CE (1972) Ultrastructural specialization at the host-pathogen interface in rust-infected flax. Protoplasma 74: 271–305
22. Luttrell ES (1974) Parasitism of fungi on vascular plants. Mycologia 66: 1–15
23. Mason DL (1973) Host-parasite relations in the Gibberidea disease of *Helianthus strumosus*. Mycologia 65: 1158–1170
24. McKeen WE, Svircev AM (1981) Early development of *Peronospora tabacina* in the *Nicotiana tabacum* leaf. Can J Plant Pathol 3: 145–158
25. Mendgen K, Heitefuss R (1975) Micro-autoradiographic studies on host-parasite interactions I. The infection of *Phaseolus vulgaris* with tritium-labelled uredospores of *Uromyces phaseoli*. Arch Microbiol 105: 193–199
26. Mercer PC, Wood RKS, Greenwood AD (1971) Initial infection of *Phaseolus vulgaris* by *Colletotrichum lindemuthianum*. In: Preece TF, Dickinson CH (eds) Ecology of leaf surface microorganisms. Academic Press, London, pp 381–390
27. Mercer PC, Wood RKS, Greenwood AD (1975) Ultrastructure of the parasitism of *Phaseolus vulgaris* by *Colletotrichum lindemuthianum*. Physiol Plant Pathol 5: 203–214
28. O'Connell RJ (1987) Absence of a specialized interface between intracellular hyphae of *Colletotrichum lindemuthianum* and cells of *Phaseolus vulgaris*. New Phytol 107: 725–734
29. O'Connell RJ, Bailey JA (1986) Cellular interactions between *Phaseolus vulgaris* and the hemibiotrophic fungus *Colletotrichum lindemuthianum*. In: Bailey J (ed) Biology and molecular biology of plant-pathogen interactions NATO ASI Series, Vol H1. Springer, Berlin Heidelberg New York Tokyo, pp 39–48
30. O'Connell RJ, Bailey JA, Richmond DV (1985) Cytology and physiology of infection of *Phaseolus vulgaris* by *Colletotrichum lindemuthianum*. Physiol Plant Pathol 27: 75–98

31. O'Connell RJ, Bailey JA, Vose IR, Lamb CJ (1986) Immunogold labeling of fungal antigens in cells of *Phaseolus vulgaris* infected by *Colletotrichum lindemuthianum*. Physiol Mol Plant Pathol 28: 99–105
32. Pierre RE, Millar RL (1965) Histology of pathogen-suscept relationship of *Stemphylium botryosum* and alfalfa. Phytopathology 55: 909–914
33. Roland JC, Vian B (1981) Use of purified endopolygalacturonase for a topochemical study of elongating cell walls at the ultrastructural level. J Cell Sci 48: 333–343
34. Spencer-Phillips PTN, Gay JL (1980) Electron microscope autoradiography of ^{14}C photosynthate distribution at the haustorium-host interface in powdery mildew of *Pisum sativum*. Protoplasma 103: 131–154
35. Tiffany LH (1951) Delayed sporulation of *Colletotrichum* on soybean. Phytopathology 41: 975–985
36. Whipps JM, Cooke RC (1978) Comparative physiology of *Albugo tragopogonis*-infected and *Puccinia lagenophorae*-infected plants of *Senecio squalidus* L. New Phytol 81: 307–319
37. Wijesundera RLC (1984) Cell wall degrading enzymes of *Colletotrichum lindemuthianum*: their role in the development of anthracnose of bean. PhD Thesis, University of Bristol
38. Wijesundera RLC, Bailey JA, Byrde RJW, Fielding AH (1989) Cell wall degrading enzymes of *Colletotrichum lindemuthianum*: their role in the development of bean anthracnose. Physiol Mol Plant Pathol 34: 403–413
39. Wolkow PM, Sisler HD, Vigil EL (1983) Effect of inhibitors of mélanin biosynthesis on structure and function of appressoria of *Colletotrichum lindemuthianum*. Physiol Plant Pathol 22: 55–71
40. Woods AM, Gay JL (1983) Evidence for a neckband delimiting structural and physiological regions of the host plasmamembrane associated with haustoria of *Albugo candida*. Physiol Plant Pathol 23: 73–88
41. Xuei XL, Järlfors U, Kuć J (1988) Ultrastructural changes associated with induced systemic resistance of cucumber to disease: host response and development of *Colletotrichum lagenarium* in systematically protected leaves. Can J Bot 66: 1028–1038

Chapter 17

Hitoshi Kunoh[1], Ralph L. Nicholson[2], and
Issei Kobayashi[1]

Extracellular Materials of Fungal Structures: Their Significance at Prepenetration Stages of Infection

1 Introduction

Historically, mechanisms of fungal infection have attracted plant pathologists, and our understanding of infection processes has been increased significantly for many disease interactions. Fungi generally undergo a series of distinct morphological changes before attempting host penetration. To date, most studies on infection processes deal with penetration and post-penetration stages of fungal development. However, several observations suggest that phenomena associated with initial stages of fungal contact on host surfaces play a critical role in host recognition, fungal differentiation, and success of infection [1, 13, 20, 26, 37, 38, 44, 64, 65, 73–76]. Since infection processes may be initiated soon after contact, prepenetration phenomena necessary for infection can easily be overlooked. One of these early prepenetration events is adhesion. The mechanisms of fungal adhesion are not as well understood as are those for bacterial adhesion [76]. Nicholson and Epstein [51] reviewed adhesion by plant pathogenic fungi and discussed mechanisms known to be involved in adhesion, as well as the broader topic of extracellular fungal matrices. The purpose of this review is to address cytological aspects of extracellular materials associated with fungal adhesion, and the subject of recognition between fungal and host cells.

2 Fungal Adhesion to Substrata

2.1 Adhesive Materials of Zoospores

Zoospores become firmly attached to any substratum on which they settle, and adhesion is strong enough to resist extensive washing [23]. An electron-opaque,

[1] Laboratory of Plant Pathology, Faculty of Bioresources, Mie University, Tsu, 514, Japan
[2] Department of Botany and Plant Pathology, Purdue University, West Lafayette, IN 47907, USA

flocculent, adhesive substance was found at the junction of encysted zoospores of *Plasmopara halstedii* and *Olpidium brassicae* and their host cell walls [67, 71]. When zoospores of *Phytophthora palmivora* attached to a plastic surface, the cyst was anchored by partially secreted, finely granular contents of a peripheral vesicle present in the zoospore, suggesting that this was an initial step in adhesion [62]. In mature cysts adhesive appeared compacted into an electron-dense, amorphous deposit binding the cyst to the plastic surface. Electron microscopy of the adhesion zone of lysed cysts revealed that cyst wall microfibrils were present with an amorphous adhesive material, suggesting that adhesion and microfibril formation occurred concurrently [62].

2.2 Adhesive Materials from Spores

Spores of several fungi are known to adhere to contact surfaces upon deposition, and prior to germination. In some cases adhesion appears associated with a mucilaginous material on the spore surface. Gold and Mendgen [16] demonstrated that mature basidiospores of *Uromyces appendiculatus* were covered with a fibrillar mucilaginous material appearing to have a role in adhesion. In *Candida albicans*, an animal pathogen, a fibrillar outer layer is evident around spores [8, 9, 41, 68]. Douglas [9] suggested a correlation between adhesion and virulence of *C. albicans*, and concluded that enhanced adhesion resulted from production of an additional fibrillar-floccular layer on the cell surface. Similar observations have been made for nematophagous fungi. Bird and McKay [5] demonstrated that a mucilagenous fibrillar material exuded from appendages of conidia of *Dilophospora alopecuri* is necessary for conidial adhesion to nematodes. Saikawa [61] demonstrated that conidia of *Meria coniospora* have a knob-like structure at the conidial apex that is covered with a mucous layer composed of radiating fibrils. Attachment of conidia to a nematode's integument is mediated by these mucous fibrils. Hamer et al. [19] demonstrated release of an extracellular fibrous matrix from the spore apex of *Magnaporthe grisea* serving as an adhesive.

Figs. 1–6. Conidia of *Erysiphe graminis* fixed and frozen at intervals after inoculation on a cellophane membrane. Bars = 5 μm except 1 μm for *insets.* **1** Conidia fixed 30 min after inoculation. Note the halo of material deposited in a zone around the conidia. **2** Conidium fixed 30 min after inoculation and moved by micromanipulation to reveal presence of an extensive deposit of material beneath the conidium and at the conidium-cellophene interface. The margin of the material appears floculent. **3** Conidia fixed 90 min after inoculation. Note the presence of a floculent deposit at the conidum-cellophane interface **4** Conidia in a chain projecting from the surface of a barley leaf. Note the reticulate network of the conidium surface which is interspersed with bent protrusions (*inset*). **5** Single conidium frozen at 10 min after inoculation. The reticulate network of the native, unstimulated conidium surface is not visible (*inset*). **6** Single conidium frozen 30 min after inoculation. There is no evidence of a reticulate network and the surface is granular and covered with globular bodies (*inset*)

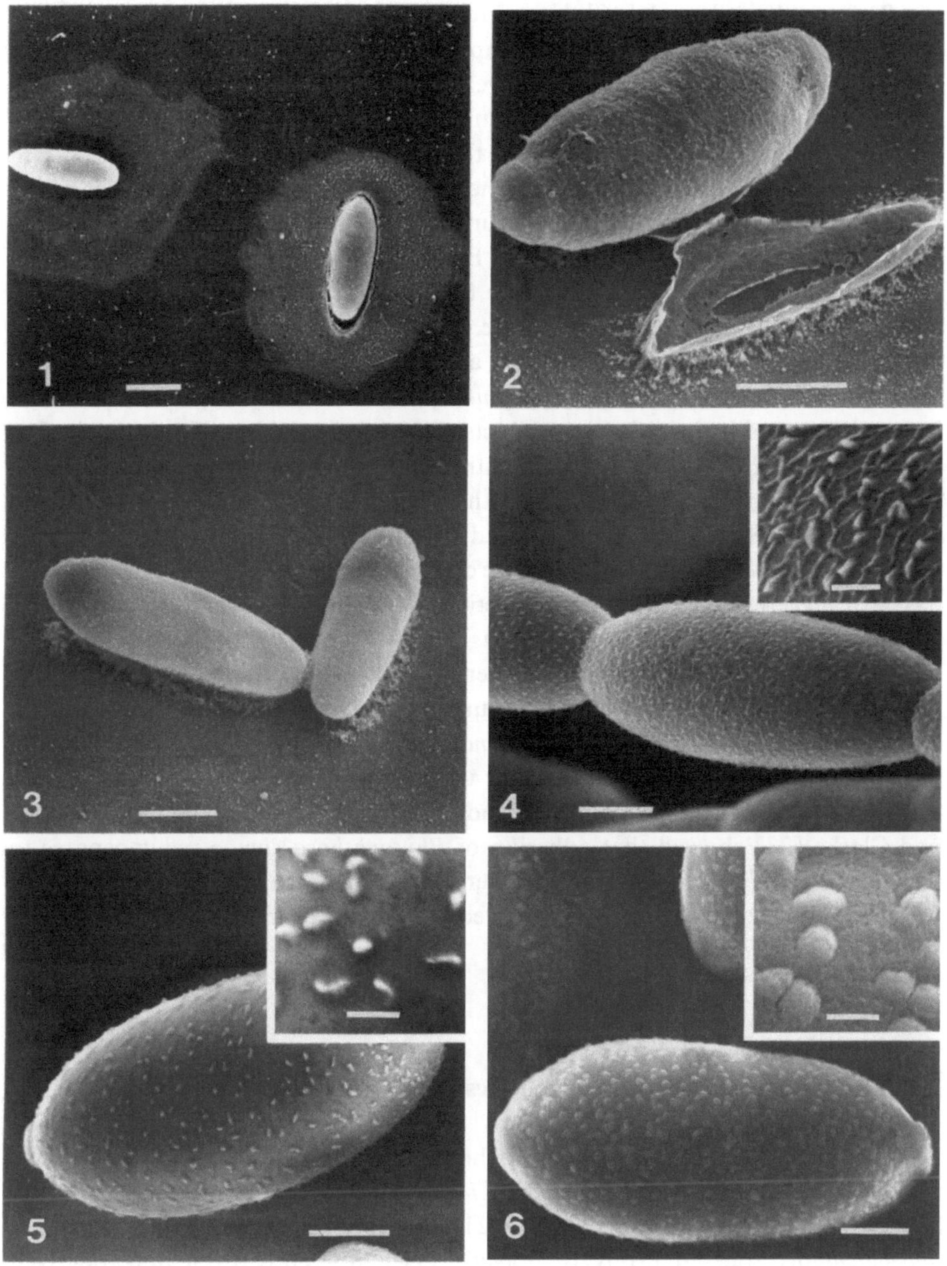
1
2
3
4
5
6

Recently, Kunoh et al. [36] showed release of a liquid from conidia of *Erysiphe graminis* stimulated by contact with cellophane or leaf surfaces. The liquid formed a film on the conidium surface and flowed outward from the conidium onto the contact surface, where it appeared fluocculant (Fig. 1). When conidia were detached from a cellophane surface, an extensive, thickened layer of material was deposited (Fig. 2). Whether this material is an adhesive is unknown, but it is present at the interface of the conidium and contact surface (Fig. 3). Importantly, the deposition occurred before conidial germination. Evans et al. [14, 15] reported a similar phenomenon in *Bipolaris maydis* where fibrillar material appears at the conidial apex before emergence of the germ tube.

When spores land on plant surfaces, adhesion may differ from that on artificial membranes. Spores of the fungus *Zygophiala jamaicensis*, causal agent of flyspeck of fruits, are encased in degradation products of the host cuticle, and it is thought that the degeneration of wax crystals and cutin is mediated by enzymes secreted from ungerminated spores [47, 49]. Thus, the cuticle itself may function in fungal adhesion. Nicholson et al. [52, 53] showed that spores of *Colletotrichum graminicola* are produced in a water-soluble mucilage containing nonspecific esterase activity. It has been suggested that the esterase functions in the degradation of the host cuticle [50] in a manner similar to *Z. jamaicensis* but involvement of cuticular products in adhesion apparently does not occur. Others have shown apparent adhesive material at the interface of spore walls and host surfaces of *Cladosporium cucumerinum* [55] and *Gymnosporangium juniperi-virginianae* [45], but the origin and timing of deposition of these secreted materials is unknown.

A role for plant mucilages in fungal adhesion has been reported [17, 23]. Hinch and Clarke [23] showed that adhesion of *Phytophthora cinnamomi* zoospores to corn roots failed when root surfaces were oxidized, when enzymatic removal of fucosyl residues of root mucilage occurred, or when root mucilage was pretreated with a fucose-binding lectin.

2.3 Adhesive Materials from Germ Tubes, Appressoria, and Hyphae

The close adhesion of germ tubes to host surfaces may be essential for morphological development of the germling [74] and the ability of germ tubes to adhere is well documented. Epstein et al. [12] showed that germlings of *Uromyces appendiculatus* adhered firmly to polyethylene, apparently by a fungal-derived protein-containing extracellular material. Germlings treated with pronase lacked extracellular material, and it was assumed that a protein was required for adhesion [12, 13]. Similar mucilage-like materials have been reported in fungi prior to appressorial formation [2, 20, 21, 28, 42, 44, 55]. Gold and Mendgen [16] showed a fibrillar exudate extending laterally on the epidermis 2 μm away from *U. appendiculatus* germ tubes and appressoria. Pringle [59] reported germinating conidia of *Bipolaris sorokiniana* were bound to glass by an adhesive, possibly containing a positively charged galactosamine polymer.

Adhesive materials have also been demonstrated with hyphae on various substrata [20, 40, 46, 55]. Hau and Rush [20] showed a glue-like adhesive outside cell walls of *Helminthosporium oryzae* hyphae and that the material was composed of an electron-dense, amorphous substance. Bonfante-Fasolo et al. [7] demonstrated an abundant fibrillar network of material surrounding hyphal walls of the mycorrhizal fungus *Pezizella ericae* and showed that the material binds specifically to Concanavalin A, suggesting mediation by mannose or glucose residues on the hyphal surface. Noninfective strains of the fungus lacked extracellular mucilage, and it was concluded that establishment of symbiosis was related to the presence of the extracellular substance [6]. Barak et al. [2] demonstrated a multilayered polysaccharide sheath around hyphae of *Sclerotium rolfsii*, and suggested that it was essential for binding of *Trichoderma hamatum*, a mycoparasite.

Intercellular hyphae also adhere to the surface of tissues within host plants. Using SEM images of complementary fractures, Beckett and Porter [4] showed adhesive pads between intercellular hyphae of *Uromyces viciae-fabae* and bean cell surfaces. Similarly, Pring [58] showed adhesive material below the haustorial mother cell of *U. phaseoli* on French bean mesophyll cells.

Appressoria adhere to host surfaces and often bind to artificial substrata [11]. Light and electron microscopy has revealed fibrillar or granular material around appressoria of many fungi [1, 4, 10, 16, 20, 24, 28, 43, 44, 46, 57, 58, 69, 73]. The adhesives occur not only on the contact side of the appressorium but may cover the whole surface [1, 44. 46]. In other fungi adhesives are mainly found at the contact surface [10, 20, 28, 43]. Hirooka et al. [24] found that the entire surface of *Pyricularia oryzae* appear covered with granular adhesive, and on cellophane the adhesive extended outward from the appressorium. It is likely that the amount of adhesive produced and the site of secretion vary depending on the contact surface (see Howard et al., Chap. 19, this Vol).

2.4 Release of Enzymes from Spores and Related Structures

A conspicuous change in surface morphology of *Erysiphe graminis* conidia occurs during the first minutes after contact stimulus [36]. Unstimulated conidia exhibit a distinct pattern of reticulation interspersed with spine-like protrusions (Fig. 4). Within 10 min of contact with either a barley leaf or with cellophane, the reticulate network becomes less distinct and eventually disappears (Fig. 5). These changes occur through deposition of a layer of material that covers the conidium like a film. The film is of sufficient thickness to cover the basal portion of the spine-like protrusions, and by 30 min the film developed to the extent that the protrusions appeared globose (Fig. 6). The events occur in the same temporal sequence regardless of the nature of the contact surface. Such a rapid change of the spore surface is a previously unreported phenomenon.

The liquid film released from *E. graminis* was subsequently shown by Nicholson et al. [54] to contain esterase enzyme activity. Ungerminated conidia released esterase immediately after contact stimulation and in two stages: the first within 2 min of contact, and the second between 10 and 15 min of contact. The release of esterase activity was correlated with changes in conidial surface morphology and cuticular degradation of the leaf surface immediately adjacent to the conidium [36]. Subsequently, it was shown that the second phase of esterase release is temperature- and cycloheximide-sensitive, suggesting that esterase released in the first phase is constitutive [34]. It is possible that the first phase is associated with overcoming the hydrophobic nature of conidia [39], and the second phase with host cuticle degradation. In support of this hypothesis is the fact that conidia are initially hydrophobic and float on water. However, conidia incubated for only 2 min after surface contact are readily suspended in water. The phenomenon represents a classic tropism response to contact stimulation as described by Wynn [74].

The release of esterase from *E. graminis* resembles zoospore encystment in which contact leads to release of glycoproteins involved in cyst adhesion [3, 18]. These events have been interpreted as steps in the infection process and suggest that enzyme release and apparent dissolution of host cuticle are required for fungal development [36, 54]. The process is not unlike calcium-induced secretion of peptidase from encysted zoospores of the animal parasitic fungus *Aphanomyces astaci* which appears to erode crayfish cuticle [56].

Degradation of host cuticle occurs below germ tubes and hyphae of various fungi [20, 24, 31–33, 35, 47–49], indicating that secretion of fungal cutinase has occurred. Emmett and Parberry [11] suggest that degradation of plant cell wall components by fungal enzymes is a prerequisite for penetration. Staub et al. [66] demonstrated that *Erysiphe graminis* dissolves cuticular components of barley leaves and of cucumber but that *E. chicoracearum*, a nonpathogen of barley, was unable to alter the barley leaf surface. They suggest that dissolution of wax crystals occurring beneath *E. graminis* on barley may be critical for direct penetration by this fungus. In contrast to the observation with *E. chicoracearum*, Kunoh et al. [31] showed that *E. pisi* was able to erode the barley cuticle beneath appressoria.

The importance of surface cuticular components and topography is only now beginning to be understood; however, the significance of wax crystals to fungal germination and differentiation has been recognized for some time [37, 47–49, 64]. Wynn [74] showed that germlings of *Puccinia sorghi* on waxless corn leaves grew away from the surface and failed to recognize stomates. Similarly, germlings of *P. coronata* were bound to the leaf apparently by a combination of fungal secretions and dissolved host wax [74] similar to the situation for *Z. jamaicensis* [47–49]. Yang and Ellingboe [75] suggested that appressorium formation by *E. graminis* on cereals was stimulated by the physical structure of epicuticular wax crystals. However, it is possible that if wax crystals and other cuticle components cannot be dissolved, close contact of the appressorium with the surface is impossible and penetration is prevented, a view supported by Nicholson [50]. Thus, the release of degradative enzymes may be essential to penetration.

3 Mechanism of Secretion of Extracellular Materials from Fungal Structures

Hamer et al. [19] demonstrated spore tip mucilage in the periplasmic space of dormant conidia of *Magnaporthe grisea* that was released upon conidium hydration. They suggested that when conidia are hydrated, mucilage is released allowing for attachment of conidia to hydrophobic leaf surfaces. Evans et al. [14, 15] reported a fibrillar mucilage present on hyphal tips and germ tubes of *Bipolaris maydis*. In older, damaged hyphae, excrescences were present that seemed to arise from damage to the pellicle. The material resembled the fibrillar material present on young hyphae. It is possible that if this is an adhesive it may be stored in a space between the pellicle and the cell wall and be released when necessary.

The mechanism for secretion of adhesives is well documented for Oomycete zoospores. Unencysted zoospores of *Phytophthora* have peripheral vesicles containing granular material [18, 22, 25, 60, 62] that is released upon encystment, suggesting that vesicle excretion upon contact cements the zoospore to the contact surface. Early in encystment the vesicles fuse with the plasmalemma and discharge their contents, forming an electron-dense zone between the cell and the solid surface. According to Gubler and Hardham [18] there are two peripheral vesicles (large and small) in zoospores of *Phytophthora cinnamomi*. Fungal adhesion involves secretion of glycoproteins by exocytosis of small vesicles. Sing and Bartnicki-Garcia [62] showed that peripheral vesicles contain glycoprotein, presumed to be involved in adhesion, with receptor sites for Concanavalin A.

Cutinase gene expression in *Fusarium solani* has been studied extensively [29, 30, 72]. Woloshuk and Kolattukudy [72] proposed that a spore senses contact with the plant by production of small amounts of cutin monomers hydrolyzed by low levels of cutinase carried by the spore. Their hypothesis may also be applicable to release of esterase by *E. graminis* observed by Nicholson et al. [54]. We have recently demonstrated that released esterase-active protein erodes the barley cuticle [34]. A contact stimulus triggers the two-stage release of the enzyme [36, 54], which may overcome the hydrophobic nature of the conidial wall [70]. The degradative products of wax substances or the first stage of esterase release may induce de novo synthesis of the second stage of release. In the sense, protrusions on the conidial wall may not be ornamental, but may play a role as sensors of the contact stimulus.

4 Significance of Extracellular Materials in Fungal Infection

A traditional view of fungal attachment is that spores become entrapped and that adhesion does not occur before germination and differentiation. However, as emphasized by Nicholson [50], fungi take more active roles in adhesion than previously assumed: the process may begin upon host contact before appressorial attachment. Adhesion may be crucial to fungal dispersal, to avoid the forces of splashing water [76] and wind [19]. Another function may be to anchor the spore to a suitable substratum that is subsequently recognized as a suitable infection court. Gold and Mendgen [16] suggested multiple roles for fungal exudates in penetration, including attachment to and sealing off of the penetration site, protection from desiccation, and as a reservoir for "penetration enzymes."

Adhesion is also suggested to be associated with physiological changes in host cells. Smith and Cruickshank [63] found that conidia of *Monilinia fructicola* adhered rapidly to pea pods and this sustained plant contact was necessary for the synthesis of the phytoalexin pisatin. Recently, Kobayashi et al. [27] demonstrated that when appressoria of the inappropriate pathogen *E. pisi* attach to barley coleoptile cells, normal host cell cyclosis is altered. These host cells become highly resistant to penetration by the normally compatible pathogen *E. graminis*. They suggested that a signal (possibly an elicitor) from the inappropriate pathogen appressorium of *E. pisi* is recognized by barley coleoptile cells.

As discussed above, release of enzymes or adhesives from spores or germlings may be crucial for differentiation and growth of fungal germlings, and may be associated with host recognition. Further study of events in prepenetration may provide interesting clues to unsolved questions in fungus-plant interactions.

Acknowledgement. This is article no. 94 of the Laboratory of Plant Pathology, Mie University and article no. 12119 of the Purdue University Agricultural Experiment Station. Supported in part by Grants-in-Aid for Co-operative Research Nos. 60304022 (1985, 1986), 62304015 (1987, 1988) and 01304014 (1989, 1990) from the Ministry of Education, Science and Culture of Japan, and by a Fellowship for Research in Japan to R.L. Nicholson from the Japanese Society for the Promotion of Science (1987).

References

1. Akai S, Ishida N (1968) An electron microscopic observation on the germination of conidia of *Collectotrichum lagenarium.* Mycopathol Mycol Appl 34: 337–345
2. Barak R, Elad Y, Mirelman D, Chet I (1985) Lectins: A possible basis for specific recognition in the interaction of *Trichoderma* and *Sclerotium rolfsii.* Phytopathology 75: 458–462
3. Bartnicki-Garcia S, Hemmes DE (1976) Some aspects of the form and function of oomycete spores. In: Hess WM, Weber DJ (eds) The fungus spore. Form and function. John Wiley, New York, pp 593

4. Beckett A, Porter R (1988) The use of complementary fractures and low-temperature scanning electron microscopy to study hyphal-host cell surface adhesion between *Uromyces vicae-fabae* and *Vicia faba*. Can J Bot 66: 645–652
5. Bird AF, McKay AC (1987) Adhesion of conidia of the fungus *Dilophospora alopecuri* to the cuticle of the nematode *Anguina agrostis*, the vector in annual ryegrass toxicity. Int J Parasito 17: 1239–1247
6. Bonfante-Fasolo P, Perotto S (1986) Visualization of surface sugar residues in mycorrhizal ericoid fungi by fluorescein conjugated lectins. Symbiosis 1: 269–288
7. Bonfante-Fasolo P, Perotto S, Testa B, Faccio A (1987) Ultrastructural localization of cell surface sugar residues in ericoid mycorrhizal fungi by gold-labeled lectins. Protoplasma 139: 25–35
8. Brawner DL, Cutler JE (1986) Ultrastructural and biochemical studies of two dynamically expressed cell surface determinants on *Candida albicans*. Infect Immun 51: 327–336
9. Douglas LJ (1985) Adhesion of pathogenic *Candida* species to host surfaces. Microbiol Sci 2: 243–247
10. Edwards HH, Allen PJ (1970) A fine-structure study of the primary infection process during infection of barley by *Erysiphe graminis* f. sp. *hordei*. Phytopathology 60: 1504–1509
11. Emmett RW, Parbery DG (1975) Appressoria. Annu Rev Phytopathol 13: 147–167
12. Epstein L, Laccetti L, Staples RC, Hoch HC (1987) Cell-substratum adhesive protein involved in surface contact responses of the bean rust fungus. Physiol Mol Plant Pathol 30: 373–388
13. Epstein L, Laccetti L, Staples RC, Hoch HC, Hoose WA (1985) Extracellular proteins associated with induction of differentiation in bean rust uredospore germlings. Phytopathology 75: 1073–1076
14. Evans RC, Stempen H, Stewart SJ (1981) Development of hyphal sheaths in *Bipolaris maydis* race T. Can J Bot 59: 453–459
15. Evans RC, Stempen H, Frasca P (1982) Evidence for a two-layered sheath on germ tubes of three species of *Biopolaris*. Phytopathology 72: 804–807
16. Gold RE, Mendgen K (1984) Cytology of basidiospore germination, penetration, and early colonization of *Phaseolus vulgaris* by *Uromyces appendiculatus* var. *appendiculatus*. Can J Bot 62: 1989–2002
17. Gould J, Northcote DH (1986) Cell-cell recognition of host surfaces by pathogens: The adsorption of maize (*Zea mays*) root mucilage by surfaces of pathogenic fungi. Biochem J 233: 395–405
18. Gubler F, Hardham AR (1988) Secretion of adhesive material during encystment of *Phytophthora cinnamomi* zoospores, characterized by immunogold labelling with monoclonal antibodies to components of peripheral vesicles. J Cell Sci 90: 225–235
19. Hamer JE, Howard RJ, Chumley FG, Valent B (1988) A mechanism for surface attachment in spores of a plant pathogenic fungus. Science 239: 288–292
20. Hau FC, Rush MC (1982) Preinfectional interactions between *Helminthosporium oryzae* and resistant and susceptible rice plants. Phytopathology 72: 285–292
21. Hawker LE, Hendy RJ (1963) An electron-microscope study of germination of conidia of *Botrytis cinerea*. J Gen Microbiol 33: 43–46
22. Hemmes DE, Hohl HR (1971) Ultrastructural aspects of encystation and cyst germination in *Phytophthora parasitica*. J Cell Sci 9: 175–191
23. Hinch JM, Clarke AE (1980) Adhesion of fungal zoospores to root surfaces is mediated by carbohydrate determinants of the root slime. Physiol Plant Pathol 16: 303–307
24. Hirooka T, Miyagi Y, Araki F, Kunoh H, Ishizaki H (1982) The effect of isoprothiolane on the emergence of infection pegs from appressoria of *Pyricularia oryzae*. Pestic Sci 13: 379–386
25. Ho HH, Zachariah K, Hickman CJ (1968) The ultrastructure of zoospores of *Phytophthora megasperma* var. *sojae*. Can J Bot 46: 37–41
26. Kim WK, Reisener HJ (1988) Evidence for binding between host proteins and pathogen wall components in the wheat stem rust system from affinity protein blotting technique. Can J Bot 66: 1702–1706

27. Kobayashi I, Komura T, Sakamoto Y, Yamaoka N, Kunoh H (1991) Recognition of pathogen and nonpathogen by barley coleoptile cells (I). Cytoplasmic responses of coleoptile cells to a nonpathogen, *Erysiphe pisi* prior to its penetration. Physiol Mol Plant Pathol (in press)
28. Koch E, Hoppe HH (1988) Development of infection structures by the direct-penetrating soybean rust fungus (*Phakopsora pachyrhizi* Syd.) on artificial membranes. J Phytopathol 122: 232–244
29. Kolattukudy PE (1985) Enzymatic penetration of the plant cuticle by fungal pathogens. Annu Rev Phytopathol 23: 223–250
30. Kolattukudy PE (1980) Biopolyester membranes of plants: Cutin and suberin. Science 208: 990–1000
31. Kunoh H, Aist JR, Hayashimoto A (1985) The occurrence of cytoplasmic aggregates induced by *Erysiphe pisi* in barley coleoptile cells before the host cell walls are penetrated. Physiol Plant Pathol 26: 199–207
32. Kunoh H, Kohno M, Tashiro S, Ishizaki H (1980) Ultrastructural studies of powdery mildew of mulberry caused by *Phyllactinia moricola* (Henn P) Homma. Fitopatol Bras 5: 11–20
33. Kunoh H, Kohno M, Tashiro S, Ishizaki H (1979) Studies of the powdery-mildew fungus, *Leveillula taurica*, on green pepper. II. Light and electron microscopic observation of the infection process. Can J Bot 57: 2501–2508
34. Kunoh H, Nicholson RL, Yoshioka H, Yamaoka N, Kobayashi I (1990) Preparation of the infection court by *Erysiphe graminis*: Degradation of the host cuticle. Physiol Mol Plant Pathol 36: 397–407
35. Kunoh H, Tsuzuki T, Ishizaki H (1978) Cytological studies of early stages of powdery mildew in barley and wheat. IV. Direct ingress from superficial primary germ tubes and appressoria of *Erysiphe graminis hordei* on barley leaves. Physiol Plant Pathol 13: 327–333
36. Kunoh H, Yamaoka N, Yoshioka H, Nicholson RL (1988) Preparation of the infection court by *Erysiphe graminis*. I. Contact-mediated changes in morphology of the conidium surface. Exp Mycol 12: 325–335
37. Lewis BG, Day JR (1972) Behaviour of urediospore germ-tubes of *Puccinia graminis tritici* in relation to the fine structure of wheat leaf surfaces. Trans Br Mycol Soc 58: 139–145
38. Longman D, Callow JA (1987) Specific saccharide residues are involved in the recognition of plant root surfaces by zoospores of *Pythium aphanidermatum* Physiol Mol Plant Pathol 30: 139–150
39. Lumbroso E, Fischbeck G, Wahl I (1982) Infection of barley with conidia suspensions of *Erysiphe graminis* f. sp. *hordei*. Phytopathol Z 104: 222–233
40. Manocha MS (1984) Cell surface characteristics of *Mortierella* species and their interaction with a mycoparasite. Can J Microbiol 30: 290–298
41. McCourtie J, Douglas LJ (1985) Extracellular polymer of *Candida albicans*: Isolation, analysis and role in adhesion. J Gen Microbiol 131: 495–503
42. McKeen WE (1974) Mode of penetration of epidermal cell walls of *Vicia faba* by *Botrytis cinerea*. Phytopathology 64: 461–467
43. McKeen WE, Rimmer SR (1973) Initial penetration process in powdery mildew infection of susceptible barley leaves. Phytopathology 63: 1049–1053
44. Metzler B (1982) Untersuchungen an Heterobasidiomyceten (23): Basidiosporenkeimung und Infektionsvorgang beim Birnengitterrost. Phytopathol Z 103: 126–138
45. Mims CW, Richardson EA (1989) Ultrastructure of appressorium development by basidiospore germlings of the rust fungus *Gymnosporangium juniperi-virginianae*, Protoplasma 148: 111–119
46. Murray GM, Maxwell DP (1975) Penetration of *Zea mays* by *Helminthosporium carbonum*. Can J Bot 53: 2872–2883
47. Nasu H, Kunoh H (1986) Scanning electron microscopy of flyspeck of grape. Ann Phytopathol Soc Jpn 52: 466–474
48. Nasu H, Kunoh H (1987) Scanning electron microscopy of flyspeck of apple, pear, Japanese persimmon, plum, Chinese quince, and pawpaw. Plant Dis 71: 361–364

49. Nasu H, Hatamoto M, Kunoh H (1986) Behavior of causal fungus and process of lesion on flyspeck-affected berries of grape. Ann Phytopathol Soc Jpn 52: 445–452
50. Nicholson RL (1984) Adhesion of fungi to the plant cuticle. In: Roberts DW, Aist JR (eds) Infection processes in fungi. Rockefeller Foundation, New York, pp 74–89
51. Nicholson RL, Epstein L (1990) Adhesion of fungi to the plant surface: prerequisite for pathogenesis. In: Cole GT, Hoch HC (eds) Fungal spore and disease initiation in plants and animals. Plenum, New York (in press)
52. Nicholson RL, Moraes WBC (1980) Survival of *Colletotrichum graminicola*: Importance of the spore matrix. Phytopathology 70: 255–261
53. Nicholson RL, Butler LG, Asquith TN (1986) Glycoproteins from *Colletotrichum graminicola* that bind phenols: Implications for survival and virulence of phytopathogenic fungi. Phytopathology 76: 1315–1318
54. Nicholson RL, Yoshioka H, Yamaoka N, Kunoh H (1988) Preparation of the infection court by *Erysiphe graminis*. II. Release of esterase enzyme from conidia in response to a contact stimulus. Exp Mycol 12: 336–349
55. Paus F, Raa J (1973) An electron microscope study of infection and disease development in cucumber hypocotyls inoculated with *Cladosporium cucumerinum*. Physiol Plant Pathol 3: 461–464
56. Persson M, Soderhall K (1986) $CaCl_2$-induced germination and peptidase secretion in *Aphanomyces astaci*. Exp Mycol 10: 205–213
57. Politis DJ, Wheeler H (1973) Ultrastructural study of penetration of maize leaves by *Colletotrichum graminicola*. Physiol Plant Pathol 3: 465–471
58. Pring RJ (1980) A fine-structural study of infection of leaves of *Phaseolus vulgaris* by uredospores of *Uromyces phaseoli*. Physiol Plant Pathol 17: 269–276
59. Pringle RB (1981) Nonspecific adhesion of *Bipolaris sorokiniana* sporelings. Can J Plant Pathol 3: 9–11
60. Reiche RE (1969) Fine structure of *Phytophthora parasitica* zoospores. Mycologia 61: 30–51
61. Saikawa M (1982) An electron microscope study of *Meria coniospora*, an endozoic nematophagous Hyphomycete. Can J Bot 60: 2019–2023
62. Sing VO, Bartnicki-Garcia S (1975) Adhesion of *Phytophthora palmivora* zoospores: electron microscopy of cell attachment and cyst wall fibril formation. J Cell Sci 18: 123–132
63. Smith MM, Cruickshank IAM (1987) Dynamics of conidial adhesion and elicitor uptake in relation to pisatin accumulation in aqueous droplets on the endocarp of pea pods. Physiol Mol Plant Pathol 31: 315–324
64. Staples RC, Macko V (1980) Formation of infection structures as a recognition response in fungi. Exp Mycol 4: 2–16
65. Staples RC, Grambow HJ, Hoch HC, Wynn WK (1983) Contact with membrane groves induces wheat stem rust uredospore germlings to differentiate appressoria but not vesicles. Phytopathology 73: 1436–1439
66. Staub T, Dahmen H, Schwinn FJ (1974) Light and scanning electron microscopy of cucumber and barley powdery mildew on host and nonhost plants. Phytopathology 64: 364–372
67. Temmink JHM, Campbell RN (1969) The ultrastructure of *Olpidium brassicae*. III. Infection of host roots. Can J Bot 47: 421–424
68. Tronchin G, Poulain D, Vernes A (1984) Cytochemical and ultrastructural studies of *Candida albicans* III. Evidence for modifications of the cell wall coat during adherence to human buccal epithelial cells. Arch Microbiol 139: 221–224
69. Uchiyama T, Ogasawara N, Nanba Y, Ito H (1979) Conidial germination and appressorial formation of the plant pathogenic fungi on the coverglass or cellophane coated with various lipid components of plant leaf waxes. Agric Biol Chem 43: 383–384
70. Weete JD (1984) Spore lipid and surface wax composition and metabolism during germination. In: Roberts DW, Aist JR (eds) Infection processes of fungi. The Rockefeller Foundation, New York, pp 118–125

71. Wetje G, Littlefield LJ, Zimmer DE (1979) Ultrastructure of compatible and incompatible reactions of sunflower to *Plasmopara halstedii*. Can J Bot 57: 315–323
72. Woloshuk CP, Kolattukudy PE (1986) Mechanism by which contact with plant cuticle triggers cutinase gene expression in the spores of *Fusarium solani* f. sp. *pisi*. Proc Natl Acad Sci USA 83: 1704–1708
73. Wynn WK (1976) Appressorium formation over stomates by the bean rust fungus: Response to a surface contact stimulus. Phytopathology 66: 136–146
74. Wynn WK (1981) Tropic and taxic responses of pathogens to plants. Annu Rev Phytopathol 19: 237–255
75. Yang SL, Ellingboe AH (1972) Cuticle layer as a determining factor for the formation of mature appressoria of *Erysiphe graminis* on wheat and barley. Phytopathology 62: 708–714
76. Young DH, Kaus H (1984) Adhesion of *Colletotrichum lindemuthianum* spores to *Phaseolus vulgaris* hypocotyls and to polystylene. Appl Environ Microbiol 47: 616–619

Chapter 18

D. E. Harder and J. Chong

Rust Haustoria

1 Introduction

The obligately biotrophic terrestrial plant pathogens such as the rust, common downy mildew, and powdery mildew fungi associate closely with their hosts by means of haustoria. Questions that come to mind are the possible role(s) of haustoria in establishing a biotrophic relationship or in nutrient absorption/transfer, and their evolutionary significance. Fundamental to these questions are an appreciation of the fine structure of haustoria and their host relationships. This chapter will attempt to describe rust haustoria based on modern methodology and draw functional and physiologic implications where possible.

1.1 Methodology and Haustorial Structure

The elucidation of haustorical structure to the present time has occurred in three distinct eras. The earliest known light microscopic illustrations were of downy and powdery mildew haustoria by DeBary [16], followed by Ward [54], Smith [52], Rice [44], Ruttle and Fraser [49], and the numerous classical papers by Allen RF in the 1930's. The next phase saw the application of electron microscopy to the study of haustoria, the pioneering work being done by Moore and McAlear [41], Ehrlich and Ehrlich [17–19], Bracker [3], Berlin and Bowen [2], and Peyton and Bowen [43]. Electron microscopy resulted in major advances in our understanding of the structure and host relationships of haustoria. However, it has become apparent that the processing and staining methods generally used for electron microscopy provided only a portion of the information, and may have mis-represented a number of structural details. The third and most recent phase has been the application of cytochemical procedures and cryoprocessing to electron microscopy. No doubt what we are now doing will in the future be seen as imprecise, and we will see the next

Table 1. Interpretation of the composition of component parts of haustoria of rust and downy mildew fungi from cytochemical data[a]

Structure	Fungal pathogen group[b] and component[c]: Dikaryotic rust																		Monokaryotic rust							Downy mildew								
	A a	A b	B a	B b	C a b	D a b	E a	E b	F a b	G a	G b	H c	I a b	J a b e	K b c	L a b	M a b	N a	B b	B d	D b	E b	G b d	H d	I b d	B e f	C f	F e f	G f	H e	I e f	L e	L f	M e f
Haustorium																																		
Proximal neck wall																																		
Outer layer	+	+	+	−	+	−				−	+		+			−		+	+	+	+	+			−			+			−	+	+	
Inner layer	−		+	+	+	−	+	+			−		+			−		+	+	+	+	+			−			+			−	+	+	
Distal neck wall																																		
Outer layer	+		−	−	+					+	+		+			−		+	+	+	+	+			−	−		+			+	−		
Inner layer	+		+	+	+	−		+			−		+			−		+	+	+	+	+			−	−		+			+	−	+	
Neck ring	−	−	−	−	−	−	−	−	−	−	−	−	−	+	−	−																		
Young body wall																																		
Outer layer	+		−		+	−				+			+			−			+	+	+				−			+			−	−	+	
Inner layer	+		+			−							+			−			+	+	+				−			+			−	−	+	

Mature body wall																																		
Outer layer	−		−	−	+	+				+	+		+			−		−	+	+	+				−			+			−	−	+	
Inner layer	−		+	+	+	+					−		+			−		−	+	+	+				−			+			−	−	+	
Extrahaustorial matrix																																		
Proximal region	+	+	+	+	−	−	+		+	+	+		−			+	+	+	+	+		−			−	+	−	+	+		+	−	+	+
Distal region	+	+	+	+	−	−	+		+	+	+		−			+	+	+	+	−		+			+	+	−	+	+		+	−	+	+
Extrahaustorial membrane			+	+							−	−	+		−								−	−	−	+				−	−			
Host plasmalèmma			+	+							+	+	+		+								+	+	+	+				+	+			
	A	A	B	B	C	D	E	E	F	G	G	H	I	J	K	L	M	N	B	B	D	E	G	H	I	B	C	F	G	H	I	L	L	M

[a] References: [7, 8, 10–14, 23, 30, 32, 47, 51, 56, 57].

[b] a = *Puccinia graminis* f. sp. *tritici*; b = *P. coronata* f. sp. *avenae*; c = *Uromyces appendiculatus*; d = *P. poarum*; e = *Uromyces vignae*; f = *Albugo candida*; g = *Peronospora pisi*.

[c] Components are judged from cytochemical procedures. It must be recognized that presence or absence is not absolute, since most components in biological tissues are in complex form or may be physically inaccessible to reagents. Also, different methods may result in different interpretations, and various components such as specific sugars or variously linked polysaccharides, either free or complexed, may result in overlap in staining.
+ = component present; − = component known to be absent or reduced in concentration; blank space = no data.
A = glycoprotein containing α-linked glucose and/or mannose; B = polysaccharide with vicinal OH, not known to be complexed; C = lipid, may be unsaturated lipoprotein; D = chitin; E = polysaccharide with α-linked glucose and/or mannose; F = same as E but α-linked; G = glycoprotein containing sugars with vicinal OH; H = ATPase activity; I = PACP positive staining; J = iron/phosphorus; K = membrane sterol; L = cellulose; M = protein, usually complexed; N = galactose.

phase, employing more advanced methods. The third and current phase is the main concern of this paper.

1.2 Cytochemical Procedures

The standard methods of electron microscopy, due both to broad and limited specificities, may reveal only part of the actual structure, may obscure details of structure, or may distort the structures that are revealed. There has also been variability in fine structure in the hands of different workers that use the same methods. Experience has shown that a wide array of procedures applied to the same object is most helpful to reveal details and to establish the accuracy of fine structure, where different structural components have different chemical composition. Space does not premit a full description of the available information, therefore some data will be tabulated and a few selected structures will be discussed in more detail.

A wide range of cytochemical procedures is available, a number of which have been applied to the study of haustoria. Briefly, they are: (1) Periodate-thiocarbohydrazide-silver proteinate (PA-TCH-SP) [53] reveals polysaccharides which contain vicinal hydroxyls. A control treatment where the periodate oxidation step is omitted may reveal unsaturated lipids or endogenous aldehyde groups. (2) Periodate-chromate-phosphotungstate (PACP) [48] stains plant plasmamembranes and has been shown to highlight portions of fungal walls [14]. (3) Specific extractive methods (enzyme, solvent) resulting in loss of staining of particular components. (4) Detection and localization of enzyme activity. (5) Energy dispersive X-ray analysis to detect most mineral elements. (6) Binding of heavy metal-conjugated lectins, the most commonly used of which are concanavalin A (ConA), which binds to α-linked glucose or mannose, and wheat germ agglutinin, which binds to substances containing oligomers of N-acetylglucosamine. Various other lectins have been used to a lesser extent, among these a new and useful lectin, arabinogalactan protein [47], which is specific for β-linked sugars. Treatment with the polyene antibotic filipin followed by freeze-fracturing has been used to locate and quantify membrane sterols [23]. Most of the above procedures, although useful singly, may provide additional or corroborating information when used in various combinations. For more detailed application of these methods, see [7–14, 28, 29, 47, 48, 51, 53, 55–57].

The cytochemical procedures have most effectively been used to analyze the structure and composition of fungal walls, the haustorial neck ring, the extrahaustorial matrix, and the extrahaustorial membrane. In Table 1 are given the structural components of the haustoria of rusts and several other biotrophic fungi, and their chemical composition as indicated by cytochemistry. The structural relationships are shown in Fig. 1. Table 1 and Fig. 1 provide the basis for most of the discussion in this chapter, and may be referred to for specific data.

2 Host Cell Penetration

One of the more useful applications of cytochemistry has been to elucidate the details of the host cell penetration region in dikaryotic rust infections. Combinations of procedures have shown the fungal walls in this region to be highly differentiated, with distinct changes in structure and composition occurring in a sequence beginning at the haustorium mother cell (HMC) wall, the HMC wall at the penetration site, then through the host wall, to the haustorial neck wall. With conventional procedures it had not been certain that fungal wall material was distinct from host wall material through the penetration region, but unique fungal wall material, although often tenuous, has clearly been shown with appropriate procedures (Figs. 2–7). Alteration of host wall material at the penetration site is also revealed by cellulase treatment (Fig. 8).

The structural wall adaptations are readily apparent, but there is little information to indicate the reasons for their existence. It is tempting to call the modifications "specialized", but specialization implies a functional adaptation, and there are few obvious reasons for the unique wall structures. The extra wall layers of the HMC [12] may contain turgor pressures developed in the HMC during penetration, or they could provide materials used during penetration. In particular, the HMC wall immediately around the penetration site is highly complex, with the development of the structure within the fungal wall, termed the *intra-wall* penetration torus [13]. The stem of the torus passes through the host wall, or through a collar adjacent to host wall, if formed [13]. Presumably such a structure is related to the process of penetration, but it is absent in the monokaryotic stage of the same fungus, or in other fungi where host cell penetration appears equally efficient. The haustorial neck wall begins abruptly at the inner surface of the host wall (Fig. 2) or collar, and is sharply differentiated cytochemically from the HMC wall, fungal wall material through the penetration site, and the haustorial body wall [13]. There is also differentiation of different portions of the neck wall, and between different rust fungal species [7, 11, 13] (Figs. 9, 10). Neck wall differentiation has also been shown for a completely unrelated biotrophic fungus, *Albugo candida* [14, 56]. Again, it is not clear what functional implications such wall "specialization" may have, although there could be some relation to host/fungal recognition (see below).

3 The Neck Ring

Bands or rings have been observed around the necks of dikaryotic rust fungal haustoria from the earlier days of light microscopy. Current analytical electron microscopic techniques have dramatically helped to elucidate their structure, composition, and probable function. The neck ring may appear as one or at times as

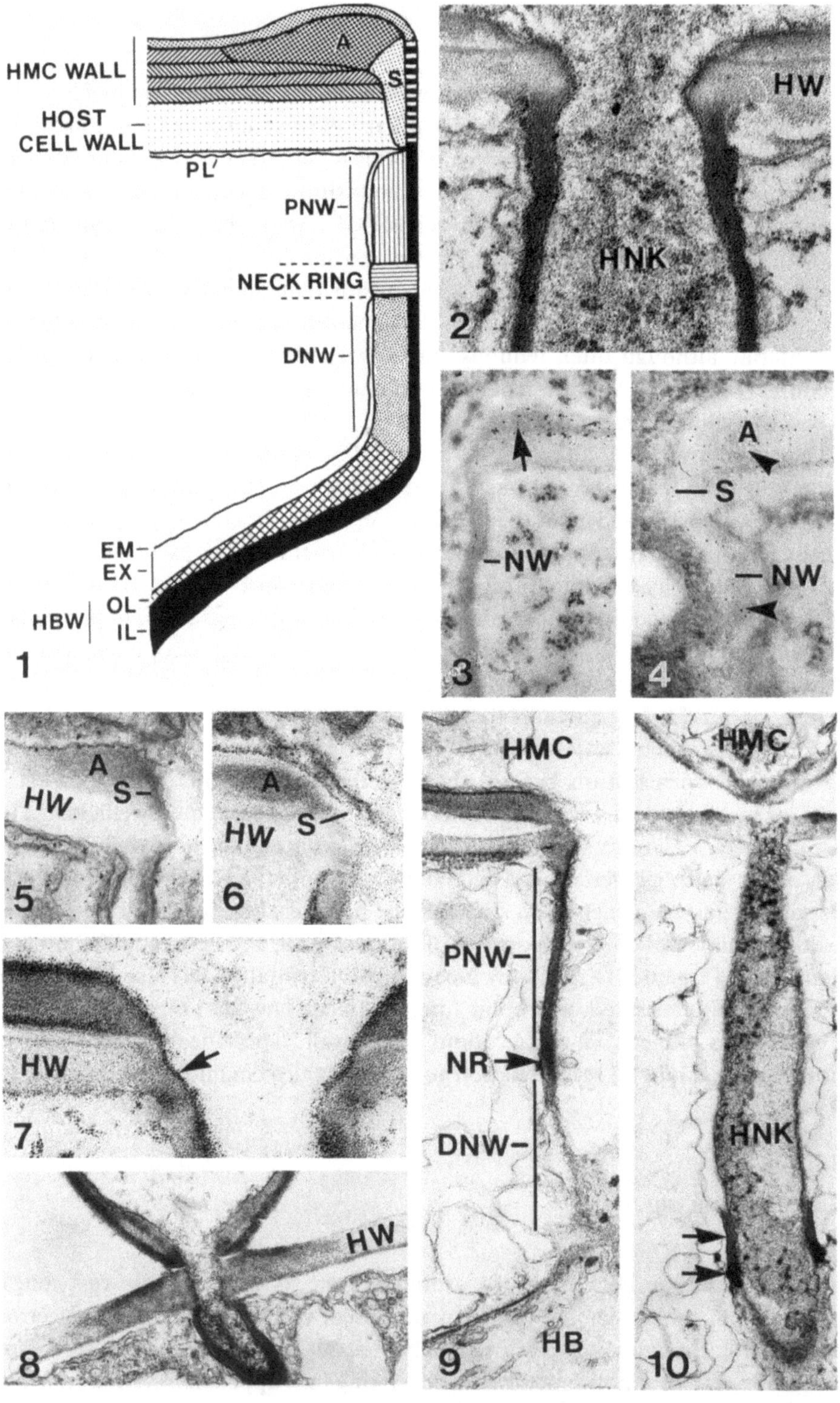
A
S
HMC WALL
HOST
CELL WALL
PL
PNW
NECK RING
DNW
EM
EX
OL
HBW
IL
1
HW
HNK
2
NW
3
A
S
NW
4
A
HW
S
5
A
HW
S
6
HW
7
HW
8
HMC
PNW
NR
DNW
HB
9
HMC
HNK
10

two bands (Fig. 10), depending on the rust species. Whether one or two bands are present, they have been shown by energy-dispersive X-ray analysis to have a high mineral content. One band is acid-labile and is rich in iron and phosphorous, in a ratio indicating iron pyrophosphate. This type of band is present [7, 8, 30] in both single and double neck rings, and the evidence to date indicates that it is conserved across the dikaryotic Uredinales. Where a second band is present, it is composed mainly of silicon, as has been shown for *Puccinia coronata* [7, 8] and *Uromyces vignae* [30]. It is not certain whether the silicon band forms an integral part of the neck ring or simply represents a fortuitous accumulation of silicon at this site. The latter is a possibility since there is extensive accumulation of silicon in the walls of haustorium mother cells [9, 11] and along the proximal portion of the neck wall in later stages of infection "unpublished". However, in *P. coronto* the silicon band appears to form early in haustorial development, somewhat earlier than the iron/phosphorous band. The extrahaustorial membrane (EM) is also shown to adhere tightly to the silicon band after protease treatment [8], indicating a probable neck ring function. The neck ring is also the location of the change in properties between the host plasmalemma and EM (see below). Changes in the composition of the neck wall in the neck ring region is some species may be involved in some additional, but as yet largely undefined factors of differentiation [30], and may have important physiologic functions (see below). A concentration of peroxidase activity has been reported in the neck wall in the neck ring region of young haustoria of *U. appendiculatus* [36].

◀

Figs. 1–10. 1 A diagramatic representation of the relationship of haustorial walls of *Puccinia graminis* f. sp. *tritici*, from the HMC, through the penetration region to the haustorial neck and body walls. The *various wall shadings* represent differences in wall composition (see Table 1 for description). Not drawn to scale. **2** The penetration region of *Puccinia coronata*, as shown by conventional processing. Glut/OsO4, UA/PbC, × 50 000. [21] **3–8** The penetration region of *Puccinia graminis* f. sp. *tritici* with various cytochemical treatments. **3** WGA binding to localize chitin. Strong binding occurs to the HMC wall (*arrow*), but not to the neck wall. Glut, WGA, 5% UA, × 31 000. **4** AGP binding specific for β-linked sugars. There is binding (*arrow heads*) to the annuals and stem of the IPT and to the neck wall. Glut, AGP, 5% UA, × 48 300. **5** A section fixed only with OsO4 and unstained. The stem of the IPT, by its greater osmiophilia, is distinguished from the annulus. × 21 600. [13] **6** Control treatment for the PA-TCH-SP method where the PA had been omitted. The annuals of the IPT, by its heavier staining, is differentiated from the stem. GLUT/OsO4, TCH-SP, × 37 000. [13]. **7** The PA-TCH-SP procedure to stain sugars with vicinal hydroxyls. A densely stained layer continuous with the inner layer of the HMC wall occurs through the penetration site. GLUT, PA-TCH-SP, × 67 000. [13] **8** Cellulase treatment, *Puccinia graminia* f. sp. *tritici*. The host wall in the vicinity of the penetration site stains more darkly than more distal regions. GLUT/OsO4, cellulase, UA/PbC. × 22 200. **9, 10** The neck walls of *Puccinia graminis* f. sp. *tritici* (**9**) and *P. coronata* (**10**) after protease treatment. The proximal neck wall of *P. graminis* is unaffected by protease, whereas the entire neck wall of *P. coronata* has been digested. **9** OsO_4, protease, PA-TCH-SP, × 34 300 [13]. **10** GLUT/OsO_4, protease, UA/PbC, × 37 500

4 The Extrahaustorial Matrix

The extrahaustorial matrix (EX) around haustoria of biotrophic fungi has long been subject to conjecture and varied opinion. The matrix has at various times been considered to be derived from the fungal wall, to be of host origin, or an artifact of preparation procedures [22, 35]. A variety of processing procedures have shown the matrix and haustorial wall to be clearly differentiated from each other (Figs. 11–13). The matrix in compatible interactions contains a variety of substances, including α- and β-linked sugars, probably largely occurring as polysaccharides (e.g., cellulose) or glycoproteins [11, 13, 15, 47]. Freeze-substitution shows the EX to be a structurally uniform and easily recognized entity [33] (Figs. 12, 13). Since cryoprocessing presumably results in minimal loss or alteration of tissue components, the EX is unlikely to be an artifact of processing for electron microscopy. The appearance of the EX varies consideraby with the age of the haustorium and is affected by the level of compatibility or incompatibility between host and fungus, becoming wider and more densely stained with advancing age or with imcompatibility. The question of a fungal or host origin of the EX has not been resolved. Several lines of evidence, e.g., apparent presence of cellulose [11, 13] or lack of radioactive label when infected with tritium-labeled urediospores [39], indicate contribution by the host. Conversely, detection of β-linked sugars suggests a fungal origin for these. There are no reasons to assign either a host or fungal origin for the numerous other substances found in the EX.

Matrices are formed around both the haustoria of dikaryotic infections and the intracellular hyphae of monokaryotic infections. However, the matrices of the two types of infections differ from one another, with those of the haploid infections displaying somewhat more host cell wall-like characteristics than those in dikaryotic infections [11, 21, 31]. The formation of host wall-like matrices in *Physopella zea* may be suppressed by the formation of dikaryotic haustoria [31], which indicates that the dikaryotic infections specifically alter host cell metabolism.

▶

Figs. 11–20. 11–13 The EX after different treatments. **11, 12** are *Puccinia gramins* f. sp. *tritici* with PA-TCH-SP staining. **11** GLUT, chloroform/methanol extraction to remove lipids, PA-TCH-SP, × 22 700. **12** freeze-substituted, PA-TCH-SP, × 40 000. **13** *P. striiformis* after high pressure freezing and freeze-substitution, UA/PbC staining, × 40 000. (Courtesy Dr. K. Mendgen). **14–20** Associations of host membranes with haustoria. **14** A young haustorium of *Puccinia graminis* f. sp. *tritici*. Note the extensive arrays of ER associated with the haustorium, × 18 200. **15** Organized interconnected tubular complexes (*arrows*), derived from the host ER, associated with a haustorium of *P. graminis*, × 29 800. [24]. **16** More randomized cytoplasmic tubules in oat cells infected by *P. coronata*, × 22 700. **17** Association of host tubules with the EM (*arrow*), stained with PACP. The EM has stained, but the tubules are relatively unstained. *P. graminis*, × 38 900. **18** Similar to **17**, but *P. coronata*, × 42 000. **19** Direct continuity between the EM and host membranes (*arrow head*) after freeze substitution. *P. striiformis*, × 40 000. (Courtesy Dr. K. Mendgen). **20** Association of rough host ER (*arrow*) with the EM in a monokaryotic infection of *P. coronata*, × 50 000

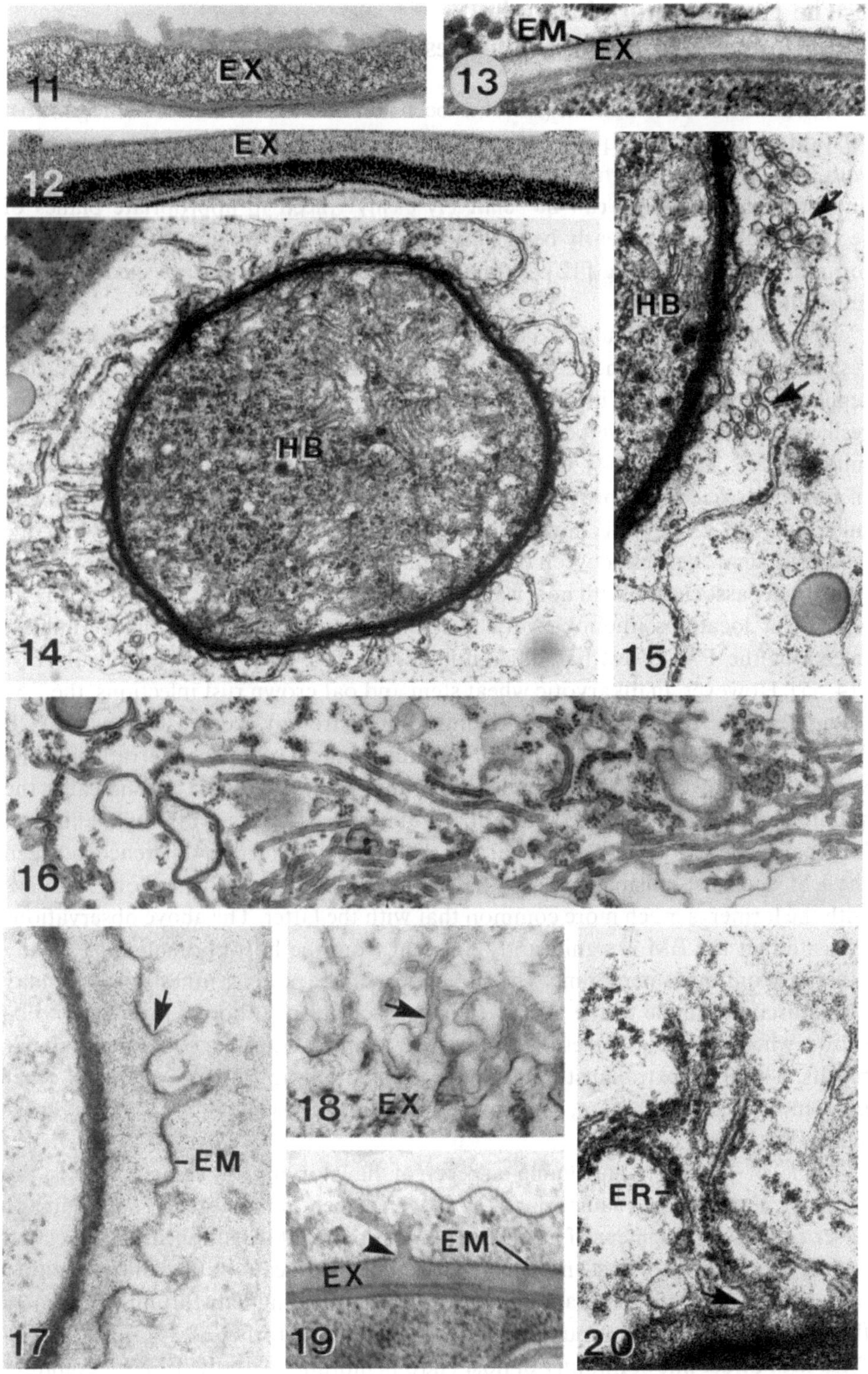
11
EX
13
EM
EX
12
EX
14
HB
15
HB
16
17
EM
18
EX
19
EX
EM
20
ER

5 The Extrahaustorial Membrane and Associated Host Membranes

An extrahaustorial membrane (EM) has been identified in all haustorial fungal/host relationships. The main questions regarding the EM have concerned its continuity with, and differentiation from, the host plasmalemma. Continuity between the EM and host plasmalemma can be relatively easily traced in appropriate planes of section, and has most clearly been shown for a young rust haustorium, before the formation of a neck ring [12]. Similar continuity has been established for most fungal/host associations.

The EM has been shown to differ from the host plasmalemma in a number of respects. Physical differences such as changes in surface characteristics in freeze-etch replicas (including loss of intramembrane particles), increased thickness, strength, or resilience of the EM have been well documented for both the rusts and powdery mildews [3, 6, 20, 24, 33, 34]. Chemical changes that have been documented are loss of some polysaccharide or ATPase activity in both rust [11, 13, 51, 57] and downy mildew [14, 32, 55, 56] infections (Table 1). In both rust and downy mildew fungi, the transition in properties between the EM and host plasmalemma occurs where there is no longer association with host wall-like material, or at the haustorial neck ring its equivalent locations (monokaryotic rust or *Albugo candida*). In downy mildew infections the PACP test has differentiated the host plasmalemma and the EM [14, 55]. However, in dikaryotic wheat stem and oat crown rust infections, the EM resembles the plasmalemma in being quite strongly contrasted by PACP (Figs. 17, 18). Staining with the PACP technique may be somewhat erratic, but the weaker staining of associated endomembranes in Figs. 17 and 18 indicates the validity of this result for the dikaryotic cereals rusts, although different host/pathogen associations may vary in this respect. A further difference between the EM and host-plasmalemma is that direct continuity of host-endomembranes with the former is much more common that with the latter. The above observations indicate that the EM is significantly different from the host plasmalemma or any other host or fungal membrane. The EM appears to be newly synthesized and it may be a sturcture unique to the host/pathogen association, although it remains to be shown whether abiotic forces (such as penetration by a needle) could have similar results. The term extrahaustorial membrane is more appropriate than the other frequently used designation, the “invaginated host plasmalemma”, and the latter should no longer be used.

One of the more intriguing aspects of haustorial/host associations is the physical relationship of haustoria to the host endomembrane system. To a greater or lesser extent, development of haustoria of most biotrophic pathogens is associated with reorganization of host endomembranes. The modifications take the form either of concentrations of membrane profiles near the invading haustorium (Fig. 14) or the formation of membranous complexes near the haustoria [22, 24] (Figs. 15, 16). Sites with direct line continuity of host endomembranes with the EM are common

(Figs. 17–19). Association of haustoria with organized host endomembrane complexes appears to be more prominent in dikaryotic cereal rust infections than corresponding monokaryotic infections, although continuity between the EM and host endoplasmic reticulum (ER) may be found in the latter (Fig. 20). In the dikaryotic infections the EM is normally continuous with smooth ER, but in the monokaryotic infections the EM is continuous with rough ER (Fig. 20). These observations raise important questions regarding the functional significance of reorganization of the host-endomembrane system and the role of the pathogen in this process. The intimacy of the EM/host endomembrane associations has clearly been shown by the continuity of these membranes in both rust and for downy mildew infections [2]. Any possible contents and/or metabolic activities of the modified host endomembrane system are directly accessed by the haustoria through the EX. Some evidence as to the role of the pathogen in reorganization of the host endomembrane system has been obtained by comparing the formation of the complexes in cells of wheat or oats infected by the same fungal species, *Puccinia graminis*, or in oats infected by an unrelated species, *P. coronata. Puccinia graminis* in wheat or oats induces similar modifications whereas *P. coronata* in oats induces a much different form of membrane complex (cf. Figs. 15, 16). These results indicate that the pathogen has a major influence in modifying the structure of the host protoplast to a form that is unique to the host/pathogen association, and should have important physiological implications.

6 Physiology and Function

Since the review by Bushnell [4], little further direct evidence has been obtained to show that role(s) rust haustoria have in the host/pathogen relationship. Some interpretations can be drawn with reasonable certainty from structural differentiation and comparison with other biotrophs where function is better understood. A successful biotrophic pathogen must presumably "instruct" the plant to recognize it as "friendly", to prevent the plant's normal defense reactions. Several lines of evidence indicate that dikaryotic rust haustoria are specifically involved in establishing longer-term compatible interactions between host and fungus. In the temperature-sensitive *Sr6/sr6* wheat/*P. graminis* association, those cells in which compatible interactions were established at higher temperatures, where resistance is inactivated, remain compatible after lowering temperatures, where resistance again should be activated [50]. Maturation of the first-formed haustorium of *Uromyces appendiculatus* rendered the invaded and neighboring bean cells unable to respond with resistance to the normally incompatible fungus *U. vignae* [26]. The host-like matrices around intracellular hyphae of *Physopella zeae* were suppressed when there was concomitant development of dikaryotic haustoria of the same fungus [31]. This also suggests that haustoria have a metabolic role in establishing longer-term compatible associations with their host cells.

Since incompatibility in most dikaryotic host/rust interactions becomes evident only after formation of the first haustorium, it has been assumed that the recognition reactions occur sometime during early contact with the host protoplast. That penetration of the host wall has an immediate effect on the host is shown by loss of the ability of cellulase to digest the host wall around the penetration site (Fig. 8). However, this may have little to do with recognition. The initial contact with the host protoplast is between the host plasmalemma and penetration peg, the wall of which develops into the haustorial neck wall. The haustorial neck wall is unique in composition compared to walls at all other growth stages of the fungus, and it's apparent glycoprotein composition indicates it as a prime candidate for a recognitional function. This may be a possible role. However, it has also been shown that in monokaryotic infections of the autoecious rust *Uromyces vignae*, where there is no comparably differentiated neck wall, that resistance reactions are initiated similarly in the monokaryotic and dikaryotic stages [28]. Thus the role of any particular haustorial structure in initiating and maintaining a host relationship is not certain. What is clear is that both host and pathogen contribute to the initiation and maintainance of a compatible association. Evidence has now been obtained [29] that without a host protoplast, haustoria are not fully elaborated or differentiated, suggesting a contributory role by the host, although it must be recognized that any contribution by the host may be induced by the fungus. Likewise, the modification of the host endomembrane system, induced by the fungus and which may be important in contributing to long-term compatibility, is a role apparently contributed by, or at least mediated by, haustoria.

Despite the prominence and chemical complexity of the EX, there is no clear indication of any particular function(s) for the matrix. No doubt the matrix is significant in fungal/host associations since a matrix of some form occurs in virtually all such associations known. It has usually been stated that the EX forms the immediate interface between the haustorium and host protoplast. However, it is also possible that the EM, because of its unique properties and association with modified host membranes, may function as a bridging structure between the two organisms. In so far as the haustorium may function as an absorbing organ, all materials must pass through the matrix. The EX is also continuous with the matrix of the modified host endomembrane complexes. Cellular endomembrane systems have important functions in biochemical synthesis and metaboli control [42,46], indicating that synthesis/control functions may be associated with the matrix.

6.1 Possible Role(s) in Nutrient Absorption

A nutritional role for haustoria could involve several dimensions; general absorption of nutrients, selective absorption of nutrients, enhanced efficiency of absorption, biosynthesis of required substances with raw materials obtained from the host, or control of nutrient flow.

Several lines of evidence, although indirect, strongly indicate a nutrient absorption role for rust fungal haustoria. Autoradiographic studies have shown the transfer of substances between host cells and rust fungi [37, 38], but the precise nature of the transfer has been difficult to verify. A comparison of monokaryotic and dikaryotic life cycle phases shows that intercellular growth in the monokaryotic phase is relatively more profuse, and the monokaryotic intracellular structures are relatively less prominent [1, 45] and are much simpler structurally [22]. For most heteroeceous macrocyclic rusts, the uredial phase must support profuse sporulation over a relatively long period, placing great demands on the host, and requiring a stable and efficient system of transfer to the fungus.

One of the more convincing arguments for a nutritional role based on structure is the nature of the haustorial neck ring. The neck ring forms the junction between the host plasmalemma and the EM, and these membranes are not separable from the neck ring with physical or enzymatic treatments [8]. The evidence indicates that the neck ring functions similarly to the Casparian strip in root endodermal cells [25] in controlling apoplastic flow, thus forcing substances symplastically through the haustorium. In this respect the neck ring would appear to have a role to direct the flow of substances, enhancing the efficiency of haustoria as putative absorbing organs.

Significantly, dissimilar biotrophic fungi have evolved similar haustorial features, presumably with similar functions. A neck ring of a possible functional equivalent has been identified respectively in *Erysiphe* [20] and *Albugo* [14, 56]. Further useful comparisons with other biotrophs are the similarities of the extrahaustorial matrices and similar modifications of the EM. Another feature of rust haustoria, and to some extent of powdery mildews [20], is the orientation of mitochondria around the periphery of haustoria, adjacent to the haustorial wall [11]. This implies enhancement of energy metabolism near the haustorial-host interface. The haustoria of *Erysiphe* have a well-demonstrated nutritional function [5, 6, 40] and the evidence indicates a similar role for rust haustoria.

In summary, haustoria appear to be a remarkable adaptation of biotrophic fungi to accomplish a variety of complex tasks. In terms of the evolution of parasitism being reflected in the development of specialized structures [27], the complex and apparently "specialized" components of dikaryotic rust haustoria appear to be involved in the establishment and maintainance of a successful biotrophic relationship, although how this is accomplished is not yet clear. The degree of structural uniformity of the dikaryotic haustoria across a large number of unrelated rust fungal genera, and indeed similarities in completely unrelated taxa, indicate the evolutionary success of these intracellular structures.

Acknowledgments. The generosity of the Deutsche Forschungsgemeinschaft for their financial support is gratefully acknowledged. Drs. M.C. Heath and R. Rohringer are thanked for their critical reading of the earlier drafts of this manuscript.

References

1. Al-Khesraji TO, Losel DM (1980) Intracellular structures of *Puccinia poarum* on its alternate hosts. Trans Br Mycol Soc 75: 397–411
2. Berlin JD, Bowen CC (1964) The host-parasite interface of *Albugo candida* on *Raphanus sativus*. Am J Bot 51: 445–450
3. Bracker CE (1968) Ultrastructure of the haustorial apparatus of *Erysiphe graminis* and its relationship to the epidermal cell of barley. Phytopathology 58: 12–30
4. Bushnell WR (1972) Physiology of fungal haustoria. Annu Rev Phytopathol 10: 151–176
5. Bushnell WR, Mendgen K, Liu Z (1987) Accumulation of potentiometric and other dyes in haustoria of *Erysiphe graminis* in living host cells. Physiol Mol Plant Pathol 31: 237–250
6. Bushnell WR, Gay JL (1978) Accumulation of solutes in relation to the structure and function of haustoria in powdery mildew. In: Spencer DM (ed) The powdery mildews. Academic Press, London, pp 184
7. Chong JYN (1981) Ontogeny of mono- and dikaryotic haustoria of *Puccinia coronata avenae*: ultrastructure, cytochemistry and electron-probe X-ray analysis. PhD. dissertation, Dept Botany, Univ Manitoba, Winnipeg
8. Chong J, Harder DE (1980) Ultrastructure of haustorium development in *Puccinia coronata avenae*. I. Cytochemistry and electron probe X-ray analysis of the haustorial neck ring. Can J Bot 58: 2496–2505
9. Chong J, Harder DE (1982) Ultrastructure of haustorium development in *Puccinia coronata avenae:* some host responses. Phytopathology 72: 1527–1533
10. Chong J, Harder DE (1982) Ultrastructure of haustorium development in *Puccinia coronata* f. sp. *avenae*: cytochemistry and energy dispersive X-ray analysis of the haustorial mother cells. Phytopathology 72: 1518–1526
11. Chong J, Harder DE, Rohringer R (1981) Ontogeny of mono- and dikaryotic rust haustoria: cytochemical and ultrastructural studies. Phytopathology 71: 975–983
12. Chong J, Harder DE, Rohringer R (1985) Cytochemical studies on *Puccinia graminis* f. sp. *tritici* in a compatible wheat host. I. Walls of intercellular hyphal cells and haustorium mother cells. Can J Bot 63: 1713–1724
13. Chong J, Harder DE, Rohringer R (1986) Cytochemical studies on *Puccinia graminis* f. sp. *tritici* in a compatible wheat host. II. Haustorium mother cell walls at the host cell penetration site, haustorial walls, and the extrahaustorial matrix. Can J Bot 64: 2561–2575
14. Coffey MD (1983) Cytochemical specialization at the haustorial interface of a biotrophic fungal parasite, *Albugo candida*. Can J Bot 61: 2004–2014
15. Coffey MD, Wilson UE (1983) An ultrastructural study of the late blight fungus *Phytophthora infestans* and its interaction with the foliage of two potato cultivars possessing different levels of general (field) resistance. Can J Bot 61: 2669–2685
16. DeBary A (1863) Recherches sur le dévelopement de quelques champignons parasites. Ann Sci Nat Bot IV 20: 5–148
17. Ehrlich MA, Ehrlich HG (1963) Electron microscopy of the host-parasite relationships in stem rust of wheat. Am J Bot 50: 123–130
18. Ehrlich MA, Ehrlich (1963) Electron microscopy of the sheath surrounding the haustorium of *Erysiphe graminis*. Phytopathology 53: 1378–1380
19. Ehrlich MA, Ehrlich HG (1966) Ultrastructure of the hyphae and haustoria of *Phytophthora infestans* and hyphae of *P. parasitica*. Can J Bot 44: 1495–1503
20. Gil F, Gay JL (1977) Ultrastructural and physiological properties of the host interfacial components of haustoria of *Erysiphe pisi* in vivo and in vitro. Physiol Plant Pathol 10: 1–12
21. Harder DE (1978) Comparative ultrastructure of the haustoria in uredial and pycnial infections of *Puccinia coronata avenae*. Can J Bot 56: 214–224
22. Harder DE, Chong J (1984) Structure and physiology of haustoria. In: Bushnell WR, Roelfs AP (eds) The cereal rusts, Vol I. Academic Press, New York, pp 431

23. Harder DE, Mendgen K (1982) Filipin-sterol complexes in bean rust- and oat crown-rust fungal/plant interactions. Protoplasma 112: 46–54
24. Harder DE, Rohringer R, Samborski DJ, Kim WK, Chong J (1978) Elecrton microscopy of susceptible and resistant near isogenic (*sr6/Sr6*) lines of wheat infected by *Puccinia graminis tritici*. I. The host-pathogen interface in the compatible (sr6/P6) interaction. Can J Bot 56: 2955–2956
25. Heath MC (1976) Ultrastructural and functional similarity of the haustorial neckband of rust fungi and the Casparian strip of vascular plants. Can J Bot 54: 2484–2489
26. Heath MC (1983) Relationship between developmental stage of the bean rust fungus and increased susceptibility of surrounding bean tissue to the cowpea rust fungus. Physiol Plant Pathol 22: 45–50
27. Heath MC (1987) Evolution of parasitism in the fungi. In: Rayner ADM, Brasier CM, Moore D (eds) Evolutionary biology of the fungi. Cambridge University Press, Cambridge, pp 149
28. Heath MC (1989) A comparison of fungal growth and plant responses in cowpea and bean cultivars inoculated with urediospores or basidiospores of the cowpea rust fungus. Physiol Mol Plant Pathol 34: 415–426
29. Heath MC (1990) Cytological and cytochemical features of the haustorium of the cowpea rust fungus, *Uromyces vignae*, formed in the absence of a living host cell. II. Electron microscopy. Can J Bot 68: 278–287
30. Heath MC, Allen FHE (1985) Morphology, element composition, and response to acids and oxidizing agents of the haustorial neckband of *Uromyces phaseoli* var. *vignae*. Can J Bot 63: 463–473
31. Heath MC, Bonde MR (1983) Ultrastructural observations of the rust fungus *Physopella zeae* in *Zea Mays*. Can J Bot 61: 2231–2242
32. Hickey EL, Coffey MD (1978) A cytochemical investigation of the host-parasite interface in *Pisum sativum* infected by the downy mildew fungus *Peronospora pisi*. Protoplasma 97: 201–220
33. Knauf GM, Welter K, Muller M, Mendgen K (1989) The haustorial host-parasite interface in rust-infected bean leaves after high pressure freezing. Physiol Mol Plant Pathol 34: 519–530
34. Littlefield LJ, Bracker CE (1972) Ultrastructural specialization of the host-pathogen interface in rust-infected flax. Protoplasma 74: 271–305
35. Littlefield LJ, Heath MC (1979) Ultrastructure of rust fungi. Academic Press, New York
36. Mendgen K (1975) Ultrastructural demonstration of different peroxidase activities during bean rust infection process. Physiol Plant Pathol 6: 275–282
37. Mendgen K (1977) Reduced lysine uptake by bean rust haustoria in a resistant reaction. Naturwissenschaften 64: 438
38. Mendgen K (1979) Microautoradiographic studies on host-parasite interactions. II. The exchange of 3H-lysine between *Uromyces phaseoli* and *Phaseolus vulgaris*. Arch Microbiol 123: 129–135
39. Mendgen K, Heitefuss R (1975) Micro-autoradiographic studies on host-parasite interactions. I. The infection of *Phaseolus vulgaris* with tritium labeled uredospores of *Uromyces phaseoli*. Arch Microbiol 105: 193–199
40. Mendgen K, Nass P (1988) The activity of powdery-mildew haustoria after feeding the host cells with different sugars, as measured with a potentiometric cyanine dye. Planta 174: 283–288
41. Moore RT, McAlear JH (1961) Fine structure of mycota. 8. On the aecidial stage of *Uromyces caladii*. Phytopathol Z 42: 297–304
42. Morré DJ, Mollenhauer HH (1974) The endomembrane concept: a functional integration of endoplasmic reticulum and Golgi apparatus. In: Robards AB (ed) Dynamic aspects of plant ultrastructure. McGraw-Hill, New York, pp 81
43. Peyton GA, Bowen CC (1963) The host-parasite interface of *Peronospora manshurica* on Glycine max. Am J Bot 50: 787–797
44. Rice MA (1927) The haustoria of certain rusts and the relation between host and pathogen. Bull Torrey Bot Club 54: 63–153

45. Rijkenberg FHJ, Truter S (1973) Haustoria and intracellular hyphae in the rusts. Phytopathology 63: 281–286
46. Robinson DG (1985) Plant membranes. Endo- and plasma membranes of plant cells. Wiley, New York, pp 331
47. Rohringer R, Chong J, Gillespie R, Harder DE (1989) Gold-conjugated arabinogalactan-protein and other lectins as ultrastructural probes for the wheat/stem rust complex. Histochemistry 91: 383–393
48. Roland JC, Lembi CA, Morré DJ (1972) Phosphotungstic acid-chromic acid as a selective electron dense stain for plasma membranes of plant cells. Stain Technol 47: 195–200
49. Ruttle ML, Frasser WP (1927) A cytological study of *Puccinia coronata* Cda. on Banner and Cowra 35 oats. Univ Calif Berkely Publ Bot 14: 21–72
50. Samborski DJ, Kim WK, Rohringer R, Howes NK, Baker RJ (1977) Histological studies on host-cell necrosis conditioned by the *Sr6* gene for resistance to wheat stem rust. Can J Bot 55: 1445–1452
51. Spencer–Phillips PTN, Gay JL (1981) Domains of ATPase in plasma membranes and transport through infected plant cells. New Phytol 89: 393–400
52. Smith G (1900) The haustoria of the Erysipheae. Bot Gaz 29: 153–184
53. Thiéry JP (1967) Mise en evidence des polysaccharides sur coupes fines en microscopie électronique. J Microsc (Paris) 6: 987–1018
54. Ward HM (1882) On the morphology of *Hemileia vastatrix*. Berk Br Quart J Microsp Sci 22: 1–11
55. Woods AM, Didehvar F, Gay JL, Mansfield JW (1988) Modification of the host plasmalemma in haustorial infections of *Lactuca sativa* by *Bremia lactucae*. Physiol Mol Plant Pathol 33: 299–310
56. Woods AM, Gay JL (1983) Evidence for a neckband delimiting structural regions of the host plasma membrane associated with haustoria of *Albugo candida*. Physiol Plant Pathol 23: 73–88
57. Woods AM, Gay JL (1987) The interface between haustoria of *Puccinia poarum* (monokaryon) and *Tussilago farfara*. Physiol Mol Plant Pathol 30: 167–185

Abbreviations Used in Figures and Legends. A annulus; *AGP* arabinogalactan protein lectin; *EM* extrahaustorial membrane; *GLUT* glutaraldehyde; *DNW* distal neck wall; *ER* endoplasmic reticulum; *EX* extrahaustorial matrix; *HB* haustorial body; *HBW* haustorial body wall; *HMC* haustorium mother cell; *HNK* haustorial neck; *HW* host wall; *IL* inner layer; *NR* neck ring; *OL* outer layer; *OsO4* osmium tetroxide; *PACP* periodate-chromate-phosphotungstate; *PbC* lead citrate; *PA-TCH-SP* periodate-thiocarbohydrazide-silver proteinate; *PL* plasmalemma; *PNW* proximal neck wall; *S* stem; *UA* uranyl acetate; *WGA* wheat germ agglutinin

Chapter 19

Richard J. Howard, Timothy M. Bourett and Margaret A. Ferrari

Infection by Magnaporthe: An In Vitro Analysis

1 Introduction

Rice blast disease, described over 300 years ago, has been the subject of research in plant pathology labs throughout the world [17]. The fungus *Magnaporthe grisea* (anamorph, *Pyricularia* sp.), an ascomycete, has been of considerable interest because of its global agronomic impact in inciting blast disease and because it is recognized as one of the most widely disseminated foliar pathogens. In recent years *M. grisea* has become more valuable for investigation because it can be genetically manipulated and characterized [5, 7, 9, 15, 18, 23–25]. For these reasons we have undertaken the task of redefining the early events of host-pathogen interaction: inoculation, pre-penetration and penetration, all prerequisites of infection. This chapter briefly summarizes several new and significant findings of an in vitro analysis concerning conidial adhesion, appressorium development, and the mechanism of penetration. Many of these new findings can be ascribed to our use of freeze substitution [12] for sample preparation.

2 Attachment of Conidia

Mature conidia not exposed to water prior to or during preparation for electron microscopy exhibit a periplasmic compartment at the conidial apex (Fig. 1 and inset). This compartment contains *spore tip mucilage* (STM), the fungal adhesive responsible for nonspecific attachment of conidia to diverse surfaces, including Teflon [8]. STM is released from conidia, presumably as a result of hydration-expansion, through a break in outer cell walls (Fig. 2). The resulting globule of STM binds concanavalin A (Fig. 3) suggesting that α-mannoside is a component of this adhesive. Conidia of *M. grisea* thus appear equipped for immediate and persistent attachment to the host surface upon first contact, even under water.

DuPont Experimental Station, P.O. Box 80402, Wilmington, DE 19880-0402, USA

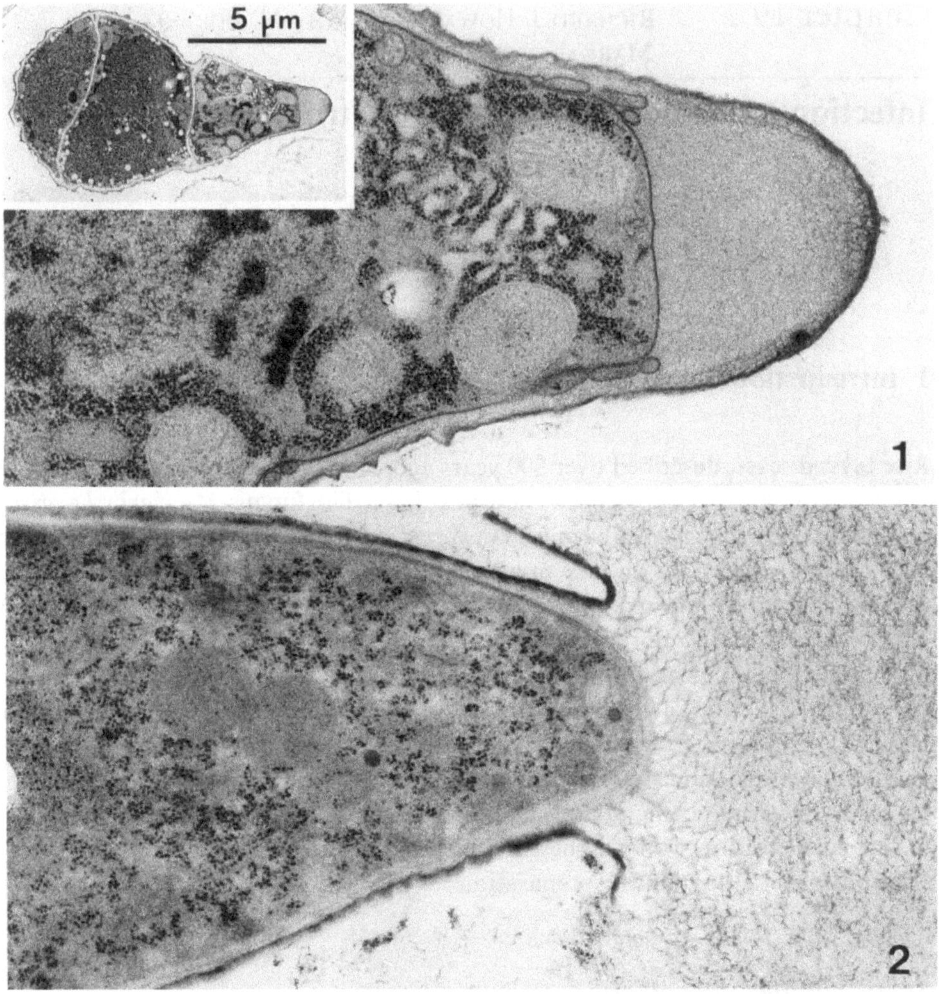
5 µm
1
2
1 µm
3

3 Appressorium Formation

Conidia were composed of two or three (predominantly three) uninucleate cells and germinated within 3 h in water. A single germ tube usually grew only a short distance across the substratum surface before initiating appressorium development. The germ tube apex swelled as a result of an apparent change in the pattern of cell wall deposition/cell expansion (Figs. 4–6). The nucleus of the germinated conidial cell divided: one daughter remained in the conidium and the other migrated into the incipient appressorium. As the apical cell volume grew, the germ tube was lifted off the substratum (Fig. 4). The nascent appressorium remained firmly attached. At this stage (Fig. 6) rough endoplasmic reticulum, smooth elements of endomembrane cisternae of two different morphologies, lipid bodies and mitochondria were found in high numbers within the cytoplasm (Fig. 6). Numerous cytoplasmic vesicles appeared near the substratum interface (Figs. 4, 5), an apparent region of increasing appressorium surface area. It remains unclear whether new cell wall was required during this cell expansion since a morphologically distinctive, but minimal, wall region was found adjacent to these vesicles. The thicker cell wall covering other areas of the forming appressorium exhibited the same structure as that of the germ tube (Fig. 4). Cell wall material along the bottom of the cell was difficult to discern especially as the appressorium matured. When the appressorium appeared fully expanded no chitinous cell wall could be seen (Fig. 7) or detected cytochemically (Fig. 8) in the area adjacent to the substratum. This specialized area of contact has been termed the *appressorium pore* [3].

4 Melanization

As the appressorium became fully expanded a septum formed across the germ tube, usually about 4 h after conidia were first hydrated. A special wall layer containing pentaketide-derived melanin then formed around the appressorium (Fig. 9). Melanin was responsible for a pigmented appearance of the appressorium under the light microscope. This layer was identified by comparing appressorium wall structure in genetically defined melanized wild-type and melanin-deficient mutant strains (cf. Fig. 10). Melanin was deposited over the appressorium side of the septum

◀

Figs. 1–3. Storage (**1**) and release (**2, 3**) of the spore tip mucilage (STM) responsible for surface attachment of *M. grisea* conidia. Dry-harvested and cryofixed conidia exhibited a periplasmic deposit near their apex (*inset*). Hydration expansion of the stored STM caused the outer cell wall to break, resulting in an extracellular globule capable of glueing conidia to Teflon. STM could be labeled in thin section using concanavalin A-colloidal gold conjugates (**3**): hapten blocking experiments indicated a high mannose content. **1–3** Same magnification (**1, 2** from [8]. Copyright 1988 by the AAAS)

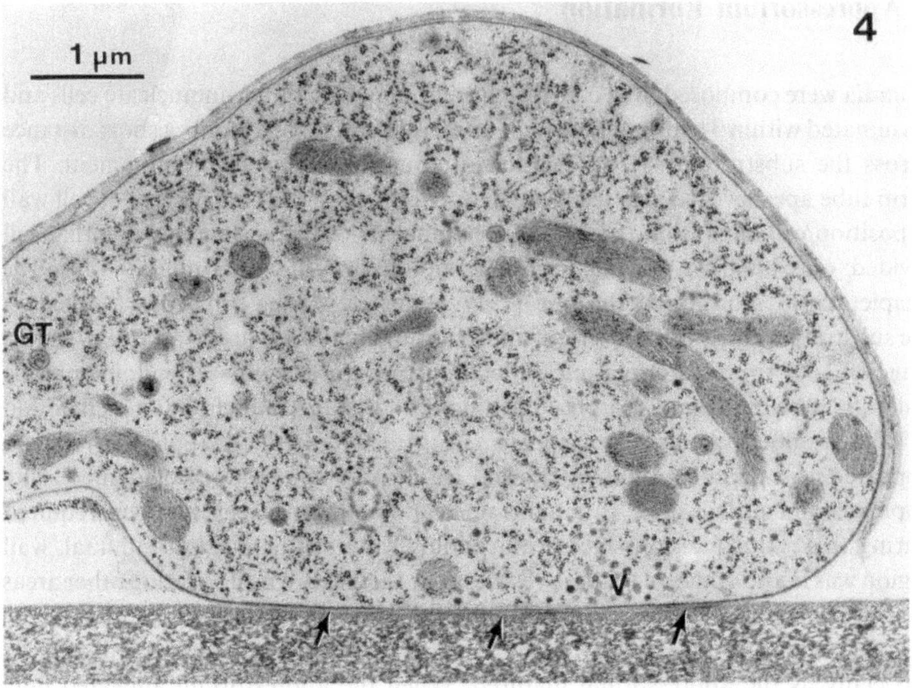
4
1 µm
GT
V

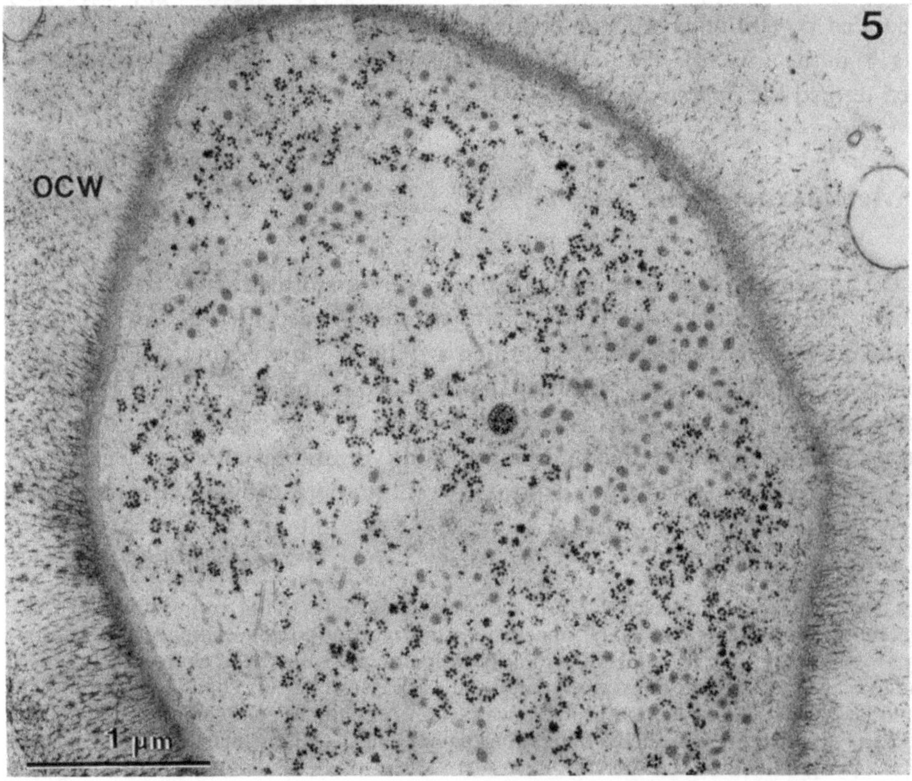
5
OCW
1 µm

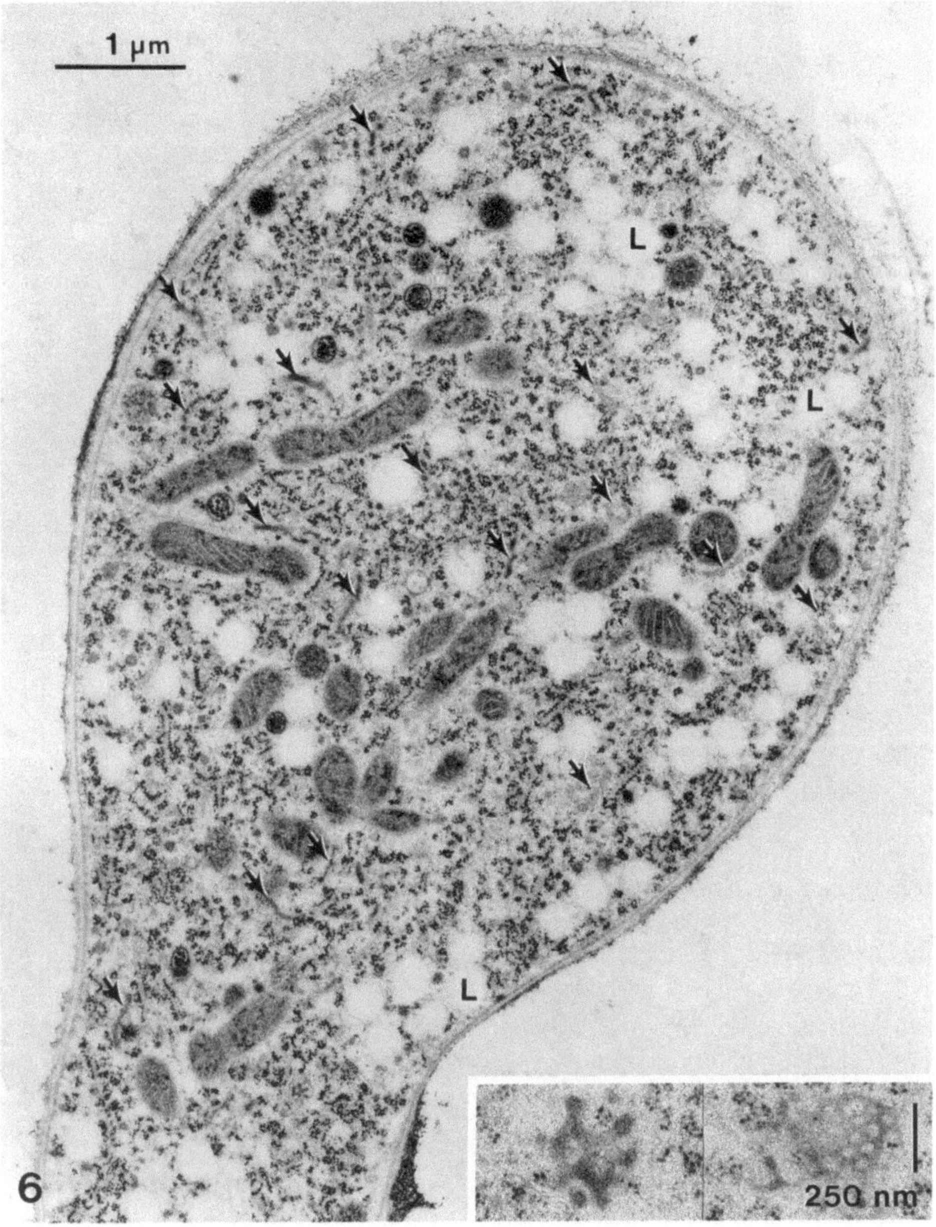

Fig. 6. The expanding region of this germ tube contains many profiles of smooth membrane cisternae (*arrows*) and lipid bodies (*L*). *Insets* depict two cisterna morphologies

◄

Figs. 4 and **5.** Expansion of the germ tube apex gives rise to an appressorium, seen here sectioned perpendicular (**4**) and parallel (**5**) to the substratum. **4** Judging from its shape, and from the location of numerous cytoplasmic vesicles (*V*), the nascent appressorium expands along its substratum interface. Note the distinctive, thin cell wall in this region of contact with the substratum (*arrows*), compared with cell wall of the germ tube (*GT*) and upper regions of the appressorium. **5** A section cut close to the substratum showing many vesicles in the region of presumed expansion. A wide layer of extracellular matrix forms a new outer layer of cell wall (*OCW*)

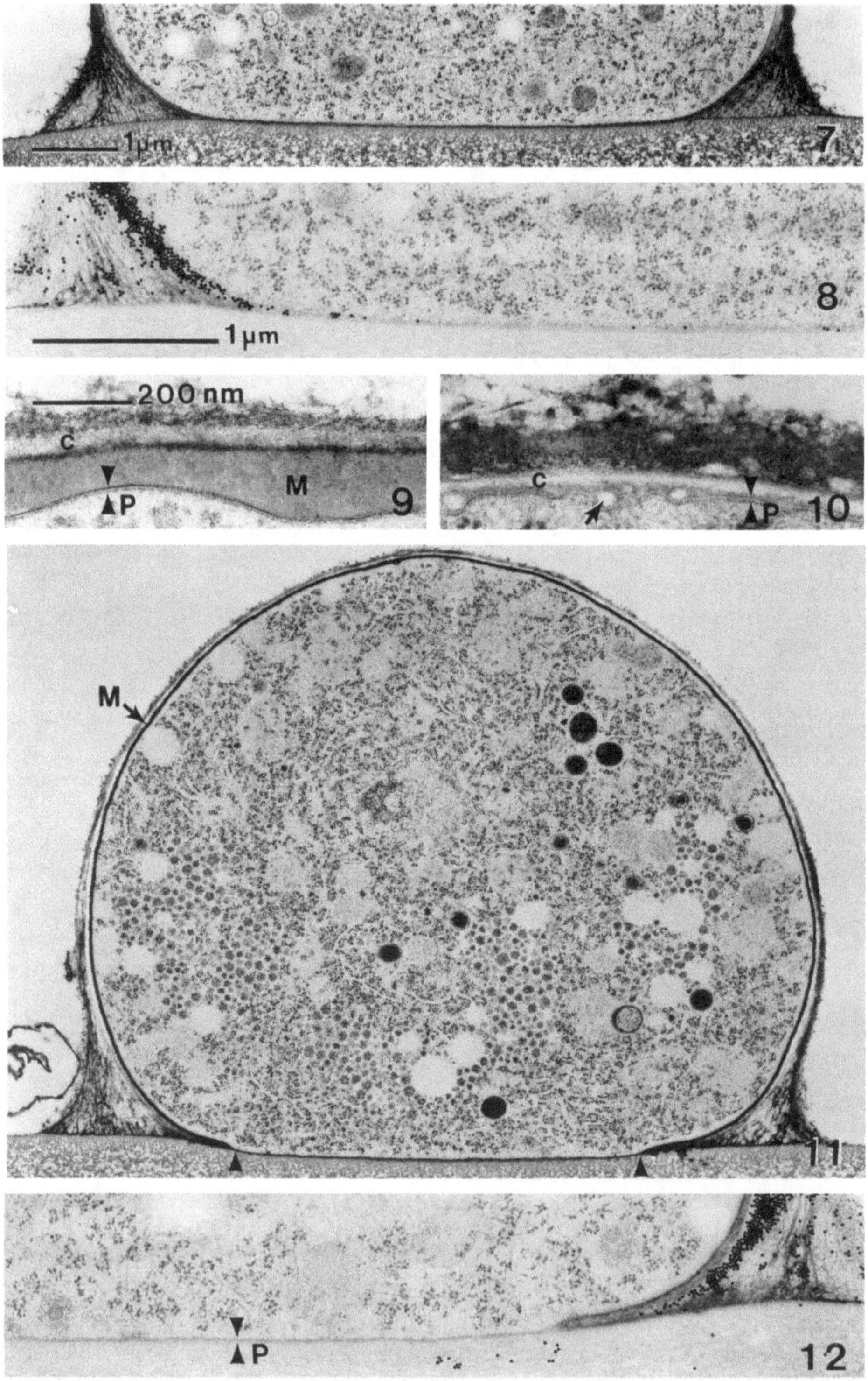
1µm
7
1 µm
8
200 nm
C
M
P
9
C
P
10
M
11
P
12

as well as around the entire appressorium, except across the pore (Fig. 11). The appressorium pore was thus retained as a specialized wall-less interface with the substratum (Fig. 12).

Following melanization the appressorium was considered structurally complete (Fig. 13) as smooth elements of endomembrane were apparently lacking (Fig. 11). However, 20–72 additional hours were required, depending on the substratum, before the next stage of development; formation of the penetration peg. That pegs formed in vitro and penetrate the surface of various artificial substrata, including plastic coverslips [11] and Mylar, was shown clearly using scanning electron microscopy (SEM). Appressoria ca. 72-h-old were fractured by sonication [22] to reveal the fungus-substratum interface including the area indented by the fungus (Fig. 14).

5 Invasive Structures

Just prior to emergence of a penetration peg a *pore wall overlay* [3], consisting of two new nonmelanin layers of wall material, was laid down over the bottom of the appressorium and completely covered the pore (Fig. 15). The penetration peg, ca. 0.7 μm in diameter, then formed from this wall overlay (Fig. 16). A wall appeared to surround the peg, at least during early stages of formation, and was continuous with the inner layer of the wall overlay (Fig. 17). Penetration pegs grew to a depth of nearly 10 μm in cellophane. Older pegs did not exhibit a uniformly labeled chitinous wall when probed with wheat germ agglutinin (Figs. 16, 17).

When incubation continued for 1–3 more days, laterally directed structures formed from the distal terminus of pegs within cellophane (Fig. 18). These structures were distinct and were composed of cells wider than the peg, were surrounded by a clear zone within the substratum, and resembled infection hyphae that formed in epidermal cells of the host (see [3]).

◄

Figs. 7–12. **7, 8** Cell wall in the region of substratum contact was not visible (**7**), or labeled by wheat germ agglutinin-colloidal gold conjugates (**8**), in these 4-h-old unmelanized appressoria. **9, 10** Cell walls of mature wild-type (**9**) and melanin-deficient strain CP220 (**10**). A prominent wall layer is missing from the mutant strain and was thus identified as melanin (*M*). Note the periplasmic inclusions (*arrow*) in strain CP220. Chitinous layer, *c*; Plasma membrane, *P*. **10** Magnification as **9**. **11** The melanin wall layer (*M*), laid down around the entire cell except in the region of contact with the substratum, defines the *appressorium pore* (*between arrows*, cf. **7**). Magnification as in 7. **12** Only the plasma membrane (*P*), no cell wall, was visible in the pore region of mature appressoria. Use of wheat germ agglutinin-colloidal gold demonstrated a dramatically different labeling pattern over appressorium cell wall vs. the pore. Magnification as in **8**

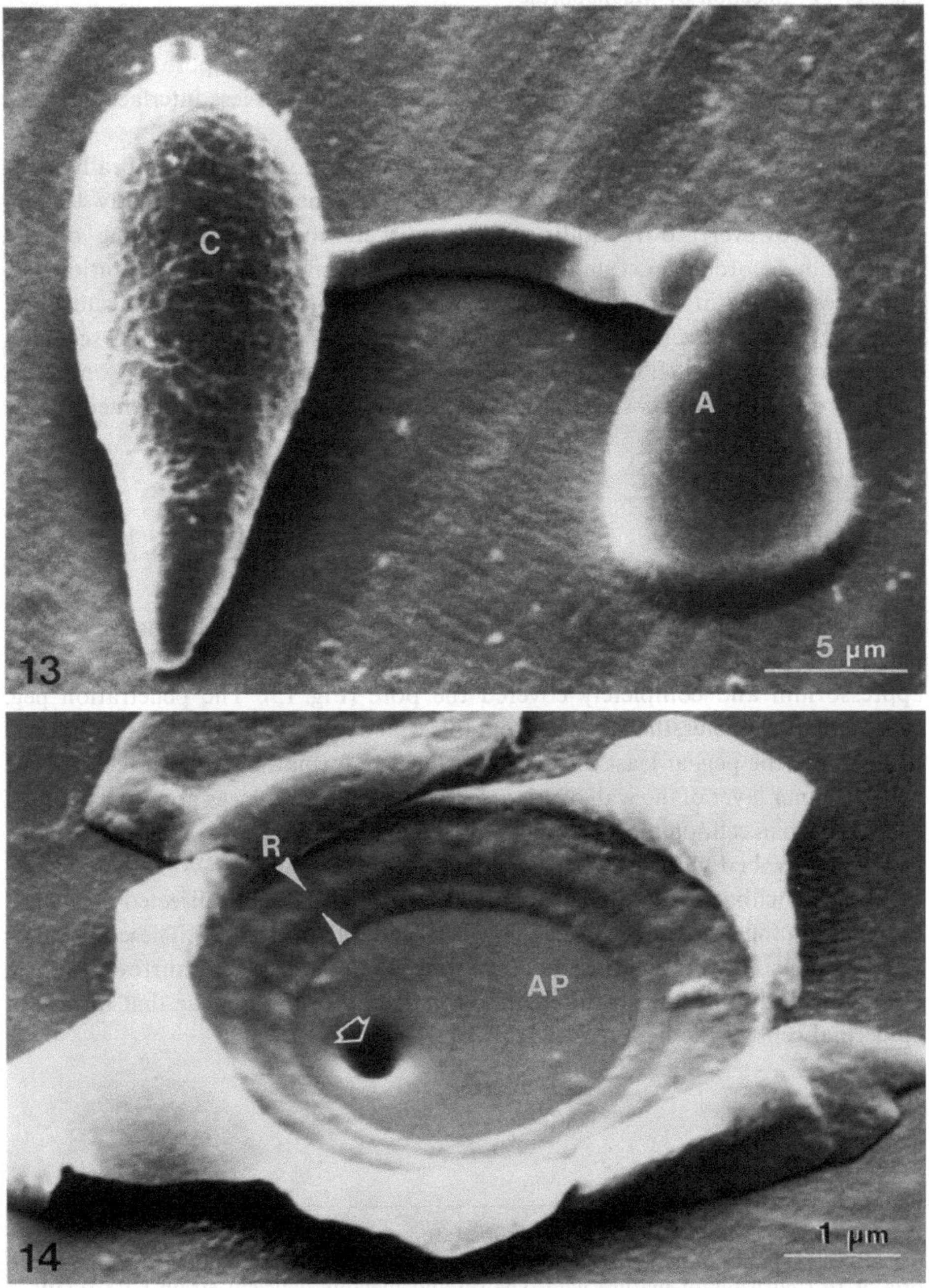

Figs. 13 and **14. 13** SEM overview of a mature appressorium (**A**) and conidium (**C**). Germination can occur from any conidial cell. The germ tube is usually no longer than the conidium before its tip begins to swell and hook. As the appressorium grows, the area of connection with the germ tube is lifted off the substratum. **14** Region of fungal-substratum attachment. The conidium, germ tube, and most of this appressorium were removed by sonication after nearly 6 days on a Mylar substratum. The appressorium cell wall fractured just above where it remains strongly attached to the Mylar in a ring configuration. Note the *pore ring* (*R*), visible as an impression in the overlying melanin layer, and the *appressorium pore* (*AP*). An indentation in the Mylar (*arrow*), caused by a penetration peg, was also exposed

6 Mechanism of Penetration

The inability of melanin-less appressoria to penetrate the host surface (see [5]) or artificial substrata [11] is a key to understanding appressorium function. A series of plasmolysis experiments demonstrated that melanin allowed appressoria to establish and maintain high internal osmotic potentials. In addition, cytorrhysis caused collapse of melanized appressoria in strongly hypertonic solutions. Melanin allowed the efflux of water but not the equilibrating influx of solute: appressoria collapsed from the negative internal pressure of plasmolysis. *Melanin was differentially permeable to water and solute* [11].

We propose that melanin acts to trap solute inside the appressorium causing water to be absorbed against the mounting osmotic gradient (Fig. 19). The resulting high internal hydrostatic pressure represents the necessary melanin-mediated element enabling penetration of the host. That pegs indented plastic coverslips composed of poly(vinyl chloride) and poly(vinyl acetate), and Mylar, poly(ethylene terephathalate), suggests that mechanical force alone was responsible for penetration since it is unlikely that the fungus produces enzymes to modify these substrata in this short time (see [20]). If we assume that the hardness of these substrata is similar to that of the host surface, we can extrapolate our results. Since the loss of melanin biosynthesis, due to either a single gene mutation or a chemical inhibitor, caused appressorial impotence and a corresponding loss of pathogenicity [5, 11], we suggest that (1) there is an essential mechanical component to penetration, and (2) enzymatic modification of the host surface may not be required for penetration of rice by *M. grisea*.

We do not wish to extrapolate our results to the penetration by other organisms of other substrata, however, as previously stated [11]. And it is with extreme caution that we equate the mechanism of penetration of artificial substrata and the host surface. We are well aware that single organisms can engage different mechanisms under different conditions. For example, penetration can be effected either via appressorium formation or by nonappressorial means (e.g., direct germ tube penetration or penetration by a peg not arising from an appressorium) as reported for *Colletotrichum trifolii* [19], *C. gloeosporioides* [4], *Venturia inaequalis* [13]; (cf. [21]), *Botrytis squamosa* [6], *Helminthosporium oryzae* [10], and *Erysiphe graminis* [16]. We cannot assume that any one species of pathogen will always employ the same mechanism of penetration and thus must exercise caution when comparing results.

7 Appressorium Function

Appressoria appeared structurally complete after melanization, ca. 8 h following the harvest of conidia. The intervening time until peg emergence is presumably required

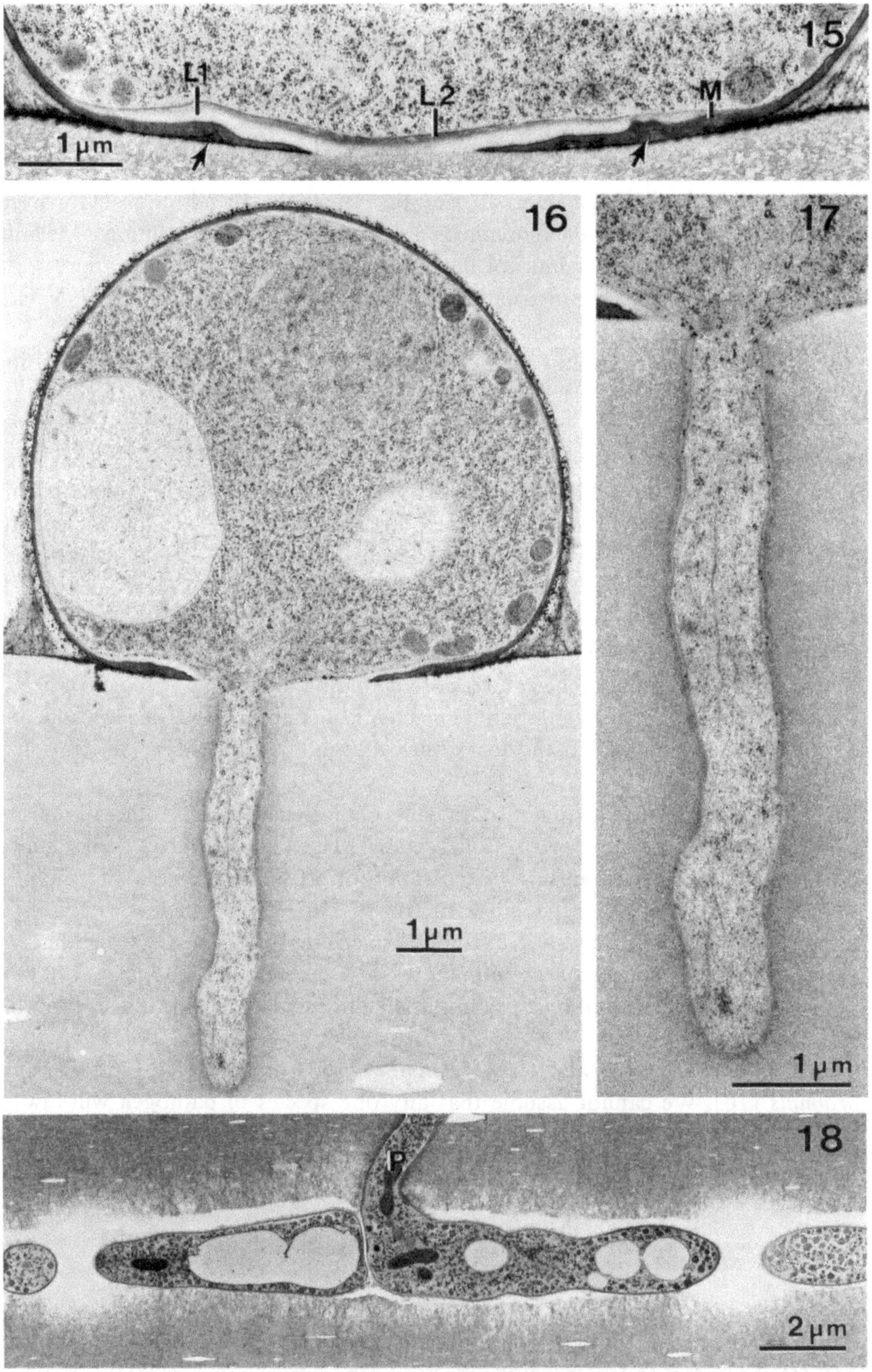
15
L1
L2
M
1 µm
16
1 µm
17
1 µm
18
P
2 µm

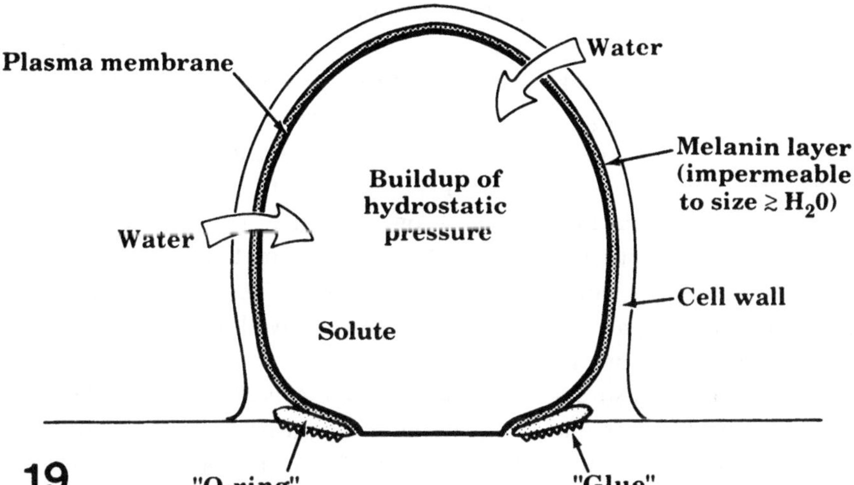

19

Fig. 19. Appressorium function. Melanin traps solute inside the appressorium but is permeable to water. As the concentration of solute is increased, more water is absorbed. The result is high internal hydrostatic pressure that subsequently serves as an essential force during penetration peg emergence. Since melanin is absent across the appressorium pore there must exist a tight seal, against the substitution, perhaps involving the action of a glue and an O-ring (i.e., the pore ring). [11]

for building sufficient hydrostatic pressure. We have described how the appressorium pore represents an apparent hole in the otherwise continuous appressorium wall. If the appressorium is to function as a device for building and applying strong mechanical pressures, then there must exist a very effective means of sealing the pore against pressure leakage. The pore wall-overlay might represent a means of sealing the pore but our observations indicate that the overlay does not form until just before penetration. Therefore, we suggest two other possible means to seal the pore, a strong glue and an O-ring (Fig. 19), and provide evidence for both.

Appressoria are commonly thought to be strongly attached to substrata, and perhaps were so termed with that function in mind. It was not surprising then that

---

Figs. 15–18. Invasive structures formed in vitro by M. grisea on cellophane membrane. **15** Just before penetration peg formation, two new cell wall layers (*L1*, *L2*), the *pore wall overlay*, are laid down covering the appressorium pore. Note the presence of material (*arrows*), under the melanin layer (*M*), probably corresponding to the pore ring. **16** Micrograph showing the distribution of wheat germ agglutinin-colloidal gold label. Note the penetration peg extending 6.4 μm into the substratum. **17** This higher magnification shows that the peg wall and pore wall overlay were continuous; however, the distribution of wheat germ agglutinin-colloidal gold labeling was not. **18** From the distal end of the peg (*P*), a multicellular structure that resembled infection hyphae formed within the cellophane. Note that the cellophane appears dissolved in the vicinity adjacent to these infection hypha-like structures, but not the peg

appressoria of *M. grisea* could not be dislodged from most substrata by any physical forces applied, including sonication for up to 2 h [11]. Sonication fractured appressorium walls but it did not disrupt attachment to substrata (Fig. 14). These results demonstrate the strength of appressorial adhesion and suggest that an adhesive substance or glue, possibly STM, could be responsible.

SEM of sonicated appressoria showed the impression of a ring of material, around the appressorium pore in the plane of contact with the substratum, visible in the overlying layer of melanin (Fig. 14; cf. [11], Fig. 8). This material was also visible in appressoria sectioned perpendicular to the substratum (Fig. 15; cf. [11], Fig. 11). We term this structure the *pore ring*, and suggest that it may play an important role in sealing the pore against pressure leakage [11].

Why is mechanical pressure necessary for peg emergence? Is it the underlying substratum that must be mechanically breached or is it (also?) the intervening pore wall overlay, as suggested by Beckett and Porter [2] and Landes and Hoffmann [14]? We have observed peg-size holes, in the absence of a peg, in the pore wall overlay of melanized *M. grisea* appressoria that were detached from Teflon (Fig. 20). This observation is consistent with the idea that a hole was enzymatically derived through the pore wall overlay and that the peg would subsequently indent the

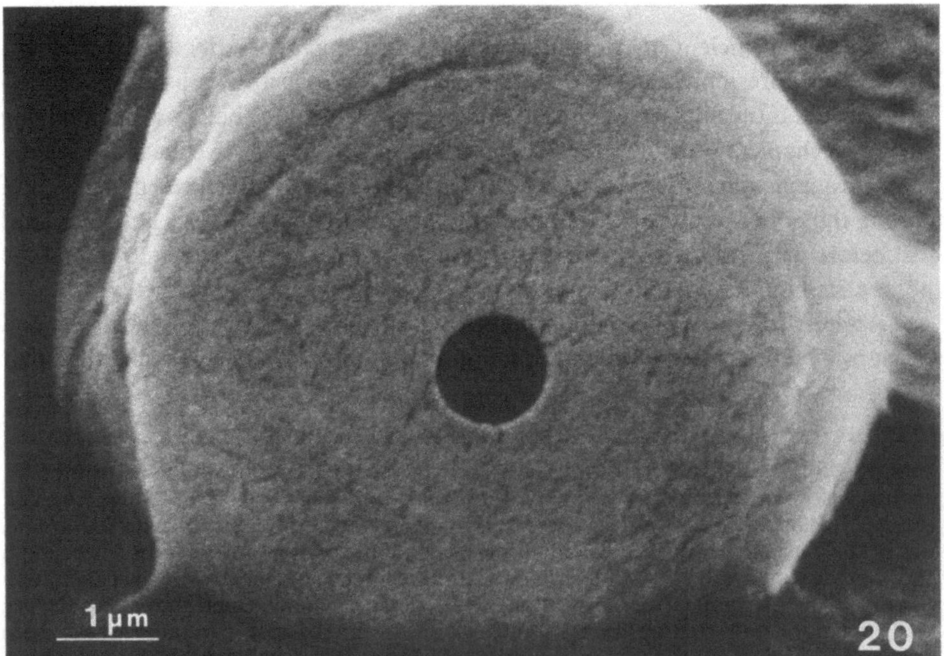

Fig. 20. SEM view of the substratum-side of an appressorium showing a hole through the pore wall overlay. This appressorium had not produced a penetration peg

substratum as a result of mechanical pressure. The pore wall overlay may be functionally similar to the wall layer that appears around the site of germ tube emergence in some conidia (see [1]).

8 Conclusions

1. Conidia of *M. grisea* have an *active* mechanism for attaching to surfaces that could play a very important role in rice blast disease.
2. Melanin acts to trap solute inside appressoria, causing water to be absorbed by the mounting osmotic gradient.
3. Penetration pegs are driven through the host surface and inert substrata by the melanin-mediated build-up of hydrostatic pressure. Enzymatic modification of these surfaces may not be required for penetration.
4. Appressorial function relies upon a level of architectural sophistication not previously appreciated.
5. Germ tube, penetration peg, and infection hypha-like structures are morphologically and functionally distinct entities. However, that this sequence of differentiation proceeded in vitro suggests that they represent a *morphogenetic unit*, triggered at some point early in development.

References

1. Aronson JM (1981) Cell wall chemistry, ultrastructure, and metabolism. In: Cole GT, Kendrick B (eds) Biology of conidial fungi, Vol 2. Academic Press, New York, pp 459
2. Beckett A, Porter R (1988) The use of complementary fractures and low-temperature scanning electron microscopy to study hyphal-host cell surface adhesion between *Uromyces viciae-fabae* and *Vicia faba*. Can J Bot 66: 645–652
3. Bourett TM, Howard RJ (1990) In vitro development of penetration structures in the rice blast fungus. *Magnaporthe grisea*. Can J Bot 68: 329–342.
4. Chau KF, Alvarez AM (1983) A histological study of anthracnose on *Carica papaya*. *Phytopathology* 73: 1113–1116
5. Chumley FG, Valent B (1990) Genetic analysis of melanin deficient nonpathogenic mutants of *Magnaporthe grisea*. Mol Plant-Microbe Interactions 3: 135–143.
6. Clark CA, Lorbeer JW (1976) Comparative histopathology of *Botrytis squamosa* and *B. cinerea* on onion leaves. Phytopathology 66: 1279–1289
7. Crawford MS, Chumley FG, Weaver CG, Valent B (1986) Characterization of the heterokaryotic and vegetative diploid phases of *Magnaporthe grisea*. Genetics 114: 1111–1129
8. Hamer JE, Howard RJ, Chumley FG, Valent B (1988) A mechanism for surface attachment in spores of a plant pathogenic fungus. Science 239: 288–290
9. Hamer JE, Valent B, Chumley FG (1989) Mutations at the *SMO* genetic locus affect the shape of diverse cell types in the rice blast fungus. Genetics 122: 351–361
10. Hau FC, Rush MC (1982) Preinfectional interactions between *Helminthosporium oryzae* and resistant and susceptible rice plants. Phytopathology 72: 285–292

11. Howard RJ, Ferrari MA (1989) Role of melanin in appressorium function. Exp Mycol 13: 403–418
12. Howard RJ, O'Donnell KL (1987) Freeze substitution of fungi for cytological analysis. Exp Mycol 11: 250–269
13. Keitt GW, Jones LK (1926) Studies of the epidemiology and control of apple scab. Wis Agric Exp Stn Res Bull 73: 1–104
14. Landes M, Hoffmann GM (1979) Zum Keimungs- und Infektionsverlauf bei *Colletotrichum lindemuthianum* auf *Phaseolus vulgaris.* Phytopathol Z 95: 259–273
15. Leung H, Borromeo ES, Brenardo MA, Notteghem JL (1988) Genetic analysis of virulence in the rice blast fungus *Magnaporthe grisea.* Phytopathology 78: 1227–1233
16. McKeen WE, Rimmer SR (1973) Initial penetration process in powdery mildew infection of susceptible barley leaves. Phytopathology 63: 1049–1053
17. Ou SH (1985) Rice Diseases, 2nd edn Commonwealth Agricultural Bureaux, Slough, UK
18. Parsons KA, Chumley FG, Valent B (1987) Genetic transformation of the fungal pathogen responsible for rice blast disease. Proc Natl Acad Sci USA 84: 4161–4165
19. Porto MDM, Grau CR, de Zoeten GA, Gaard G (1988) Histopathology of *Colletotrichum trifolii* on alfalfa. Phytopathology 78: 345–349
20. Potts JE (1984) Plastics, environmentally degradable. In: Eckroth D (ed) Encyclopedia of chemical technology, 3rd edn, Suppl Vol Wiley, New York, pp 626
21. Turner ML, MacHardy WE, Gadoury DM (1986) Germination and appressorium formation by *Venturia inaequalis* during infection of apple seedling leaves. Plant Dis 70: 658–661
22. Uchiyama T, Ogasawara N, Nanba Y, Ito. H (1979) Conidial germination and appressorial formation of the plant pathogenic fungi on the coverglass or cellophane coated with various lipid components of plant leaf waxes. Agric Biol Chem 43: 383–384
23. Valent B, Chumley FG (1987) Genetic analysis of host species specificity in *Magnaporthe grisea.* UCLA Symp Mol Cell Biol (New Series) 48: 83–93
24. Valent B, Crawford MS, Weaver CG, Chumley FG (1986) Genetic studies of pathogenicity and fertility of *Magnaporthe grisea.* Iowa State J Res 60: 569–593
25. Valent B, Farrall L, Chumley FG (1990) *Magnaporthe grisea* genes for pathogenicity and virulence identified through a series of backcrosses. Genetics (in press)

Chapter 20 Paola Bonfante-Fasolo and Silvia Perotto

Mycorrhizal and Pathogenic Fungi: Do They Share Any Features?

1 Introduction

Symbiotic interactions of higher plants and fungi range from pathogenic to mutualistic associations, depending on the nutritional strategy of the microorganism [26]. Plant pathogens drain sugars from specific autotrophic partners and express their virulent phenotype on them, producing toxins and/or enzymes, lastly causing "plant diseases" [30]. Mycorrhizal fungi, which are the most common symbiotic fungi, spend an important part of their life cycle inside the host. Like pathogenic fungi, most mycorrhizal fungi drain sugars from the host, but they counter this drawback by improving the mineral nutrition of the plant. The common response during these symbiotic associations is a growth increase of the host [23]. Therefore, beside the different responses of the colonized plant at the organismic level, the pathway of the nutrient flux represents an important difference between pathogenic and mycorrhizal associations at the cellular level. It goes from the plant towards the heterotrophic fungus following a one-way route in pathogenic interaction, while there is a two-way exchange in the mycorrhizal interaction [39].

According to Keen's [28] terminology, two types of pathogens can be recognized: one type is the nonspecialized "thug", which is necrotrophic, while the second type is represented by the "con men", which grow biotrophically and do not damage the host severely, at least in the early stages of the interaction [30]. While mycorrhizal fungi and "thugs" possess completely different strategies, an interesting parallel study can be made with biotrophic pathogenic fungi, to discover whether mycorrhizal and biotrophic pathogenic fungi share common features.

In this paper we shall exclusively consider endomycorrhizal fungi, i.e., vesicular-arbuscular and ericoid mycorrhizal fungi. In fact, major events of their infection process (such as adhesion between partners, penetration into cortical cells and triggering of host responses) can be best paralleled with those shown by biotrophic pathogenic fungi [32].

Dipartimento di Biologia Vegetale dell' Università and Centro di Studio sulla Micologia del Terreno del CNR Viale Mattioli 25, I-10125 Torino, Italy

2 The Infection Process

VAM fungi accomplish a complex morphogenesis in the presence of the host: they form appressoria, intercellular and intracellular hyphae, coils, vesicles, and highly branched structures, the so-called arbuscules (Fig. 1). All these structures seem to be formed under the control of the plant host. VAM fungi are ecologically and physiologically obligate biotrophs: they do not grow in pure culture, even though recent advances have been made in the development of growth-stimulating nutrient media. Spores germinate in axenic conditions and extensive hyphal growth with the formation of secondary spores and auxiliar cells is obtained ([24, 14], Bianciotto unpubl. results) (Fig. 2). However, as mycelium cannot be subcultured, growth is only sustained by spores, probably because of the absence of nuclear division [14]. Consequently, the infection structures do not develop in absence of the root. This is particularly important for the arbuscules, which represent the key point of the interaction from a physiological point of view. They are not formed in absence of the plant and so far we do not know which molecules or what conditions trigger the events by which the fungus ends its apical growth—typical of filamentous fungi [44]—and branches out, originating numerous growing points (Figs. 3 and 4).

It is interesting to observe that these host-controlled morphogenetical events closely resemble those shown by some rust fungi, which are not capable of differentiating typical haustoria in vitro [32].

3 Adjustment of the Fungal Cell Surface

During the infection process the symbiotic fungus develops a large contact surface area with the host cell, that provides a structural basis for exchanges in the effective mycorrhiza. All our results suggest that fungal cell surfaces with specific characteristics are necessary [9]. During the life cycle, the wall of VAM fungi shows striking modifications. It is thick, (from 10 μm in the spore to 0.50 μm in the hyphae), fibrillar, sometimes with a helicoidal arrangement of the chitin fibrils during the extraradical phase; then it becomes thinner and finally amorphous during the intraradical phase (Figs. 4 and 5), especially during the formation of the arbuscules. Throughout this phase the fungal wall is very thin, about 30 nm. The morphological changes are

Figs. 1–4. Morphological aspects of the infection process of VA mycorrhizal fungi. **1** Leek roots colonized by *Glomus versiforme*. The fungal arbuscules (*A*) mostly develop in the inner cortical cells. × 140. **2** Scanning electron micrograph of a spore of *scutellospora gregaria* germinating in vitro. × 100. **3** Scanning electron micrograph of highly branched arbuscules of *Glomus versiforme* which colonizes cortical cells in *Ginkgo biloba* roots. × 1600. **4** Transmission electron micrograph of arbuscules (*A*) colonizing *Ginkgo* cells (*H*). × 18 000

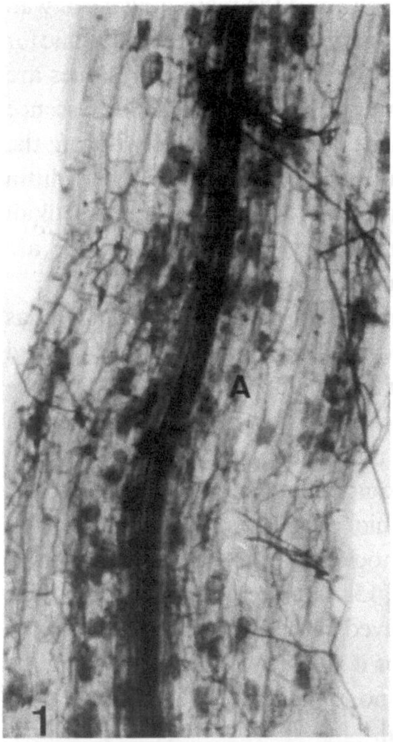
A
1

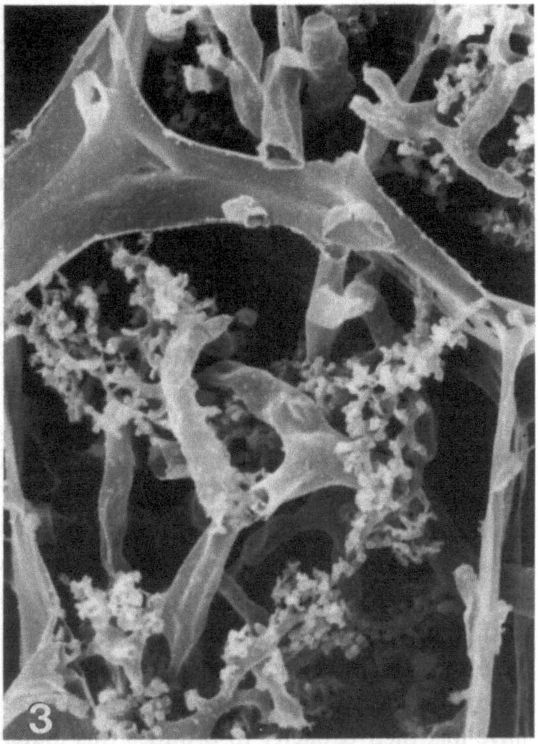
3

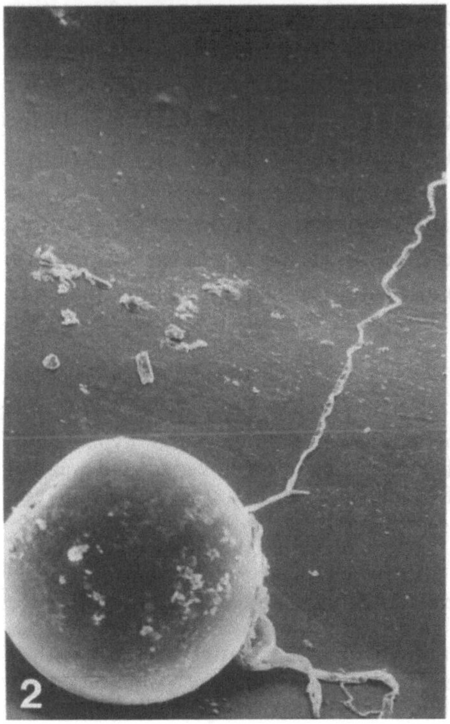
2

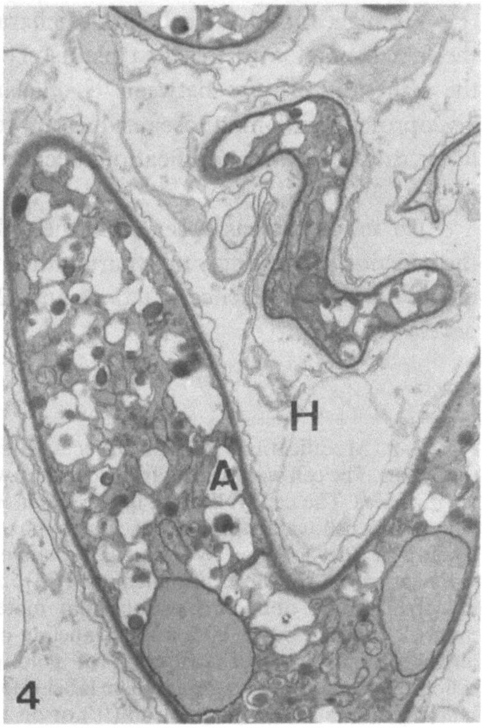
H
A
4

mirrored by different features of the cell surfaces: chitin is the main fungal cell wall component, at least in the spores [8]. When wheat germ agglutin—lectin specific for N-acetyl glucosamine (GlNAc)—is used at ultrastructural level, such residues are localized over the walls of all the fungal structures (Figs. 5 and 6). Fluorescence microscopy visualizes GlNAc residues on the entire hyphae exclusively during the arbuscular phase. During the other steps, characterized by thick walls, the chitin seems to be hidden by the presence of other components [12]. The latter are alkali soluble, sensitive to oxidizing agents and are visualized by concanavalin A: they are probably glucans covering the chitin skeleton (unpubl. results).

The symbiotic ericoid fungi change their surface characteristics drastically as well. Different sugar residues can be visualized, but in this case too, N-acetyl glucosamine residues do not extend to the outer surface of the hyphae [11], since they are covered by a fibrillar material. Such material shows a mosaic pattern: a combination of chemical, immunological, and ultrastructural investigations demonstrate the presence of soluble and insoluble components, sugars such as glucose, mannose, proteins and molecules with antigenic properties [37] (Fig. 7). The presence of an antigen cross-reacting with a monoclonal antibody raised against a cell surface mannoprotein of *Candida albicans* [15] is extremely interesting. The fibrillar material of ericoid fungi is probably involved in the adhesion process on the epidermal cell of the host, since it disappears—or it is highly reduced—during the host colonization. Moreover, ericoid strains are no longer infective [22] once they have lost their capacity to produce this material.

We suggest that the cell surface characteristics of the symbiotic fungi control the first contact with the host plants. This situation resembles that described in the adhesion process between *C. albicans* and epithelial cells, where specific glycoproteins with mannose residues are required [15] or in some rust fungi. In the latter, drastic surface modifications occur during the adhesion or the infective phases [27].

The modifications observed in the cell surface of the mycorrhizal fungi during the infection process may take part in the mechanisms (1) leading to the suppression of the host plant defence reaction, and (2) allowing the cellular compatibility to develop [12].

Figs. 5–8. Modifications of cell surfaces during symbiosis. **5** Intercellular hypha of *G. versiforme*. The cell wall is regularly labeled by wheat germ agglutinin linked to gold granules (*arrow head*). The cell wall of an intracellular BLO (*B*) is not labeled. × 58 000. **6** Thin and amorphous cell walls of the arbuscular hypha (*A*) of *Glomus versiforme*. They are regularly labeled by the complex WGA/gold. Granules do not occur in the interfacial material, surrounded by the host plsmalemma (*PL*). × 50 000. **7** Fungal isolate from an ericoid mycorrhiza with abundant extracellular material. After incubating the section with the monoclonal antibody CB25, a faint labeling of gold granules was observed at the surface (*arrows*). × 40 000. **8** Distribution of gold granules after incubating sections with a cellobiohydrolase-gold complex. The label is found exclusively over the pea walls (*HW*). No labeling occurs on the fungal wall (*W*) or on the intercellular space [IS]. × 31 000

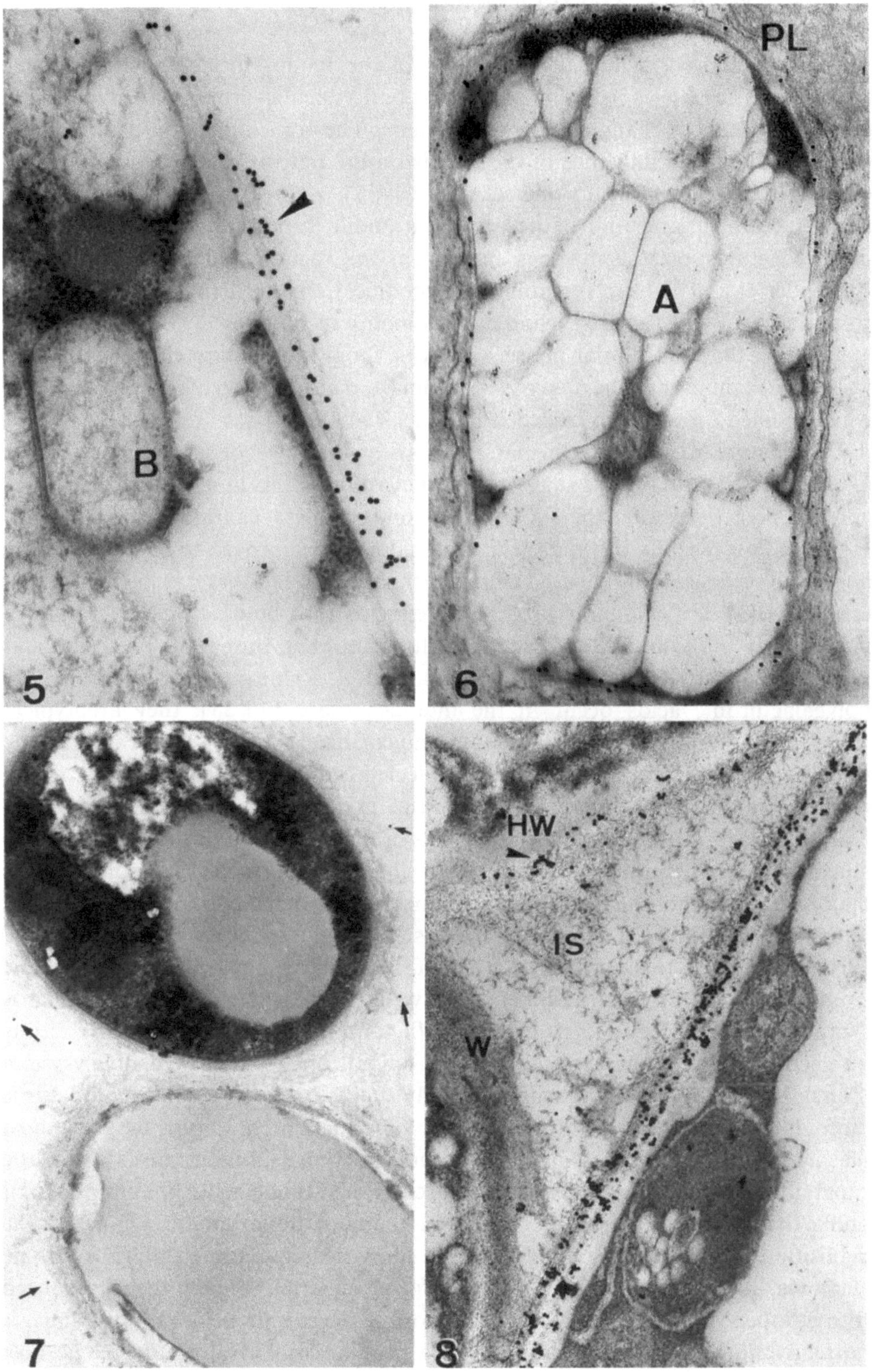
B
5
PL
A
6
7
HW
IS
W
8

4 Host Responses: The Interface Zone

The infection process and the spreading of the fungus inside the root trigger conspicuous responses in the host. Ericoid and vesicular-arbuscular mycorrhizal fungi create marked changes in the root systems. The branching pattern is modified, probably due to an influence of the fungus on the mitosis rate and consequently on the cell cycle of the colonized plants [4, 5]. At cellular level, the general ultrastructural organization of the host cell is modified and all the organelles display distinct features: plastids, nucleus and chromatin, vacuole and membranes [5, 8]. Alexander et al. [1] have quantified the modifications occurring in cortical cells colonized by VAM fungi by using morphometric techniques.

During the intracellular phase, symbiotic fungi cause a stretching of the host wall, an invagination of the host plasmalemma and the creation of an interface zone, which is filled up with the so-called interfacial material [7].

The nature of the interfacial material has been the object of many investigations: in the past a host origin was suggested on the grounds of its morphological features and of its reactivity after the PATAg reaction for polysaccharide localization [7]. However, just recently the interfacial material has been proved to be prevalently a host-produced material. We used affinity techniques and probes specific for the host cell wall [43], i.e., cellobiohydrolase, an enzyme that binds cellulose [3] and a monoclonal antibody JIM 5, reacting with pectin [42]. Immunofluorescence and gold granule distribution demonstrated that 1–4 β-glucan polymers as well as polygalacturonic acids occur in the interface space [13, 38]: they are typical components of a plant cell wall [6, 35]. In a mycorrhizal root they are localized in the peripheral cell walls of infected and noninfected cells (Fig. 8), close to penetration point where the hyphae are passing from the intercellular spaces into the cortical cells, and also around the thinner branches of the arbuscule. Residues of N-acetyl glucosamine are never detected in the intercellular space (Fig. 6), suggesting that the major component of the fungal cell wall does not participate in the creation of the interfacial material [12]. How does the host cell form this new wall-like envelope? We are suggesting two possibilities: (1) that the material may originate by a simple stretching of the original peripheral cell wall; (2) that new cell wall material is synthesized [13]. Morphometric analyses [1] and some of our preliminary results on electron micrographs, analyzed by using a Kontron AT Vidas/Videoplan system (Zeiss) to quantify density labeling, strongly suggest that the second hypothesis is more likely to be correct (unpubl. results). This implies that the new host plasmamembrane, which invaginates around the fungus, retains all its enzymatic capabilities. We can imagine the periarbuscular membrane being in a situation not unlike that of a protoplast membrane: it must contain the enzymatic complex for the deposition of cellulose-like molecules and allow the secretion of polygalacturonic acids via the Golgi apparatus. The interfacial material is likely to be produced through the synthetic activity of all, or most of the metabolic apparatus which is normally working on the deposition of a primary wall. The nature of the host perisymbiotic membrane has been the object of numerous investigations: increases in enzymatic activities such as phosphatases and ATPases could be very important in explaining the two-way exchanges [31, 21].

A similar interface also occurs at the contact area between plant cells and the intracellular hyphae of many pathogenic biotrophs, such as Oomycetes, rusts, powdery mildews, and hemibiotrophic fungi such as *Colletotrichum lindemuthianum* [36 and quoted references]. Material of variable thickness has been described as consisting of polysaccharides reacting with the PATAg reaction and rich in proteins. Micro-autoradiographic studies suggest that the material is solely of host origin [33], while other reports suggest that fungal proteins occur in the matrix [36]. Even around the pathogenic biotrophic fungi there is an extensive invagination of the host plasmalemma. However, important differences between the mycorrhizal and pathogenic systems can be pointed out: the invaginated host membrane surrounding the pathogenic fungus seems to be highly modified by comparison to the normal plasmalemma. It has different affinities to stains such as PTA, a much lower sterol content, and lacks ATPase activity [32, 20]. It is suggested that all these modifications channel the one-way route of nutrients from the host towards the fungus, avoiding the flow back towards the host thanks to the presence of specialized hydrophobic structures [20].

5 Host Responses: The Defense Reactions

When a plant is challenged by microorganisms or it is subjected to treatment with an elicitor or mechanical damages, it elaborates numerous inducible defense responses. Lamb et al. [30] describe a long list of such defenses, including synthesis of phytoalexins, reinforcement of cell walls by deposition of callose, lignin and related wall-bound phenols, accumulation of hydroxyproline-rich glycoproteins (HRGP's), production of proteinase inhibitors and lytic enzymes such as chitinase and glucanase.

In the last few years we studied the possibility that VAM fungi could evoke similar defense responses in their hosts: the results were neither completely negative or positive. In fact, callose is not laid down in the interfacial material, at least as seen by using anilin blue as probe, while suberin occurs in some mycorrhizal roots such as leek. Suberin is important in controlling the fungal development through the root, but it is a constitutive component also present in nonmycorrhizal roots [10]. Cell wall-bound phenols were studied in detail, with biochemical and ultrastructural techniques: the same types of phenols occur in mycorrhizal and nonmycorrhizal roots, and phenylalanine ammonia lyase does not show significant variations [17]. Interestingly, such phenols do not occur in the interfacial material. Other substances are reported to appear during fungal colonization: phytoalexins can be produced in low quantity as response to the fungus colonization [34], and some undetermined pathogenesis-related proteins have been found [18]. In our laboratory, we considered the occurrence of peroxidase activity during infection. Peroxidase is an important biochemical marker of plant-pathogen interactions, since it makes the cell wall more resistant to pathogen penetration by creating cross-links that stiffen it. Time course experiments demonstrated that there is an early peak of activity, corresponding to a stage where the fungus has not yet fully colonized the root and is mostly intercellular. When the fungus penetrates the host cell and the infection is fully established, the peroxidase activity decreases and levels at the same values of a nonmycorrhizal plant [40]. A similar early peak with subsequent decrease was also observed during the time course of a host chitinase activity in mycorrhizal roots. The parallel chitinase localization with immunogold techniques demonstrates that the enzyme occurs in the intercellular spaces, in the vacuoles or around the fungus, but never directly on the fungal wall. Hydroxyproline-rich glycoproteins (HRGP's) are major structural proteins of the plant cell [45], which usually increase in infected plants [19]. Preliminary results obtained by using an antibody against HRGP on semi-thick sections of *Ginkgo biloba*, show a regular labeling in the interfacial material, more abundant than in the peripheral wall (Vian et al. unpubl. results). In the latter examples there appears to be a modification in the expression of some enzymatic and structural proteins. This corresponds to a regulation of the interactions between the two partners and—at least in the case of peroxidase and chitinase—probably to a suppression of the host defence reaction thanks to an unknown mechanism. Induction of defence responses involves transcriptional activation of genes as part of a massive switch in the pattern of host gene expression [30]. Changes in mRNA populations in a time course of the symbiotic infection process have not yet been monitored, but the study of protein expression shows strong changes during the establishment of an ectomycorrhizal symbiosis [25].

6 Conclusions and Speculations

All the events (infection process, cell surface modifications, interfacial material, host defense reactions) show remarkable similarities and differences with those already described for biotrophic pathogenic fungi. One may suggest that there is a common background both for the penetration strategies in mycorrhizal and pathogenic fungi and for the host responses. However, mycorrhizal fungi elude plant surveillance mechanisms, suppressing their defence mechanisms and giving rise to a compatible interaction. How does this happen? We proposed the following possibilities. (1) Some pathogenic fungi possess dsRNA molecules which can be related to modifications of their virulence: a classical example of this is given by *Endothia parasitica*: strains with dsRNA molecules are hypovirulent, unlike the virulent strains lacking such molecules [2]. Our research now in progress suggests that similar molecules exist in some VAM fungi (Perotto et al. unpubl. results). It is possible that VAM fungi originated as pathogenic strains which were further attenuated by the presence of exogenous genomic molecules. (2) The different behavioral patterns of mycorrhizal and pathogenic fungi involve a completely different enzymatic equipment. Mycorrhizal fungi may be capable of penetrating a host cell without affecting its viability. Ectomycorrhizal and ericoid fungi have been demonstrated to possess cellulase and polygalacturonase activities at a much lower level than pathogenic fungi [29, 16].

In addition to this, mycorrhizal fungi mostly have to deal with the hydrophilic surfaces of roots in primary structure, while the more aggressive pathogenic fungi usually have to face the hydrophobic surfaces covered by cutin and waxes. Mycorrhizal fungi usually cross nonencrusted cell walls consisting of cellulose and pectins. In order to do this, they use low amounts of enzymes which—we may speculate—would be not so different from those used by the growing plant cell to loosen its wall. Cell wall synthesis is in fact the result of a balance between endogenous hydrolytic enzymes produced under hormone control and the stimulation of the different polysaccharide synthases [6]. Mycorrhizal fungi may produce hydrolytic enzymes at a rate low enough not to activate the reaction defenses. The low quantity of oligosugars solubilized after the glycosidic linkage breakdown may even represent a positive stimulus to produce new wall material.

We suggest that this complex strategy evolved by mycorrhizal fungi could be the result of an ancient co-existence with the host plant.

References

1. Alexander T, Meier R, Toth R, Weber HC (1988) Dynamics of arbuscule development and degeneration in mycorrhizas of *Triticum aestivum* L. and *Avena sativa* L. with reference to *Zea mays*. New Phytol 110: 363–370
2. Anagnostakis SL (1982) Biological control of chestnut blight. Science 15: 466–471
3. Berg RH, Erdos GW, Gritzali M, Brown RD Jr (1988) Enzyme-gold affinity labelling of cellulose. J Electron Microsc Tech 8: 371–379
4. Berta G, Fusconi A, Trotta A, Scannerini S (1990) Morphogenetic modifications induced by the mycorrhizal fungus *Glomus* strain E_3 on the root system of *Allium porrum* L. New Phytol 114: 207–215
5. Berta G, Sgorbati S, Soleri V, Fusconi A, Trotta A, Citterio MG, Sparvoli E, Scannerini S (1990) Chromatin structure variations in host nuclei of a VA mycorrhiza. New Phytol 114: 199–205
6. Bolwell GP (1988) Synthesis of cell wall components: aspects of control. Phytochemistry 27: 1235–1253
7. Bonfante-Fasolo P (1984) Anatomy and morphology of VA mycorrhizae. In: Powell CL, Bagyaraj DJ (eds) VA mycorrhizas. CRC Press, Boca Raton, USA, pp 5
8. Bonfante-Fasolo P (1987) Vesicular-Arbuscular Mycorrhizae: Fungus-plant interactions at the cellular level. Symbiosis 3: 249–268
9. Bonfante-Fasolo P (1988) The role of the cell-wall as a signal in mycorrhizal associations. In: Scannerini S, Smith DC, Bonfante-Fasolo P, Gianinazzi-Pearson V (eds) Cell to cell signals in plant, animal, and microbial symbiosis. Springer, Berlin Heidelberg New York Tokyo, pp 219

10. Bonfante–Fasolo P, Vian B (1989) Cell wall architecture in mycorrhizal roots of *Allium porrum* L. Ann Sci Nat Bot 13: 97–109
11. Bonfante-Fasolo P, Perotto S, Testa B, Faccio A (1987) Ultrastructural localization of cell surface sugar residues in ericoid mycorrhizal fungi by gold-labeled lectins. Protoplasma 137: 25–35
12. Bonfante-Fasolo P, Faccio A, Perotto S, Schubert A (1990) Correlation between chitin distribution and cell wall morphology in the mycorrhizal fungus *Glomus versiforme*. Mycol Res 94: 157–165
13. Bonfante-Fasolo P, Vian B, Perotto S, Faccio A, Knox A (1990) Cellulose and pectin localization in roots of mycorrhizal *Allium porrum:* labelling continuity between host cell wall and interfacial material. Planta 180: 537–547
14. Burggraaf JP, Beringer JE (1989) Absence of nuclear DNA synthesis in vesicular-arbuscular mycorrhizal fungi during in vitro development. New Phytol 111: 25–33
15. Cassone A, Torosantucci A, Boccanera M, Pellegrini G, Palma C, Malavasi F (1988) Production and characterization of a monoclonal antibody to a cell surface, glucomannoprotein constituent of *Candida albicans* and other pathogenic *Candida* species. J Med Microbiol 27: 233–238
16. Cervone F, Castoria R, Spanu P, Bonfante-Fasolo P (1988) Pectinolytic activity in some ericoid mycorrhizal fungi. Trans Br Mycol Soc 91: 537–539
17. Codignola A, Verotta L, Maffei M, Spanu P, Scannerini S, Bonfante-Fasolo P (1989) Cell wall bound phenols in roots of vesicular-arbuscular mycorrhizal plants. New Phytol 112: 221–228
18. Dumas E, Gianinazzi-Pearson V, Gianinazzi S (1990) Production of new soluble proteins during VA endomycorrhiza formation. In: 2nd European Symposium on Mycorrhizae Aug 14–20 1988 Prague, Agricultural Ecosystems and Environments 29: 111–114
19. Esquerre-Tugaye MT, Lafitte C, Mazau D, Toppan A, Touze A (1979) Cell surfaces in plant microorganism interactions. II. Evidence for the accumulation of hydroxyproline-rich glycoproteins in the cell wall of diseased plants as a defense mechanisms. Plant Physiol 64: 320–326
20. Gay JL, Salzberg A, Woods AM (1987) Dynamic experimental evidences for the plasmamembrane ATPase domain hypothesis of haustorial transport and for ionic coupling of the haustorium of *Erysiphe graminis* to the host cell (*Hordeum vulgare*). New Phytol 107: 541–548
21. Gianinazzi-Pearson V, Gianinazzi S (1988) Morphological integration and functional compatibility between symbionts in vesicular arbuscular endomycorrhizal associations. In: Scannerini S, Smith DC, Bonfante-Fasolo P, Gianinazzi-Pearson V (eds) Cell to cell signals in plant, animal, and microbial symbiosis. Springer, Berlin Heidelberg New York Tokyo, pp 73
22. Gianinazzi-Pearson V, Bonfante–Fasolo P, Dexheimer J (1986) Ultrastructural studies of surface interactions during adhesion and infection by ericoid endomycorrhizal fungi. In: Lugtenberg B (ed) Recognition in microbe plant symbiotic and pathogenic interactions. NATO ASI Series Vol H4, Springer, Berlin Heidelberg New York Tokyo, pp 273
23. Harley JL, Smith SE (1983) Mycorrhizal symbiosis, Academic Press, London
24. Hepper CM (1979) Germination and growth of *Glomus caledonius* spores: the effects of inhibitors and nutrients. Soil Biol Biochem 11: 269–277
25. Hilbert JL, Martin F (1988) Regulation of gene expression in ectomycorrhizas. I. Protein changes and the presence of ectomycorrhiza specific polypeptides in the *Pisolithus-Eucalyptus* symbiosis. New Phytol 110: 339–346
26. Jeffries P (1987) Pathways for the exchange of materials in mycoparasitic and plant-fungal interactions. In: Pegg GF, Ayres PG (eds) Fungal infection of plants. Cambridge University Press, Cambridge, pp 735
27. Kapooria RG, Mendgen K (1985) Infection structures and their surface changes during differentiation in *Uromyces fabae*. Phytopathol Z 113: 317–323
28. Keen NT (1986) Pathogenic strategies for fungi. In: Lugtenberg B (ed) Recognition in

microbe plant symbiotic and pathogenic interactions. NATO ASI Series Vol H4, Springer, Berlin Heidelberg New York Tokyo, pp 171

29. Keon JPR, Byrde RJW, Cooper RM (1987) Some aspects of fungal enzymes that degrade plant cell walls. In: Pegg GF, Ayres PG (eds) Fungal infection of plants. Cambridge University Press, Cambridge, pp 133
30. Lamb CJ, Lawton MA, Dron M, Dixon RA (1989) Signals and transduction mechanisms for activation of plant defenses against microbial attack. Cell 56: 215–224
31. Marx C, Dexheimer J, Gianinazzi-Pearson V, Gianinazzi S (1982) Enzymatic studies on the metabolism of vesicular arbuscular mycorrhizas. VI. Ultracytoenzymological evidence (ATPase) for active transfer processes in the host-arbuscule interface. New Phytol 90: 37–43
32. Mendgen K, Schneider A, Sterk M, Fink W (1988) The differentiation of infection structures as a result of recognition events between some biotrophic parasites and their hosts. J Phytopathol 123: 259–272
33. Mendgen K, Heitefuss R (1975) Microautoradiographic studies on host-parasite interactions. I. The infection of *Phaseolus vulgaris* with tritium-labelled uredospores of *Uromyces phaseoli.* Arch Microbiol 105: 193–199
34. Morandi D, Bailey JA, Gianinazzi-Pearson V (1984) Isoflavonoid accumulation in soybean roots infected with vesicular-arbuscular mycorrhizal fungi. Physiol Plant Pathol 24: 357–364
35. McNeil M, Darvill GA, Fry CS, Albersheim P (1984) Structure and function of the primary cell wall of plants. Annu Rev Biochem 53: 625–663
36. O'Connell RJ, Bailey JA (1986) Cellular interactions between *Phaseolus vulgaris* and the hemibiotrophic fungus *Colletotrichum lindemuthianum.* In: Bailey JA (ed) Biology and molecular biology of plant-pathogen interactions, NATO ASI Series, Vol H1 Springer, Berlin Heidelberg New York Tokyo, pp 39
37. Perotto S, Faccio A, Malavasi F, Bonfante-Fasolo P (1988) Detection of cell surface molecules in mycorrhizal fungi by using monoclonal antibodies. Giorn Bot Ital 122: 60–61
38. Perotto S, VandenBosch KA, Brewin NJ, Faccio A, Knox JP, Bonfante–Fasolo P (1990) Modifications of the host cell wall during root colonization by Rhizobia and VAM fungi. In: Nardon P, Gianinazzi-Pearson V, Grenier AM, Margulis L., Smith DC (eds) Endocytobiology IV, pp 115
39. Smith SE, Gianinazzi-Pearson V (1988) Physiological interactions between symbionts in vesicular-arbuscular mycorrhizal plants. Annu Rev Plant Physiol Plant Mol Biol 39: 221–244
40. Spanu P, Bonfante-Fasolo P (1988) Cell-wall bound peroxidase activity in roots of mycorrhizal *Allium porrum.* New Phytol 109: 119–124
41. Spanu P, Boller T, Ludwig A, Wiemken A, Faccio A, Bonfante-Fasolo P (1989) Chitinase in roots of mycorrhizal *Allium porrum*: regulation and localization. Planta 177: 447–455
42. VandenBosch KA, Bradley DJ, Knox JP, Perotto S, Butcher GW, Brewin NJ (1989) Common components of the infection thread matrix and the intercellular space identified by immunocytochemical analysis of pea nodules and uninfected roots. EMBO J 8: 335–342
43. Vian B (1986) Ultrastructural localization of carbohydrates. Recent developments in cytochemistry and affinity methods. In: Bailey JA (ed) Biology and molecular biology of plant-pathogen interactions, NATO ASI Series, Vol H1, Springer, Berlin Heidelberg New York Tokyo, pp 49
44. Wessels JGH (1986) Cell wall synthesis in apical hyphal growth. Int Rev Cytol 104: 37–79
45. Wilson LG, Fry LC (1986) Extensin—a major cell wall glycoprotein. Plant Cell Environ 9: 239–260

Chapter 21 ROSMARIE HONEGGER

Haustoria-Like Structures and Hydrophobic Cell Wall Surface Layers in Lichens

1 Lichen Mycobionts and Photobionts

The polyphyletic, taxonomically very heterogenous group of lichen-forming fungi comprises about 21% of all fungi (±13250 spp. or 46% of all ascomycetes, ±50 spp./0.3% of basidiomycetes, and ±200 spp./1.2% of imperfect fungi [20]). Lichen-forming fungi are nutritionally specialized, but otherwise normal representatives of their subclasses, and there are no reasons for separating them taxonomically from nonlichenized taxa. In only about 2% of lichens has the photobiont ever been determined at the species level [52], and very little is known about the range of compatible photobionts of individual lichen mycobionts. A large number of lichen-forming fungi seem to be moderately specific (several algal species of a genus are acceptable partners), but highly selective towards their photobionts (compatible photobionts are often very rare in natural ecosystems, and the most common taxa of free-living algae are not acceptable partners [16]). Unicellular and filamentous Charophyceae, Chlorophyceae and Ulvophyceae [34, 48] and/or nitrogen-fixing cyanobacteria (altogether about 100 spp.) have been identified as lichen photobionts [52]. The lichen symbiosis is not inheritable, but numerous taxa disperse very efficiently by means of vegetative symbiotic propagules.

2 The Lichen Symbiosis

In contrast to mycorrhizal, plant pathogenic, or other endophytic fungi which inhabit or associate with multicellular host structures, most lichen mycobionts are exhabitants of their minute photobionts. A wide range of mycobiont-photobiont interactions have been observed in lichens (summaries: [26, 28, 31]). Correlations between thallus morphology, type of mycobiont–photobiont relationship, and the

---

Institute of Plant Biology, University of Zürich, Zollikerstr. 107 CH-8008 Zürich, Switzerland

structure and composition of the photobiont cell wall have been described in lichens with green algal photobionts [26, 28, 31, 46, 51].

Primitive lichen mycobionts are not capable of differentiating an internally stratified thallus with various types of plectenchymata and pseudoparenchymata. They ensheath or overgrow photobiont cells or colonies at the sites where these occur in the aposymbiotic state, the resulting thallus being a crust-like structure on or within the substratum. In many of the primitive crustose lichen species the mycobiont forms finger-like intracellular haustoria (Fig. 1) unless the photobiont cell wall contains an outermost layer of enzymatically nondegradable sporopollenin [7, 25, 26, 33]. Such finger-like intracellular haustoria may bear structural similarity with the finger-like fungal protrusions in the rhizodermal cells of monotropoid mycorrhizas [14], but they are likely to be functionally quite different, since the achlorophyllous, C-heterotrophic Monotropaceae depend on water, nutrient, and carbohydrate supply by their mycorrhizal partners. More than 50% of all lichen-forming fungi (representatives of 18 orders of asco- and basidiomycetes [20]) form relatively primitive, crustose thalli. This taxonomically very heterogenous group of lichen mycobionts and their likewise taxonomically heterogenous photobionts [52, 32] are the "testing ground" of nutritionally specialized fungi which associate as biotrophs with cyanobacteria and/or algal cells.

Competition for space at the sites where lichen photobionts normally occur (surface or uppermost layers of bark, soil, rock, etc.) and the scarcity of some of the most frequently acquired compatible photobionts (*Trebouxia* spp.) in the aposymbiotic state in natural ecosystems were probably among the factors which imposed a strong evolutionary pressure on lichen-forming fungi to rise above the substratum by developing three-dimensional structures in which the photobiont cells are integrated and their development controlled by the mycobiont. Such squamulose, foliose, or fruticose, internally stratified vegetative structures are unique in the fungal kingdom.

Figs. 1–5. The mycobiont-photobiont interface in Lecanorales and Teloschistales (Ascomycotina) with *Trebouxia* (Pleurastrales. Ulvophyceae) photobiont. **1** Detail of a cross-section of the crustose, nonstratified thallus of *Lecanora subfuscata* Magn. (Lecanorales). Each photobiont cell (*PH*) is invaded by at least one intracellular haustorium (*ic*). *MY* mycobiont. **2–5** *Xanthoria parietina* (L.)Th.Fr. (Teloschistales). **2** Cross-section through part of the dorsiventrally organized, internally stratified thallus. The photobiont cell population (*PH*) is kept in a uniformly thick layer underneath the conglutinate upper cortex (*uc*). *m* gas-filled medullary layer; *lc* conglutinate lower cortex. **3** Intraparietal haustoria (*ip*) **4** Juvenile photobiont (*PH*) cells (autospores; three cells are visible) ensheathed by the autosporangial (mother cell) wall. The *arrow* points to an outgrowth of the intraparietal haustorial complex [formerly contacting the mother cell (autosporangium)] which is invading the mother cell wall in order to form a haustorial complex with the daughter cells. **5** Short distance shifting of photobiont (*PH*) cells (originating from the same autosporangium) by growth processes of the intraparietal haustorial complex (*ip*). *Small arrows* point to a site where the thin, mycobiont-derived wall surface layer was torn away. Preparative techniques see [27, 29]. **1** from [27]. Bar = 5 μm in all micrographs

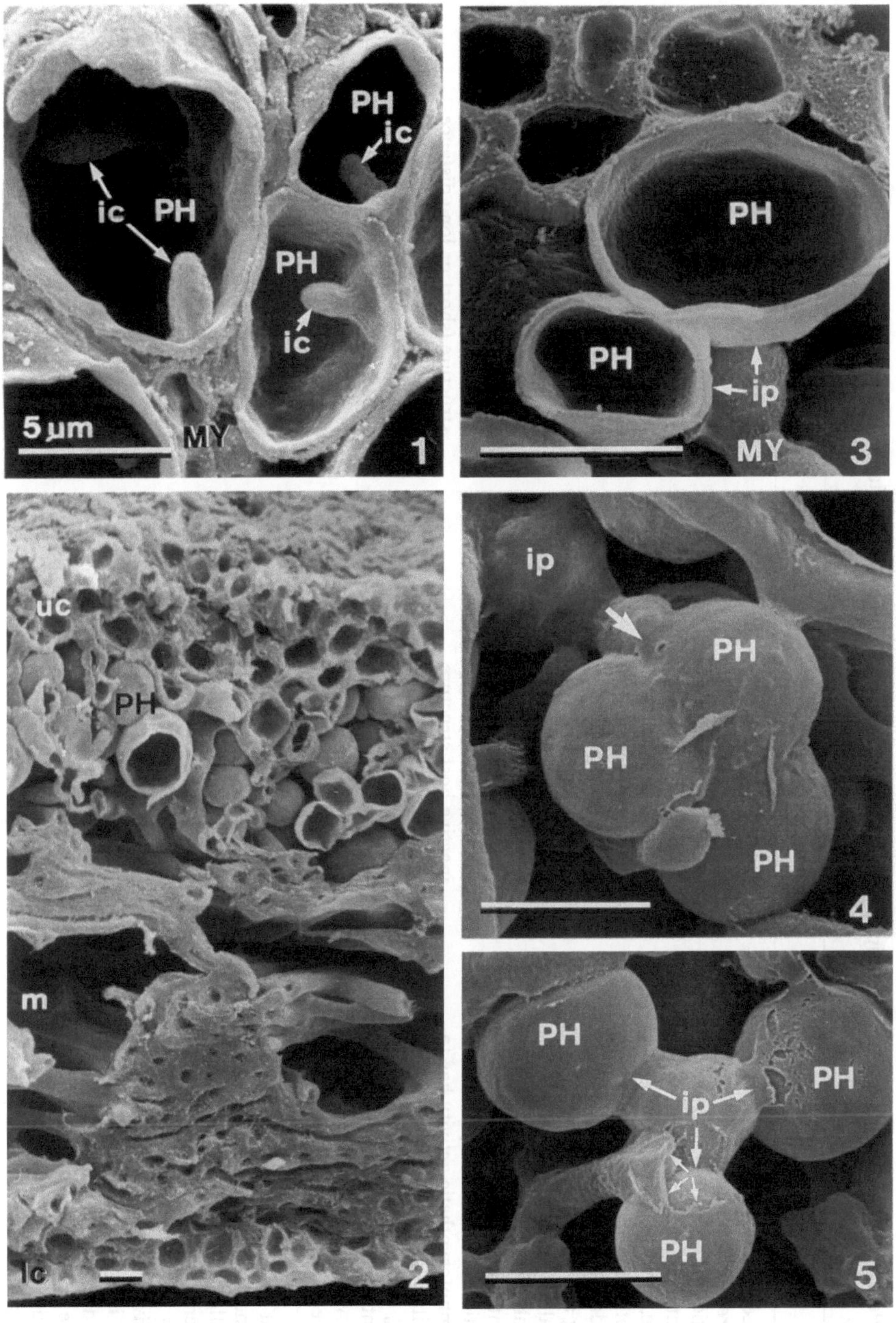
PH
ic
ic PH
PH
ic
5 µm
MY
1
PH
PH
ip
MY
3
uc
PH
m
lc
2
ip
PH
PH
PH
4
PH
ip
PH
PH
5

Table 1. Peculiarities of the symbiotic relationship in highly advanced Lecanorales and Teloschistales in comparison with ectomycorrhizae and leaf-inhabiting rusts

Lichens	Ectomycorrhizae [18, 3]	Leaf-inhabiting rusts [41]
(Foliose Lecanorales and Teloschistales with *Trebouxia* photobiont. > 1000 spp.)		
Mutualistic (+/+) relationship	Mutualistic (+/+) relationship	Parasitic (+/−) relationship
Mycobiont is an ecologically obligate, physiologically facultative biotroph	Mycobiont is an ecologically obligate or facultative, physiologically facultative biotroph	Mycobiont is an ecologically obligate and physiologically facultative biotroph
Pokilohydrous symbionts, regular wetting and drying cycles are essential for the maintenance of the symbiotic relationship	Homoiohydrous plant host	Homoiohydrous plant host
Mycobiont is exhabitant of a population of unicellular photobionts whose photosynthetic activity has to be facilitated; mycobiont amounts 85–90% of total biomass	Mycobiont is exhabitant of a multicellular, photosynthetically inactive host structure	Mycobiont is inhabitant of a multi-cellular, photosynthetically active host structure (short or long distance translocation of photosynthates to the contact site possible)
Mycobiont forms three-dimensional vegetative structure (dorsiventrally organized, internally stratified thallus, "symbiotic phenotype")	Mycobiont forms conglutinate sheath around given host structure	Mycobiont forms no three-dimensional structures in the purely vegetative part
Individual photobiont cells are positioned within the algal layer of the thallus and their growth and cell turnover controlled by the mycobiont	Growth and development of the mycorrhizal root are influenced by the fungal partner	Morphology of the host may or may not be influenced by the mycobiont
Symbiosis is not inheritable, but vegetative symbiotic propagules are formed in a large number of spp.	Symbiosis is not inheritable	Symbiosis is not inheritable
Long-lived structures (individual cells are subjected to a certain cell turnover): decades to centuries → High regulatory potential required	Relatively short-lived symbiotic structures (several months to few years	Relatively short-lived symbiotic structures (several months to few years)

The present chapter aims to focus on highly advanced lecanoralean and teloschistalean lichen mycobionts which form dorsiventrally organized, foliose, internally stratified thalli (Fig. 2) and position their photobionts of the genus (*Trebouxia* s. lat. [17] within the algal layer by means of intraparietal haustoria [25, 26, 28]. *Trebouxia* spp. are supposed to be the photobionts of 50–70% of all lichens [1]. Their cell walls are structurally comparable with the pecto-cellulosic walls of plants [25, 32]. The growth and cell turnover of the photobiont cell population of such thalli are controlled by the mycobiont, the regulatory mechanisms being unknown [21, 22, 30]. It is important to note that these photobiont cells are cut off from the external world and depend on water and nutrient supply by the mycobiont (Figs. 2, 10). In contrast to mycorrhizal fungi, which mobilize nutrients by invading large areas of soil, lichen mycobionts absorb key ions (ammonium, nitrate, phosphate) quite rapidly from rain water (atmospheric deposition, aeolic particles) [12]. Some of the peculiarities of these symbioses in comparison with ectomycorrhizae and leaf-inhabiting plant pathogens are summarized in Table 1.

3 Physiological Aspects of the Lichen Symbiosis

Lichens were the first fungus-plant symbiosis in which the chemical nature of the carbohydrates translocated from the plant cell to the fungal partner has been determined (the pioneering work of Smith and co-workers) by using the so-called "inhibition" or "isotope trapping technique" [47, 49, 50]. These experiments were designed on the basis of the following four assumptions:

—no haustoria are present in lichens;
—the apoplastic space of both photobiont and mycobiont is easily accessible for experimental studies;
—by incubating horizontally dissected thallus fragments (whose medullary layer was previously removed) on an incubation medium which contains carbohydrates the C-requirements of the mycobiont can be satisfied, and further uptake will be inhibited. Mobile photosynthates produced during a subsequent $^{14}CO_2$ pulse would therefore leak out of the apoplastic space and thus be detectable in the incubation medium.
—Until recently, water saturation was supposed to be a situation in which water would cover the wall surfaces of both symbionts within the thallus [35]; leakage fluids would therefore be easily recovered.

An acyclic polyol (its type depending upon the identity of the photobiont) was found to be the mobile carbohydrate in lichens with green algal photobionts [47, 49, 50]. *Trebouxia* spp. release ribitol, and they do produce the same photosynthate when kept under suitable conditions in axenic culture [23, 37, 43]. Translocation rates of

2–4% of total fixed C within 3 h, and 20–45% within 24h (under water saturation) were determined in the highest advanced, foliose lichens with internally stratified thallus and *Trebouxia* photobionts [47]. These very low and slow translocation rates are in contrast to ecophysiological data on these pokilohydrous symbioses (e.g., Lange and co-workers, [39]), which indicates that full photosynthetic activity is attained very rapidly after, and declines equally fast at the onset of a stress situation. Lichen mycobionts and photobionts of both arctic/alpine and xeric habitats have relatively low temperature optima, and at least sufficient air humidity is required for keeping them metabolically active. In extreme habitats such as deserts or arctic-alpine areas, photosynthesis can be performed only during a very short period of time every day [39; summaries: 36]. A 2–4% translocation of total fixed C within 3 h would be definitively insufficient to keep the system working under extreme conditions, since the photobiont cell population amounts distinctly less than 15% (usually about 10%) of the total biomass of internally stratified thalli.

4 Fine Structure of the Mycobiont-Photobiont Interface

Detailed investigations of the mycobiont-photobiont interface with combined light and electron microscope techniques have significantly improved our understanding of the biology of lichens in general, and of possible routes of translocation between the symbionts. The mycobiont-photobiont interface in internally stratified lecanoralean and teloschistalean lichens with *Trebouxia* photobiont was found to be quite different from what had been previously assumed (Table 2):

—all photobiont cells are contacted by one or several haustoria-like structures, the so-called intraparietal haustoria (summaries: [28, 31]) which had already been described as "intramembranous haustoria" in comparative LM investigations by Mameli [42], Tschermak [51] and Plessl [46]. Intraparietal haustoria may, at first sight, resemble appressoria of plant pathogens which are forming a small penetration peg (Fig. 10), but differ from those by their ontogeny and function. Intraparietal haustoria are usually formed by the tip or subapical area of a growing hypha which meets a juvenile algal cell (autospore which is still enclosed within the degrading mother cell wall; Fig. 4). Subsequently both symbionts grow and develop co-ordinately [26]. The photobiont cells may be shifted over short distances by intercalary growth processes of the haustorial apparatus (Fig. 5).

—The apoplastic space of neither symbiont is easily accessible in the medullary and algal layer since the wall surfaces of both are covered by a thin, mycobiont-derived, hydrophobic layer which is continuously spreading from the mycobiont over the wall surface of the photobiont at the contact site [25, 28, 31, 32] (Figs. 6, 8, 9). The degree of hydrophobicity of this wall surface layer varies considerably even within species of the same genus, depending on its chemical composition.

Table 2. Structural aspects of the mycobiont-photobiont relationship in highly advanced Lecanorales and Teloschistales with *Trebouxia* photobiont in comparison with ectomycorrhizae and leaf-inhabiting rusts

Lichens	Ectomycorrhizae [18, 3]	Leaf-inhabiting rusts [41]
(Foliose Lecanorales and Teloschistales with *Trebouxia* photobiont. > 1000 spp.)		
Symbiotic relationship has to be established between juvenile partners (tip area of growing hypha/autospores), co-ordinated development and growth of both symbionts at the interface	Symbiotic relationship has to be established between juvenile partners (growing hyphae/developing root)	Symbiosis can be established between growing hyphae and adult host cells
Each photobiont cell is contated by one or more intraparietal haustoria; shifting of the photobiont cells by intercalary growth processes of the haustorial complex	Wall-to-wall apposition (few intra-parietal fungal protrusions?) in the rhizodermal and cortical layers where the Hartig net is formed	Wall-to wall apposition, appressoria and intracellular haustoria in mesophyll cells (e.g., [45])
Sealing of each individual photobiont cell by a water-repellent, mycobiont-derived cell wall surface layer (important for the translocation of solutes between the symbionts and for facilitating the algal gas exchange)	Sealing of the mycorrhizal root by a conglutinate sheath, and of individual root cells by extracellular meterial released by the Hartig net	???
Mycobiont —Water, nutrients → (Passive, apoplastic)— Photobiont Mycobiont ← Ribitol (Passive or active?)— Photobiont	Mycobiont —Water, nutrients → Photobiont Mycobiont ← Carbohydrates— Photobiont	Mycobiont ← Water, nutrients— Photobiont Mycobiont ← Carbohydrates— Photobiont

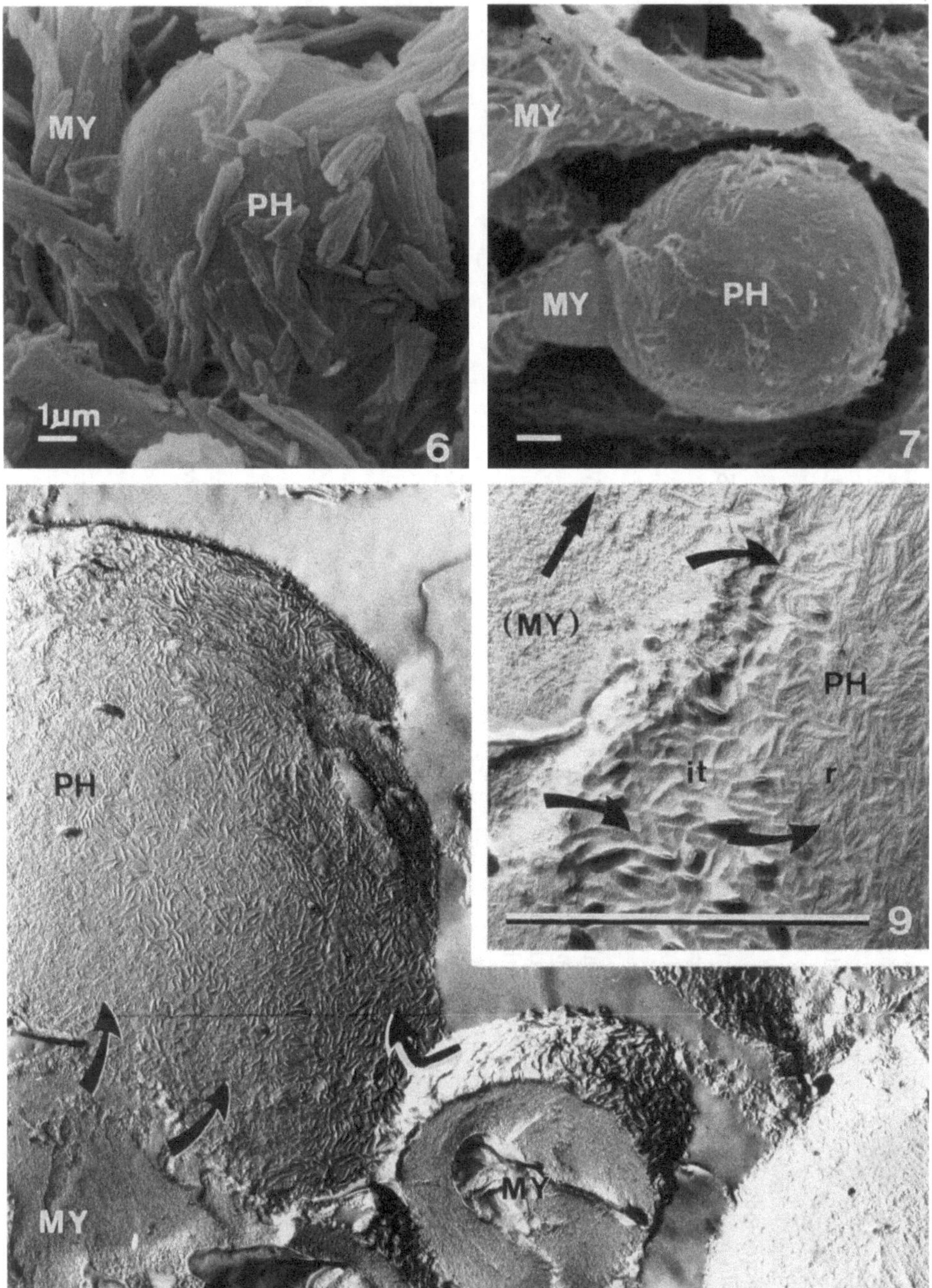
MY
PH
1µm
6
MY
MY
PH
7
(MY)
PH
it
r
9
PH
MY
MY
8

5 The Functional Morphology of the Mycobiont-Photobiont Interface

On the basis of combined ultrastructural and histochemical studies, translocation of water, nutrients, and fungal metabolites were postulated to occur apoplastically underneath the thin hydrophobic wall surface layer within the peripheral part of the fungal cell wall [26, 28, 29, 31]. The first LTSEM studies of a foliose, internally stratified lecanoralean species with *Trebouxia* photobiont in the water-saturated and dry state by A. Beckett [6] impressively support this view. No ice crystals were discovered on the cell walls of either symbiont within the algal and medullary layers indicating that these surfaces are dry even in the water-saturated state. Therefore it is likely that only a small amount of mobile carbohydrates (those which had been leaking out of cut or otherwise damaged cells) were detectable in the incubation medium in experiments using the inhibition technique. First ^{13}C NMR studies [40] indicated that translocation of polyols occurs faster than was previously calculated on the basis of inhibition technique experiments.

Mycobiont-derived, hydrophobic cell wall surface layers have been observed in numerous unrelated lichen taxa [26]. They seem to play a functional key role in the symbiotic relationship. The hyphae of the medullary and algal layers of lichens are aerial hypae developing within a gas-filled internal zone of the thallus underneath or even inside a structurally and functionally different, conglutinate peripheral zone, the cortical layer(s) (Figs. 2, 10). Like the aerial hyphae of numerous non-lichenized zygo-, asco-and basidiomycetes and conidial fungi the aerial hyphae of lichen mycobionts have a thin, outermost wall layer built up mainly by proteins and lipids (as determined with chemical methods in various types of conidia [5, 8,

◄

Figs. 6–9. Ultrastructural features of hydrophobic, mycobiont-derived cell wall surface layers in Lecanorales (Ascomycotina) with *Trebouxia* (Pleurastrales, Ulvophyceae) photobiont. **6–7** SEM of *Cetrelia olivetorum* (Nyl.)Culb. and Culb. **6** Freeze-dried (chemically untreated) specimen showing needle-shaped crystals of olivetoric acid [a *para*-depside (phenolic carboxylic acid derivative)], a secondary metabolite of the mycobiont (*MY*) embedded in a hydrophobic, mycobiont-deived cell wall surface layer which is also ensheathing the photobiont cells (*PH*). **7** Chemically fixed, dehydrated and critical point-dried specimen. The needle-shaped crystals of the phenolic secondary metabolite, and part of the lipidic fraction of the hydrophobic cell wall surface layer were lost during the preparative procedure; impressions of the (lost) crystals are visible in the proteinaceous fraction of the mycobiont-derived wall surface layer on fungal (*MY*) and algal (*PH*) cells. **8, 9** The mycobiont-photobiont interface in freeze-fracture preparations of chemically fixed, glycerinated specimens of *Cladonia macrophylla* (Schaerer) Stenh. **8** The walls of both mycobiont (*MY*) and photobiont (*PH*) are covered by the mycobiont-derived, highy hydrophobic surface layer, revealing an irregularly tessellated pattern. **9** Detail of the interface at the first contact of a growing hyphal tip with the surface of a juvenile *Trebouxia* (*PH*) cell. The tightly adhering hypha was torn away (*MY*; *large arrow* points towards the tip area); a rodlet layer (*r*) was initially spreading from the fungal hypha over the smooth (not visible) algal wall surface. Subsequently, the bundles of rodlets become obscured by an amorphous material which gives the wall surface layer an irregularly tessellated pattern (*it*). Preparative techniques see [29]. Bar = 1 μm in all micrographs

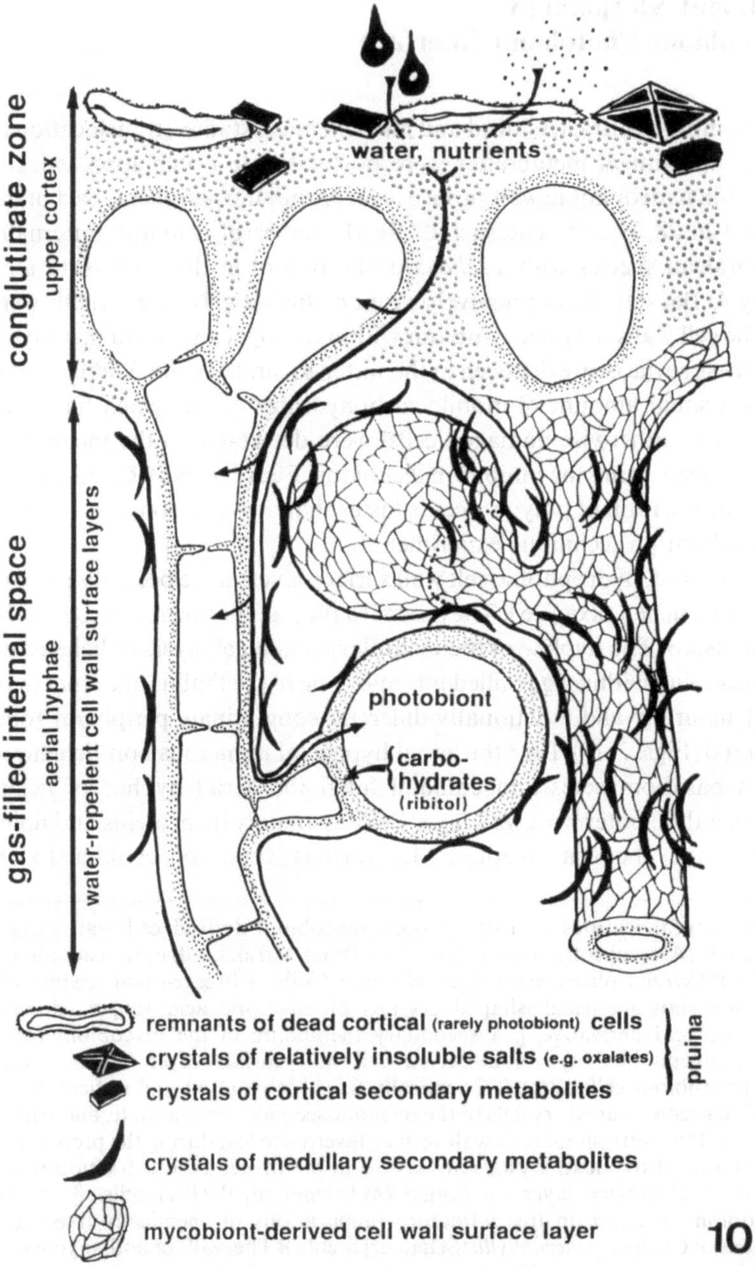

Fig. 10. Diagram illustrating the functional morphology of the mycobiont-photobiont interface in the internally stratified thalli of the Parmeliaceae (Lecanorales, Ascomycotina) with *Trebouxia* spp. (Pleurastrales, Ulvophyceae) as photobiont. Advanced Teloschistales (Ascomycotina) differ from Lecanorales by the lack of medullary secondary metabolites (the taxon-specific, bright yellow anthraquinones crystallize within the uppermost zone of the upper cortex), and by the slightly different external morphology of the haustorial complex. [25, 26, 28–32]

10, 11, 13, 19], and with histochemical techniques in lichen thalli [29, 31]). Some of the very peculiar phenolic secondary metabolites of lecanoralean lichen mycobionts crystallize within and on this wall surface layer and increase its hydrophobicity (Fig. 6) [29]. During dehydration in conventional preparative procedures for TEM and SEM these wall surface layers and adhering crystalline secondary metabolites are often partly or completely lost, depending on their chemical composition (Fig. 7). Best preservation during conventional preparative procedures (including chemical fixation and dehydration) was achieved in a lecanoralean specimen (*Cladonia caespiticia*) in which the wall surface layer is built up by insoluble protein-lipid complexes [28, 29, 31].

Although their chemical composition may vary considerably, these hydrophobic cell wall surface layers often reveal a distinct structure which can best be seen in freeze-fractured preparations. Irregularly arranged bundles of semi-crystalline rodlets were found, e.g., on hyphophores and conidia [2, 4, 5, 8, 9, 10, 11, 13, 19], on hyphae of the medullary and algal layers in peltigeralean lichen mycobionts [24, 25, 32], and in the thin pellicle which is spreading over the hymenial surface of the gills of basidiomycetes [44]. A rodlet layer which soon becomes covered by a smooth material (lipids and/or phenolics?), giving the surface layer an irregularly tessellated pattern was found in lecanoralean lichen mycobionts with internally stratified thallus and *Trebouxia* photobionts [25, 26, 29, 32] (Figs. 8, 9). This particular fungal wall surface layer spreads over the wall surface of the photobiont at the very first contact [25, 32]. Each individual photobiont cell is thus not only contacted by an intraparietal haustorium, but also sealed with a hydrophobic surface layer, an optimal situation with regard to the algal gas exchange on the one, and the canalization of the flow of solutes between the symbionts on the other hand.

Even in nonlichenized taxa, these particular wall surface layers seem to be somehow involved in water relations. Reduction of transpiratory losses of solutes from the substratum to the highly turgescent conidia-bearing cells might be their role in hyphophores. The wall surface of wind-dispersed conidia becomes hydrophobic due to this surface layer as demonstrated in a rodlet-layer-deficient mutant [4]. The rodlet layer of the pellicle on the gill surface of basidiomycetes is likely to facilitate the release of the basidiospores by forming a boundary layer between the fully turgescent tips of hymenial cells and the air space between the gills, thus keeping the hymenial surface dry (a sine qua non condition for successful basidiospore release).

Sealing of a cell or multicellular structure which donates required solutes occurs also in mycorrhizal symbioses (e.g., by a conglutinate sheath and extracellular material secreted by the Hartig net [3, 18], and in nonsymbiotic cell-cell interactions (e.g., at the pollen-stigma interface, where rapidly spreading oils of the pollen surface cover the cell surface of papillae, thus facilitating a rapid hydration of the pollen grain by canalizing the flow of solutes [15]). It could be that similar phenomena occur even in plant pathogenic associations, but there the possible routes and modes of translocation between the symbionts are distinctly more difficult to investigate than in the lichen symbiosis.

Wall surface layers which are rapidly spreading over suitable substrates are likely to be involved in some rather unspecific first adhesion phenomena which may precede subsequent specific cell-cell interactions. Examples are the rapidly spreading oils at the pollen-stigma interface [15], or the proteinaceous fraction of the erysiphalean conidial surface which is spreading over the host cuticle within half an hour [38]. This hypothesis is currently tested in axenic resyntheses of aposymbiotically cultured lichen mycobionts with a range of compatible and incompatible photobionts.

The mycobiont-photobiont interface in lichens shares many features with, but at the same time differs in many respects considerably from other fungus-plant cell interactions. Although lichens will hardly ever be considered as easy to handle systems for experimental studies on cell-cell recognition, they seem to be well suited for studying certain fundamental aspects related to translocation processes between fungal and plant host cells, since the mycobiont is the predominant partner and the mycobiont-photobiont interface is quite easily accessible for combined biochemical, histochemical, and ultrastructural studies.

References

1. Ahmadjian V (1982) Algal/fungal symbiosis. In: Round F, Chapman DJ (eds) Progress in phycological research. Elsevier, Amsterdam, Vol 1, pp 179–233
2. Aronson JA (1981) Cell wall chemistry, ultrastructure and metabolism. In: Cole GT, Kendrick B (eds) Biology of conidial fungi, Vol 2. Academic Press, New York, pp 459–507
3. Ashford AE, Peterson CA, Carpenter JL, Cairney JWG, Allayway WG (1988) Structure and permeability of the fungal sheath in the *Pisonia* mycorrhiza. Protoplasma 147: 149–161
4. Beever RE, Dempsey GM (1978) Function of rodlets on the surface of fungal spores. Nature 272: 608–610
5. Beever RE, Redgwell RJ, Dempsey GM (1979) Purification and chemical characterization of the rodlet layer of *Neurospora crassa* conidia. J Bacteriol 140: 1063–1070
6. Brown DH, Rapsch S, Beckett A, Ascaso C (1987) The effect of desiccation on cell shape in the lichen *Parmelia sulcata* Taylor. New Phytol 105: 295–299
7. Brunner U, Honegger R (1985) Chemical and ultrastructural studies on the distribution of sporopollenin-like biopolymers in six genera of lichen phycobionts. Can J Bot 63: 2221–2230
8. Cole GT, Pope LM (1981) Surface wall components of *Aspergillus niger* conidia. In: Turian G, Hohl HR (eds) The fungal spore: morphogenetic controls. Academic Press, London, pp 195–215
9. Cole GT, Samson RA (1979) Patterns of development in conidial fungi. Pitman, London
10. Cole GT, Sekiya T, Kasai R, Yokoyama T, Nozawa Y (1979) Surface ultrastructure and chemical composition of the cell walls of conidial fungi. Exp Mycol 3: 132–156
11. Cole GT, Pope LM, Huppert M, Sun SH, Starr P (1983) Ultrastructure and composition of conidial wall fractions of *Coccidioides immitis*. Exp Mycol 7: 297–318
12. Crittenden PD (1989) Nitrogen relations of mat-forming lichens. In: Boddy L, Marchant R, Read DJ (eds) Nitrogen, phosphorus and sulphur utilization by fungi. Cambridge University Press, pp 243–268
13. Dempsey GP, Beever RE (1979) Electron microscopy of the rodlet layer of *Neurospora crassa* conidia. J Bacteriol 140: 1050–1062

14. Duddridge J, Read DJ (1982) An ultrastructural analysis of the development of mycorrhizas in *Monotropa hypopitys* L. New Phytol 92: 203–214
15. Ferrari TE, Best V, More TA, Comstock P, Muhammad A, Wallace H (1985) Intercellular adhesions in the pollen-stigma system: pollen capture, grain binding and tube attachments. Am J Bot 72: 1466–1474
16. Galun M, Bubrick P (1984) Physiological interactions between the partners of the lichen symbiosis. In: Linskens HF, Heslop-Harrison J (eds) Cellular interactions. Encyclopedia of Plant Physiology. Springer, Berlin Heidelberg New York Tokyo, pp 362–401
17. Gärtner G (1985) Die Gattung *Trebouxia* PUYMALY (Chlorellales, Chlorophyceae). Arch Hydrobiol Suppl 71: 495–548
18. Harley JL, Smith SE (1983) Mycorrhizal symbiosis. Academic Press, London
19. Hashimoto T, Wu-Yuan CD, Blumenthal HD (1976) Isolation and characterization of the rodlet layer of *Trichophyton mentagrophytes* microconidial wall. J Bacteriol 127: 1543–1549
20. Hawksworth DL (1988) The fungal partner. In: Galun M (ed) Handbook of lichenology. CRC, Boca Raton FL, Vol 1, pp 35–38
21. Hill DJ (1985) Changes in photobiont dimensions and numbers during co-development of lichen symbionts. In: Brown DH (ed) Lichen physiology and cell biology. Plenum, New York, pp 303–317
22. Hill DJ (1989) The control of the cell cycle in microbial symbionts. New Phytol 112: 175–184
23. Hill DJ, Ahmadjian V (1972) Relationship between carbohydrate movement and the symbiosis in lichens with green algae. Planta 103: 267–277
24. Honegger R (1982) Cytological aspects of the triple symbiosis in *Peltigera aphthosa.* J Hattori Bot Lab 52: 379–391
25. Honegger R (1984) Cytological aspects of the mycobiont-phycobiont relationship in lichens. Haustorial types, phycobiont cell wall types, and the ultrastructure of the cell wall surface layers in some cultured and symbiotic myco- and phycobionts. Lichenologist 16: 111–127
26. Honegger R (1985a) Fine structure of different types of symbiotic relationships in lichens. In: Brown DH (ed) Lichen physiology and cell biology. Plenum, New York, pp 287–302
27. Honegger R (1985b) Scanning electron microscopy of the fungus-plant cell interface: a simple preparative technique. Trans Br Mycol Soc 84: 530–533
28. Honegger R (1986a) Ultrastructural studies in lichens. I. Haustorial types and their frequencies in a range of lichens with trebouxioid phycobionts. New Phytol 103: 785–795
29. Honegger R (1986b) Ultrastructural studies in lichens II. Mycobiont and photobiont cell wall surface layers and adhering crystalline lichen products in four Parmeliaceae. New Phytol 103: 797–808
30. Honegger R (1987) Questions about pattern formation in the algal layer of lichens with stratified (heteromerous) thalli. Bibl Lichenol 25: 59–71
31. Honegger R (1988) The functional morphology of cell-to-cell interactions in lichens. In: Scannerini S, Smith DC, Bonfante-Fasolo P, Gianinazzi-Pearson V (eds) Cell to cell signals in plant animal and microbial symbiosis. NATO ASI series, Vol H17 Springer, Berlin Heidelberg New York Tokyo, pp 39–53
32. Honegger R (1990) Surface interactions in lichens. In: Wiessner W, Robinson DG, Starr RC (eds) Experimental Phycology, Vol. 1. Cell walls and surfaces, Reproduction, Photosynthesis. Springer, Berlin Heidelberg New York Tokyo pp. 40–54
33. Honegger R, Brunner U (1981) Sporopollenin in the cell walls of *Coccomyxa* and *Myrmecia* phycobionts of various lichens: an ultrastructural and chemical investigation. Can J Bot 59: 2713–2734
34. Irvine DEG, John DM (eds) (1984) Systematics of the green algae. Academic Press, London
35. Jahns HM (1984) Morphology, Reproduction and water relations—a system of morphogenetic interactions in *Parmelia saxatilis.* Nova Hedwigia Beih 79: 715–737
36. Kershaw KA (1985) Physiological ecology of lichens. Cambridge University Press

37. Komyia T, Shibata S (1971) Polyols produced by the cultured phyco- and mycobionts of some *Ramalina* species. Phytochemistry 10: 695–699
38. Kunoh H, Yamaoka N, Yoshioka H, Nicholson R (1988) Preparation of the infection court by *Erysiphe graminis* I. Contact-mediated changes in morphology of the conidium surface. Exp Mycol 12: 325–335
39. Lange O (1969) Die funktionelle Anpassung der Flechten an die ökologischen Bedingungen arider Gebiete. Ber Dtsch Bot Ges 82: 3–22
40. Lines CEM, Ratcliffe RG, Rees TAV, Southon TE (1989) A ^{13}C NMR study of photosynthate transport and metabolism in the lichen *Xanthoria calcicola* Oxner. New Phytol 111: 447–456
41. Littlefield LJ (1981) Biology of the plant rusts. Iowa State University Press, Ames
42. Mameli E (1920) Note critiche ad alcune moderne teorie sulla natura del consortio lichenico. Atti Ist Bot Univ Pavia NS 17: 209–227
43. Maruo B, Hattori T, Takahashi H (1965) Excretion of ribitol and sucrose by green algae into the culture medium. Agric Biol Chem 29: 1084–1089
44. McLaughlin DJ (1982) Basidial and basidiospore development. In: Wells K, Wells E (eds) Basidium and basidiocarp. Springer, Berlin Heidelberg New York, pp 37–74
45. Mendgen K (1978) Der Infektionsverlauf von *Uromyces phaseoli* bei anfälligen und resistenten Bohnensorten. Phytopathol Z 93: 295–313
46. Plessl A (1963) Ueber die Beziehunge von Pilz und Alge im Flechtenthallus. Oesterr Bot Z 110: 194–269
47. Richardson DHS, Hill DJ, Smith DC (1968) Lichen Physiology XI. The role of the alga in determining the pattern of carbohydrate movement between lichen symbionts. New Phytol 67: 469–486
48. Sluiman HJ (1989) The green algal class Ulvophyceae. An ultrastructural survey and classification. Crypt Bot 1: 83–94
49. Smith DC (1978) What can lichens tell us about real fungi? Mycologia 70: 915–934
50. Smith DC, Douglas A (1987) The biology of symbiosis. Edward Arnold, London
51. Tschermak E (1941) Untersuchungen über die Beziehungen von Pilz und Alge im Flechtenthallus. Oesterr Bot Z 90: 233–307
52. Tschermak-Woess E (1988) The algal partner. In Galun M (ed) Handbook of lichenology, Vol 1. CRC Boca Raton FL, pp 39–92

Chapter 22 BURTON Y. ENDO

Ultrastructure of Nematode-Plant Interactions

1 Introduction

The light microscope was the foundation for our basic concepts of host-parasite relationships of plant parasitic nematodes in crop plants. The electron microscope has extended our knowledge of the feeding apparatus of the nematodes and its relation to secretions from esophageal glands and host response during parasitism. Much of the progress in studying host-parasite interactions is due to increased resolution of transmission and scanning electron microscopes and improved technologies of fixation and embedment of both plant and animal tissues. This chapter summarizes portions of this research in ultrastructure directed to nematode anatomy, its relation to host penetration, nematode development during feeding, and host response.

2 The Sensory System of Nematodes

In a discussion of nematode-plant interactions, it is essential to consider the anterior sensory organs of nematodes as they relate to host penetration and establishment of a feeding site. Basic anatomy and arrangement of nematode sensory organs were first described in large specimens such as *Ascaris* spp. [31, 35] and various other animal and plant parasitic species [10, 14]. It was only as recently as 1965 that nematodes were shown to have cilia in their sensory organs [34]. Since that time, cilia have been described in both animal and plant parasitic forms. Among the intensive efforts to describe sensory organs were the observations of the anterior parts of males of the root-knot nematode, *Meloidogyne incognita*, and the soybean cyst nematode, *Heterodera glycines* [1–4]. During this period, significant strides

---

U.S. Dept. of Agriculture, ARS Beltsville Agricultural Research Center, Plant Sciences Institute, Nematology Laboratory, BARC-West, Room 138, Bioscience Bldg. 011A Beltsville, MD 20705, USA

were made to utilize the saprophytic nematode, *Caenorhabditis elegans*, as a basic tool for studying biological systems. As part of the effort, the anterior sensory organs of this organism were described [70]. In a comprehensive review, the ultrastructure of sensory organs of both plant and animal parasitic forms were described and compared [49]. A major study of 22 species and 19 genera of plant parasites revealed the functional morphology of sensory organs [15]. Based on these and related observations [50], a description of sensory organ morphology and arrangement was developed, utilizing new data on the ultrastructure of sensilla [12, 13].

A possible role of anterior sensory organs and feeding sites was evident in the formation of feeding plugs derived from exudations that originated in the amphidial duct. A nerve process associated with the amphid and described in juveniles of the root-knot nematode [28, 29, 72] and soybean cyst nematode [22], is directly associated with the electron-dense mass surrounding cilia within the amphidial duct. Amphidial secretions apparently flow and solidify to plug the host cell opening after stylet removal during feeding pauses and at molt [21, 26, 64, 85].

3 Host Reactions with Ectoparasitic Species

Sensory organs apparently play a role in exploration and subsequent feeding activities of ectoparasitic nematodes. The exploration mode is apparent in observations of *Tylenchorhynchus dubius* in its contact with epidermal cells and root hairs of seedlings of *Brassica rapa* var. *silvestris*. By utilizing cine-film, feeding activity of *T. dubius* was compared to feeding activities of *Trichodorus similis* on *Nicotiana tabacum* seedlings. In cells of the host, a massive aggregation of cytoplasm and associated nuclear changes in roots and sudden death of root hairs was discussed in terms of the inefficiency of *T. similis* as a virus vector [74]. Subsequently, a feeding tube is formed from hardened saliva that enters the host cell and is anchored in place by a small plug just inside the perforated cell wall [83].

The feeding behavior and pathogenicity of *Xiphinema index* on grape roots showed that *X. index* prefers the elongation zone of growing grape roots. Histologically, the swollen tips of grapevine roots fed upon by *X. index* show a collapsed outer cortical layer under which lie the hypertrophied multinucleate cells stimulated by nematode saliva. The hypertrophied cells are filled with dense cytoplasm, sometimes devoid of vacuoles [71]. Observations of swollen root tips of *Ficus carica* 1 day after feeding by *X. index* showed necrotic cells, either singly or in small groups. These cells appear similar to cells exhibiting a hypersensitive reaction and are the cells fed upon by the nematode [86]. Each necrotic cell is surrounded by several enlarged binucleate cells with dense cytoplasm. A day later, binucleate cells become multinucleate, each having four to eight nuclei [84]. These multinucleate cells are considered to be induced by repeated mitosis without cytokinesis [78], thus resembling giant cells induced by *Meloidogyne* [36]. Wyss described, with film

[77, 86], the progressive stages of stylet insertion, feeding, and movement of *Xiphinema* into deeper-lying cells of the host meristem of *Ficus carica*. This activity, together with the stimulation of surrounding cells, provides optimal conditions for nematode feeding. After maximum protrusion, the stylet is retracted and re-entry occurs at another site of the galled root tip [79].

4 Host Reactions with Endoparasitic Species

4.1 Endoparasitic Nematodes in Compatible Hosts

Among plant parasitic nematodes, the root-knot nematode has been a subject of extensive study in terms of ultrastructural anatomy/host response in the form of tissue hyperplasia, hypertrophy, and the induction of giant cells upon which they feed [27]. Once established at the feeding site, the root-knot nematode stimulates cells to become giant cells. The morphology and physiology of these giant cells have been examined and studied extensively [5–7, 11, 16, 19, 20, 39, 40–42, 46, 52, 53, 69, 73]. The root-knot nematode induction of giant cells in the broadbean, *Vicia faba*, results in a geometric increase of chromosome numbers from 24 to 384 within giant cell nuclei; these giant cells are formed by endomitosis without cytokinesis [36, 37]. However, Bird [6] found that cell wall breakdown occurs during giant cell formation in *Vicia faba* infected with *Meloidogyne incognita*. It was noted that in epoxy embedded tissues, cytoplasmic continuity exists between the syncytium and adjacent cells. Thus it was hypothesized that syncytia (giant cells) are formed partly by the incorporation of cells whose walls were dissolved rather than by expansion of single cells. Observation by Jones and Payne [46], utilizing light and electron microscopy, re-emphasized the concept proposed by Huang and Maggenti [36, 37] that cell wall breakdown is not evident during the induction of giant cells. The first signs of giant cell formation in *Impatiens balsamina* by *Meloidogyne javanica* was the division of cells surrounding the juvenile. In the cell being fed upon by the nematode, cell plate alignment took place but cytokinesis was not successful and the result was the formation of binucleate cells. Subsequently, the numbers of nuclei in each giant cell appeared to increase by repeated mitosis with the absence of cytokinesis.

Electron micrographs revealed the proliferation of the cell walls of giant cells and syncytia in the form of plasmalemma-covered ingrowths or protuberances [36, 37, 40, 43, 44, 55]. Jones and Northcote [43] observed that these plasmalemma-associated cell wall protuberances form extensively on the walls of giant cells lying adjacent to xylem vessels and to a lesser extent next to sieve elements. The presence of protuberances opposite xylem elements and their analogy to transfer cells found in normal plant tissues led to the conclusion that giant cells and syncytia are modified transfer cells and probably account for selective flow of nutrients across plasma membranes that separate the plant vascular components and the giant cells or syncytia upon which nematodes feed and gain nutrients [32, 44, 54].

4.2 Resistant Host Reactions

Root-Knot Nematode. In contrast to the induction of giant cells in susceptible host tissue, *M. incognita* infections in resistant host cells of tomato, *Lycopersicon esculentum*, result in stimulated host cells having dense cytoplasm characterized by large numbers of ribosomes, proliferated endoplasmic reticulum, and increased stainability of cytoplasmic ground substance. Giant cell development is generally suppressed by the hypersensitive reaction [55]. In a related study, Bleve-Zacheo et al. [9] observed the ultrastructural changes in the tomato cultivar Brech after inoculation with juveniles of *M. incognita.* Similar to reactions in the cultivar Nematex [55] in the cultivar Brech a hypersensitive response results in necrotic cells near the nematode and giant cell suppression. Callose deposits were thought to form in response to nematode movement through the tissues [9].

Soybean Cyst Nematode. The discovery of Peking as a source of resistance to the soybean cyst nematode provided opportunities to study host responses to nematode infection with emphasis on resistant reactions. Initially, Ross [63] emphasized the hypersensitive reaction of Peking to invasion by the soybean cyst nematode and noted the absence of syncytia. However, later investigations showed that syncytia are induced by the soybean cyst nematode in resistant cultivars in much the same manner as observed in susceptible cultivars. The main difference is the early deterioration of the syncytium and the subsequent termination of feeding by the nematode [17, 18, 60]. The fine structure of the host-parasite relations of the soybean cyst nematode is consistent with the responses observed with light microscopy [30, 60]. Syncytia observed at 42 h after inoculation had cell walls with many perforations and cytoplasm with numerous plastids and mitochondria among proliferated endoplasmic reticulum. Within 4 days after inoculation, syncytia in these resistant roots began to deteriorate and cell wall thickenings formed. Secondary wall deposits tended to wall off the diseased areas. It was proposed that these depositions could obstruct the plasmadesmata of the syncytial walls to prevent the cell to cell movement of nutrients from the host vascular system to the syncytium. The result would be a food deficiency for the nematode or a buildup of toxic byproducts within the syncytium [60]. Superimposed on the walling-off and obstruction of plasmadesmata shown in Peking is the reaction found in the resistant cultivar Forrest, in which a necrotic layer of cells forms around the syncytium and produces a necrotic boundary between the syncytium and normal host cells. Another resistant cultivar, Bedford, differs from Forrest in that nuclear degeneration precedes cytoplasmic degradation. The chromatin is scattered throughout the cytoplasm [47].

Potato Cyst Nematode. Resistant host reaction of potato cultivars derived from *Solanum tuberosum* spp. *andigenum* to infection by the potato cyst nematode, *Globodera rostochiensis*, consisted of a hypersensitive-like reaction [59]. The cell walls of the inner cortex and stele adjacent to the invading nematode break down to form a syncytial complex, but the cells surrounding the syncytium become necrotic and are presumed to limit nutrient transfer from the plant to the nematode. The resistance mechanism appears different in cultivars derived from *Solanum vernei*. After inoculation by *Globodera pallida* and *G. rostochiensis*, the *S. vernei*-derived cultivars show a hypersensitive reaction in the cortex in the immediate vicinity of the nematode body. After movement through the cortex, the juvenile comes to rest in the inner cortex and initiates a syncytium.

The hypersensitive type of reaction around the body of the juvenile does not appear to inhibit initiation of feeding sites by the nematode and may not be part of the resistance mechanism. In contrast, the changes occurring after initiation of the feeding sites can be directly implicated in the resistance reaction. Thus transfer cell development is inhibited and the developing nematode is deprived of food or provided with a depleted food supply. The author suggest that two distinct resistance processes exist—the first directly elicited by an external property of the juvenile and the second by a salivary component of the nematode that is introduced into the cytoplasm at the feeding site. It was further suggested that the hypersensitive response enclosing the syncytium may result either from a component of the juvenile's saliva directly inducing the resistant reaction or from salivary component(s) inducing processes which eventually alter cell surfaces in such a way as to make adjacent cells react defensively. The latter concept is preferred by the authors [59].

5 Feeding Site Adaptations

5.1 Feeding Plugs

During the induction of syncytia in soybean roots infected by the soybean cyst nematode, a feeding plug is formed primarily from secretions of the amphid. The plug occurs between the stylet and outer wall of the initial syncytial cell and remains in place during the nematode secretion and food ingestion processes (Figs. 1, 2, 5). The plug remains in place after stylet withdrawal at the time of feeding and during molt and also appears to maintain the integrity of the syncytial contents during the feeding and molting periods [21]. This same type of plug formation occurs in both susceptible and resistant hosts infected by other species of cyst nematodes [64, 85].

5.2 Feeding Tubes

An integral part of the host-parasite interaction of endoparasitic nematodes are the secretions (Figs. 1, 2) synthesized in the dorsal gland and accumulated in the dorsal

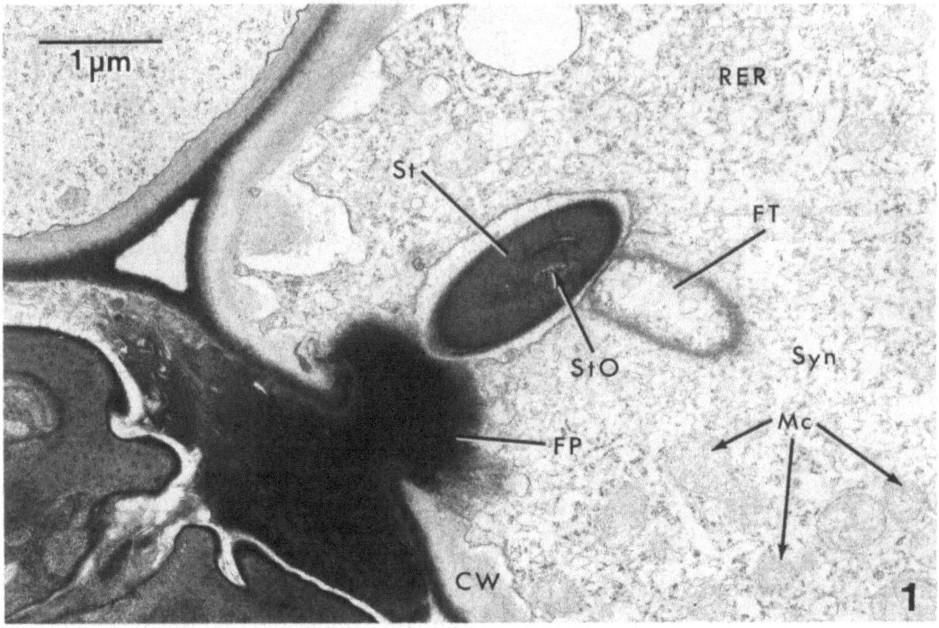

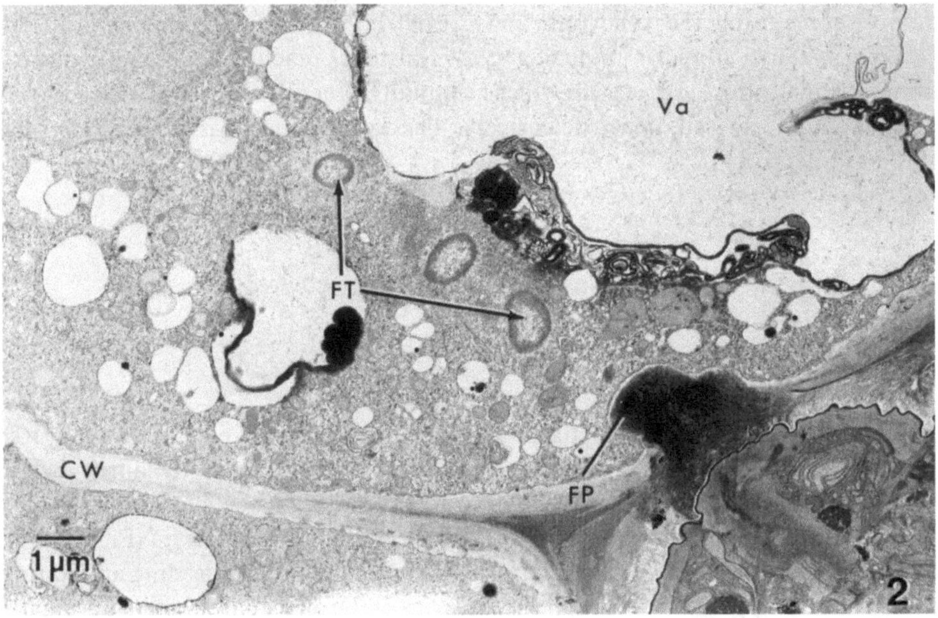

Figs. 1 and 2. 1 Longitudinal section through the lip region of a nematode 3 days after inoculation. Close to the stylet tip (*St*), cut tangentially, is a feeding tube (*FT*) with a filamentous internal matrix. The cytoplasm of the initiated syncytium (*Syn*) contains large concentrations of rough endoplasmic reticulum (*RER*) and mitochondria (*Mc*). The cell wall (*CW*) near the feeding site shows signs of proliferation. *FP* feeding plug; *StO* stylet orifice. [26]. **2** Section of syncytium stimulated by nematode. Series of ovate secretion structures in the cytoplasm that may be oblique sections through a sinuous feeding tube (*FT*) extending into the host cell. The first cell that was stimulated shows evidence of accumulation of dense cytoplasm with many ribosomes that are free or associated with ER, plastids, and mitochondria. Size of vacuoles (*Va*) depends on the cytoplasmic density. *CW* cell wall; *FP* feeding plug [26]

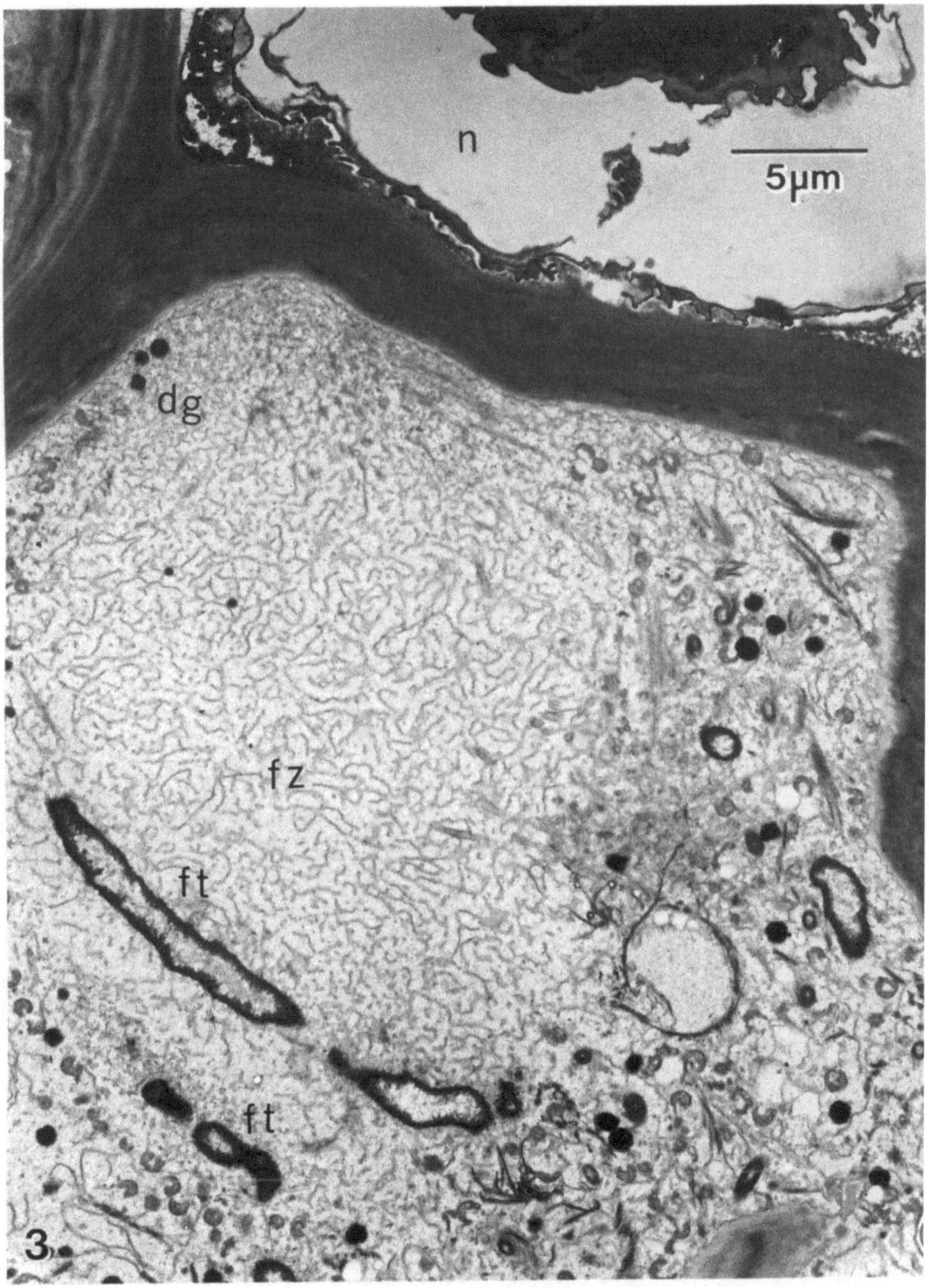

Fig. 3. Electron micrograph of the feeding site of *G. pallida* (*n*) in susceptible potato root. Feeding zone (*fz*) with uniformly smooth endoplasmic reticulum nearly free from cell organelles surrounded by electron-dense globules (*dg*). Feeding tubes (*ft*) present in feeding zone and in cytoplasm rich in organelles [64]

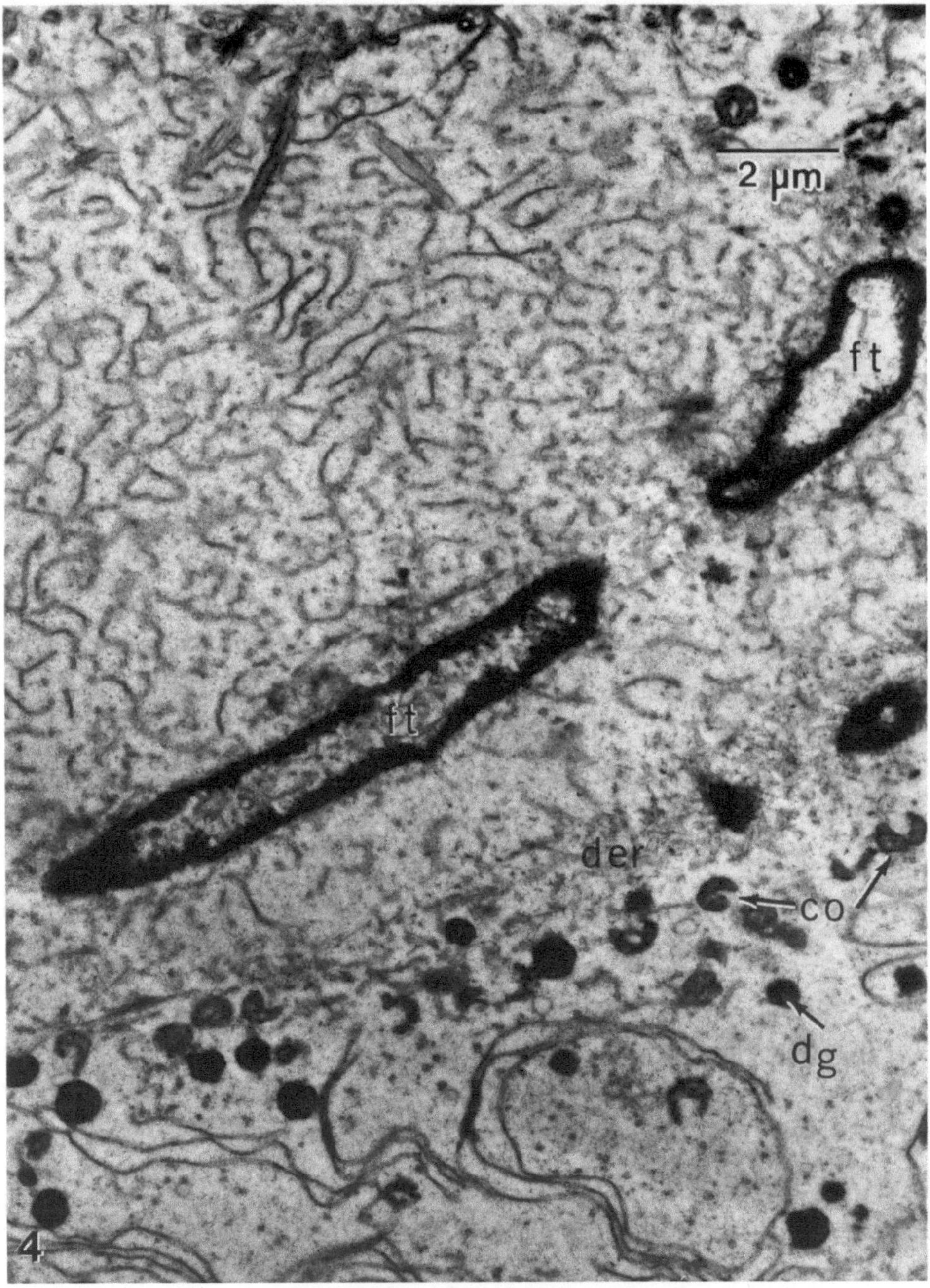

Fig. 4. Same section as Fig. 3 at higher magnification. Upper feeding tube with host ER passing through the tube, lower feeding tube surrounded by dense tubular ER (*der*); *dg* electron-dense globules; *co* cup-shaped organelles. [64]

gland ampulla before being released into the host cell as saliva [24, 26, 38, 67]. These secretions are probably the material that Nemec [51] observed in sugar beet roots infected with *Heterodera schachtii.* The filament-like proteinaceous inclusions were first thought to be mitochondria but later termed filaments. Recent investigations utilizing interference-contrast light microscopy have re-emphasized the filamentous nature of nematode secretions in syncytia. These secretions have now been termed feeding tubes and have been observed in cyst and root-knot infection and in a special continuous tubular form by reniform nematodes. Among the species that have been described to form feeding tubes in host cells are the cyst nematodes *Globodera rostochiensis, G. pallida, Heterodera schachtii, H. trifolii;* root-knot nematodes *Meloidogyne hapla* and *M. incognita* [64]; and the reniform nematode, *Rotylenchulus reniformis* [56, 58]. It was hypothesized that the feeding tube (Figs. 3, 4) was a duct into which the plant cell secreted special products which were then directed into a feeding ampulla. Thus by feeding on the contents of an ampulla within the host, the nematode was assumed to withdraw secreted products instead of regular cytoplasmic components [64]. However, the ultrastructure of the feeding site of *H. schachtii* in radish roots, *Raphanus sativus* L. var. *oleiformis*, showed direct contact between the feeding tube and the aperture of the stylet cone terminus. It was concluded that the feeding tubes are formed by hardened saliva secreted through the stylet aperture of infective juveniles during interaction with the host cytoplasm [85]. This concept has been further reinforced by video-enhanced light microscopy [81, 82]. Nutrients from the host were observed to be drawn through a feeding tube that extends into a zone of modified cytoplasm. This modified cytoplasmic zone is free of organelles and has fewer ribosomes and polyribosomes than the surrounding cytoplasm [85].

Ultrastructure of the esophagi of *Aphelenchoides blastophthorus, Ditylenchus dipsaci* and *Heterodera glycines* [23, 24, 65, 66] complemented video-enhanced light microscope observations of *H. schachtii* in rape seedlings [81, 82]. These comparisons provided an excellent base for study of the structure and function of *H. schachtii* during its feeding activity [80, 82].

5.3 Feeding Pegs

The feeding site and feeding tube of the endoparasitic species, *Rotylenchulus reniformis,* differs from those formed by cyst nematodes. Fundamental to this difference is that *R. reniformis* establishes its feeding site as an adult whereas cyst nematodes initiate syncytia as J_2 juveniles and undergo three molts while at the feeding site. Since *R. reniformis* does not molt after establishment at the feeding site, the stylet is held in place by a feeding peg (Fig. 5) consisting of cell wall material [57, 58]. The single helical feeding tube extending from the stylet and into the feeding cell is thought to act as a filter through which the nematode could derive solutes without passage of host cell particles that might block the stylet lumen [56].

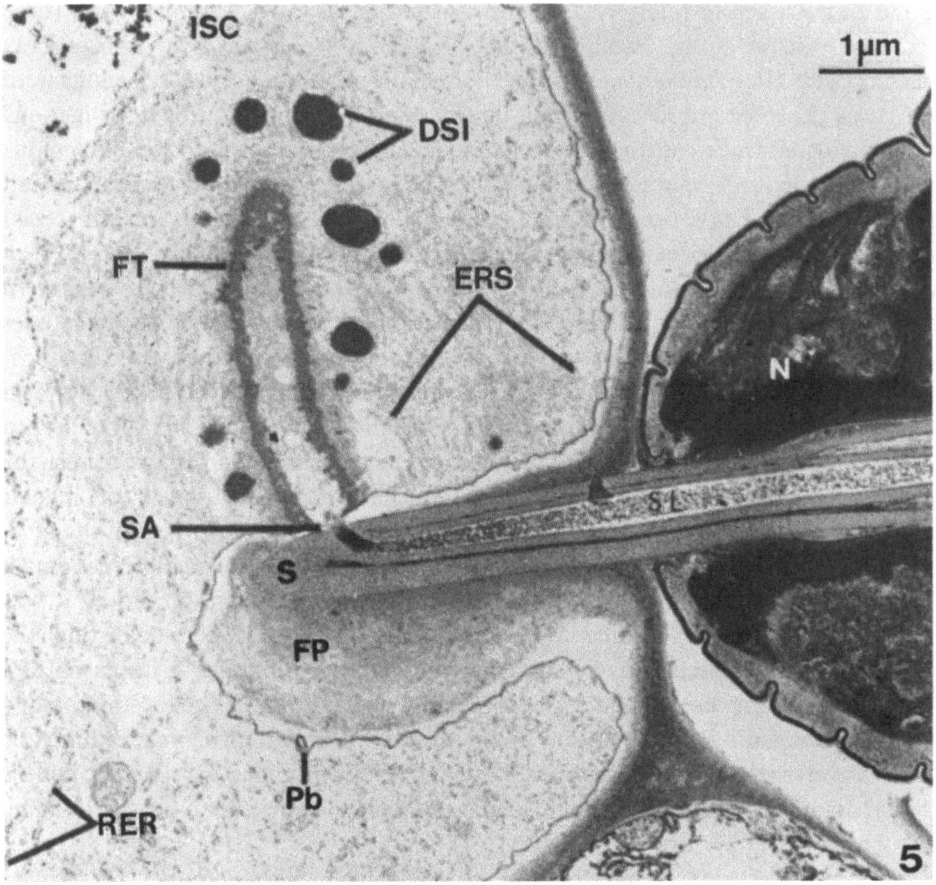

Fig. 5. Cross-section through cotton root cv. Auburn 56 3 days after inoculation with *R. reniformis*. Swirls of endoplasmic reticulum (*ERS*) are associated with the feeding tube (*FT*) near the stylet aperture (*SA*) side of the peg (*FP*). The central vacuole has been replaced by accumulations of endoplasmic reticulum, ribosomes, many small vacuoles, dense spherical inclusions (*DSI*), and other organelles. Portion of rough endoplasmic reticulum (*RER*) is scattered through the cytoplasm of the host. Initial syncytial cell (*ISC*); paramural bodies (*Pb*); stylet (*S*); stylet lumen (*SL*). [57]

6 Nematodes as Virus Vectors

The first report on the role of nematodes as vectors of viruses was made by Hewitt et al. [33], who demonstrated that the virus causing fan leaf symptoms and degenerative disease of grape was transmitted by *Xiphinema index*. Since that time, a wide range of viruses have been found to be transmitted by four genera of ectoparasitic nematodes, *Xiphinema*, *Longidorus*, *Trichodorus*, and *Paratrichodorus*. Reviews of nematode-virus interactions [48, 68] show their extensive nature. The virus retention sites of *Xiphinema* and Trichodorus are similar in that both show virus adsorption to surfaces of the esophagus. In contrast, viruses are adsorbed to the odontostyle and guiding ring of Longidorus [68]. Structural features of the feeding apparatus of virus-transmitting, trichodorid, and longidorid nematodes were described in relation to (a) cell wall exploration, (b) cell wall perforation, (c) salivation, (d) ingestion, and (e) departure from the feeding site [61, 62].

7 Conclusion

Significant progress has been made in describing the host-parasite interactions of nematodes and plants through histology and ultrastructure. Several concepts have been projected to show the nature of host susceptibility and resistance. However, there is a paucity of information on the chemical, structural, and physiological factors that control these reactions. Among the challenges in nematology is the identification of the chemical and biological properties of secretions synthesized by plant parasitic nematodes and their interactions with host tissues. Current technology of immunochemistry and labeling techniques, with heavy metal compounds such as immunogold, provide opportunities to identify the sites of synthesis and movement of secretions from the esophageal glands, and their interaction with host cells components. The increased resolution of video-enhanced interference contrast light microscopy and the electron microscope provides new opportunities for studying the host-parasite interactions of nematode-infected plants.

References

1. Baldwin JG, Hirschmann H (1973) Fine structure of cephalic sense organs in *Meloidogyne incognita* males. J Nematol 5: 285–302
2. Baldwin JG, Hirschmann H (1975) Fine structure of cephalic sense organs in *Heterodera glycines* males. J Nematol 7: 40–53
3. Baldwin JG, Hirschmann H (1976) Comparative fine structure of the stomatal region of males of *Meloidogyne incognita* and *Heterodera glycines*. J Nematol 8: 1–17
4. Baldwin JG, Hirschmann H, Triantaphyllou AC (1977) Comparative fine structure of the esophagus of males of *Heterodera glycines* and *Meloidogyne incognita*. Nematologica 23: 239–252
5. Bird AF (1968) Changes associated with parasitism in nematodes. 4. Cytochemical studies on the ampulla of the dorsal esophageal gland of *Meloidogyne javanica* and on exudations from the buccal stylet. J Parasitol 54: 879–890
6. Bird AF (1972) Cell wall breakdown during the formation of syncytia induced in plants by root-knot nematodes. Int J Parasitol 2: 431–432
7. Bird AF (1973) Observations on chromosomes and nucleoli in syncytia induced by *Meloidogyne javanica*. Physiol Plant Pathol 3: 387–391
8. Bird AF (1974) Plant response to root-knot nematode. Annu Rev Phytopathol 12: 69–85

9. Bleve-Zacheo T, Zacheo G, Melillo MT, Lamberti F (1982) Ultrastructural aspects of the hypersensitive reaction in tomato root cells resistant to *Meloidogyne incognita.* Nematol Mediterr 10: 81–90
10. Chitwood BG, Chitwood MB (1974) An introduction to nematology. University Park Press, Baltimore, Maryland, pp 334
11. Christie JR (1936) The development of root-knot nematode galls. Phytopathology 26: 1–22
12. Coomans A (1979) The anterior sensilla of nematodes. Rev Nématol 2: 259–283
13. Coomans A, De Grisse A (1981) Sensory structures. In: Zuckerman BM, Rohde RA (eds) Plant parasitic nematodes. Vol III, Academic Press, New York, pp 127–174
14. De Coninck LAP (1965) In: Grasse PO (ed) Traité de Zoologie, Tome IV, Fasc. II. Masson, Paris, France, pp 9
15. De Grisse A (1977) The ultrastructure of the nerve system in the head of 22 species of plant parasitic nematodes belonging to 19 genera (Nematoda: Tylenchida) (in Dutch). D.Sc. Thesis, University of Gent, Belgium
16. Dropkin VH (1969) Cellular responses of plants to nematode infections. Annu Rev Phytopathol 7: 101–122
17. Endo BY (1964) Penetration and development of *Heterodera glycines* in soybean roots and related anatomical changes. Phytopathology 54: 79–88
18. Endo BY (1965) Histological responses of resistant and susceptible soybean varieties, and backcross progeny to entry and development of *Heterodera glycines.* Phytopathology 55: 375–381
19. Endo BY (1971) Nematode-induced syncytia (giant cells). Host-parasite relationships of Heteroderidae. In: Zuckerman BM, Mai WF, Rhode RA (eds) Plant parasitic nematodes, Vol 2. Academic Press, New York, pp 91–117
20. Endo BY (1975) Pathogenesis of nematode-infected plants. Annu Rev Phytopathol 13: 213–238
21. Endo BY (1978) Feeding plug formation in soybean roots infected with the soybean cyst nematode. Phytopathology 68: 1022–1031
22. Endo BY (1980) Ultrastructure of the anterior neurosensory organs of the larvae of the soybean cyst nematode, *Heterodera glycines.* J Ultrastruct Res 72: 349–366
23. Endo BY (1983) Ultrastructure of the stomatal region of the juvenile stage of the soybean cyst nematode, *Heterodera glycines.* Proc Helminthol Soc Wash 50: 43–61
24. Endo BY (1984) Ultrastructure of the esophagus of larvae of the soybean cyst nematode, *Heterodera glycines.* Proc Helminthol Soc Wash 51: 1–24
25. Endo BY (1985) Ultrastructure of the head region of molting second-stage juveniles of *Heterodera glycines* with emphasis on stylet formation. J Nematol 17: 112–123
26. Endo BY (1987) Ultrastructure of the esophageal gland secretory granules in juveniles of *Heterodera glycines.* J Nematol 19: 469–483
27. Endo BY, Wergin WP (1973) Ultrastructural investigation of clover roots during early stages of infection by the root-knot nematode, *Meloidogyne incognita.* Protoplasma 78: 365–379
28. Endo BY, Wergin WP (1977) Ultrastructure of anterior sensory organs of the root-knot nematode, *Meloidogyne incognita.* J Ultrastruct Res 59: 231–249
29. Endo BY, Wergin WP (1988) Ultrastructure of the second-stage juvenile of the root-knot nematode, *Meloidogyne incognita.* Proc Helminthol Soc Wash 55: 286–316
30. Gipson I, Kim KS, Riggs RD (1971) An ultrastructural study of syncytium development in soybean roots infected with *Heterodera glycines.* Phytopathology 61: 347–353
31. Goldschmidt R (1903) Histologische Untersuchungen an Nematoden. 1. Die Sinnesorgane von *Ascaris lumbricoides* L. und *A. megalocephala* Cloqu., Zool Jahrb 18: 1–57
32. Gunning BES, Pate JS, Green LW (1970) Transfer cells in the vascular system of stems: taxonomy, association with nodes, and structure. Protoplasma 71: 147–171
33. Hewitt WB, Raski DJ, Goheen AC (1958) Nematode vector of soilborne fanleaf virus of grapevines. Phytopathology 48: 586–595
34. Hope WD (1965) A taxonomic review of the genus Thoracostoma, Marion, 1870

(Nematoda: Leptosomatidae) and a study of the histologic morphology of *Thoracostoma californicum*. Ph.D. Thesis, University of CA, Davis, USA Diss Abstr 26: 2389
35. Höppli R (1925) Über das Vorderende der Ascariden. Vergleichende histologische Untersuchungen unter besonderer Berücksichtigung der Zellkonstanzfrage. Z Zellforsch Mikrosk Anat 2: 1–68
36. Huang CS, Maggenti AR (1969a) Mitotic aberrations and nuclear changes of developing giant cells in *Vicia faba* caused by root-knot nematode, *Meloidogyne javanica*. Phytopathology 59: 447–455
37. Huang CS, Maggenti AR (1969b) Wall modifications in developing giant cells of *Vicia faba* and *Cucumis sativus* induced by root-knot nematode, *Meloidogyne javanica*. Phytopathology 59: 931–937
38. Hussey RS (1987) Secretions of esophageal glands of Tylenchida nematodes. In: Veech JA, Dickson DW (eds) Vistas in nematology. Soc of Nematologists, Inc. E. O. Painter Printing Co, De Leon Springs, Florida, pp 509
39. Jones MGK (1981) The development and function of plant cells modified by endoparasitic nematodes. In: Zuckerman BM, Rohde RA (eds) Plant parasitic nematodes, Vol 3. Academic Press, New York, pp 588
40. Jones MGK, Dropkin VH (1975) Scanning electron microscopy of syncytial transfer cells induced in roots by cyst-nematodes. Physiol Plant Pathol 7: 259–263
41. Jones MGK, Dropkin VH (1976) Scanning electron microscopy of nematode-induced giant transfer cells. Cytobios 15: 149–161
42. Jones MGK, Gunning BES (1976) Transfer cells and nematode induced giant cells in *Helianthemum*. Protoplasma 87: 273–279
43. Jones MGK, Northcote DH (1972a) Multinucleate transfer cells induced in Coleus roots by the root-knot nematode, *Meloidogyne arenaria*. Protoplasma 75: 381–395
44. Jones MGK, Northcote DH (1972b) Nematode-induced syncytium—a multinucleate transfer cell. J Cell Sci 10: 789–809
45. Jones MGK, Novacky A, Dropkin VH (1975) Transmembrane potentials of parenchyma cells and nematode-induced transfer cells. Protoplasma 85: 15–37
46. Jones MGK, Payne HL (1978) Early stages of nematode-induced giant-cell formation in roots of *Impatiens balsamina*. J Nematol 10: 70–84
47. Kim YH, Riggs RD, Kim KS (1987) Structural changes associated with resistance of soybean to *Heterodera glycines*. J Nematol 19: 177–187
48. Lamberti F, Roca F (1987) Present status of nematodes as vectors of plant viruses. In: Veech JA, Dickson DW (eds) Vistas on nematology. E. O. Painter Printing Co, De Leon Springs, Florida, pp 509
49. McLaren DJ (1976) Nematode sense organs. In: Dawes B (ed) Advances in parasitology, Vol 14. Academic Press, London, England, pp 462
50. Natasasmita S (1979) Transmission and scanning electron microscope observations of *Tylenchulus semipenetrans* juveniles (J2), males and females. Ph.D. Thesis, University of Gent, Belgium
51. Nemec B (1932) Über die Gallen von *Heterodera schachtii* auf der Zuckerrübe. Stud Plant Physiol Lab, Charles University Prague 4: 1–14
52. Orion D, Wergin WP (1982) Chloroplast differentiation in tomato root galls induced by the root-knot nematode *Meloidogyne incognita*. J Nematol 14: 77–83
53. Owens RG, Specht HN (1964) Root knot histogenesis. Contrib Boyce Thompson Inst 22: 471–489
54. Pate JS, Gunning BES (1972) Transfer cells. Annu Rev Plant Physiol 23: 173–196
55. Paulson RE, Webster JM (1970) Giant cell formation in tomato roots caused by *Meloidogyne incognita* and *Meloidogyne hapla* (Nematoda) infection. A light and electron microscope study. Can J Bot 48: 271–276
56. Razak AR, Evans AAF (1976) An intracellular tube associated with feeding by *Rotylenchulus reniformis* on cowpea root. Nematologica 22: 182–189
57. Rebois RV (1980) Ultrastructure of a feeding peg and tube associated with *Rotylenchulus reniformis* in cotton. Nematologica 26: 396–405

58. Rebois RV, Madden PA, Eldridge BJ (1975) Some ultrastructural changes induced in resistant and susceptible soybean roots following infection by *Rotylenchulus reniformis*. J Nematol 7: 122–139
59. Rice SL, Stone PR, Leadbeater BSC (1987) Changes in cell structure in roots of resistant potatoes parasitized by potato cyst nematodes. 2. Potatoes with resistance derived from *Solanum vernei*. Physiol Mol Plant Pathol 31: 1–14
60. Riggs RD, Kim KS, Gipson I (1973) Ultrastructural changes in Peking soybeans infected with *Heterodera glycines*. Phytopathology 63: 76–84
61. Robertson WM, Wyss U (1979) Observation on the ultrastructure and function of the dorsal esophageal gland cell in *Xiphinema index*. Nematologica 25: 391–396
62. Robertson WM, Wyss U (1983) Feeding processes of virus-transmitting nematodes. In: Harris KF (ed) Current topics in vector research, Vol 1. Praeger, New York, pp 271–295
63. Ross JP (1958) Host-parasite relationship of the soybean cyst nematode in resistant soybean roots. Phytopathology 48: 578–579
64. Rumpenhorst HJ (1984) Intracellular feeding tubes associated with sedentary plant parasitic nematodes. Nematologica 30: 77–85
65. Shepherd AM, Clark SA, Hooper DJ (1980) Structure of the anterior alimentary tract of *Aphelenchoides blastophthorus* (Nematoda: Tylenchida, Aphelenchina). Nematologica 26: 313–357
66. Shepherd AM, Clark SA (1983) A re-examination of the esophageal ultrastructure in *Ditylenchus dipsaci* (Nematoda, Tylenchida) with some observations on intestinal structure. Nematologica 29: 151–170
67. Sundermann CA, Hussey RS (1988) Ultrastructural cytochemistry of secretion granules of esophageal glands of *Meloidogyne incognita*. J Nematol 20: 141–149
68. Taylor CE, Brown DJF (1981) Nematode-virus interactions. In: Zuckerman B, Rohde RA (eds) Plant parasitic nematodes, Vol 3. Academic Press, New York, pp 281–299
69. Veech JA, Endo BY (1970) Comparative morphology and enzyme histochemistry in root-knot resistant and susceptible soybeans. Phytopathology 60: 896–902
70. Ward S, Thomson JN, White JG, Brenner S (1975) Electron microscopical reconstruction of the anterior sensory anatomy of the nematode *Caenorhabditis elegans*. J Comp Neurol 160: 313–337
71. Weischer B, Wyss U (1976) Feeding behaviour and pathogenicity of *Xiphinema index* on grapevine roots. Nematologica 22: 319–325
72. Wergin WP, Endo BY (1976) Ultrastructure of a neurosensory organ in a root-knot nematode. J Ultrastruct Res 56: 258–276
73. Wergin WP, Orion D (1981) Scanning electron microscope study of the root-knot nematode (*Meloidogyne incognita*) on tomato root. J Nematol 13: 358–367
74. Wyss U (1973) Feeding of *Tylenchorhynchus dubius*. Nematologica 19: 125–136
75. Wyss U (1975a) Response of root cells to the feeding by the ectoparasitic root-infesting nematode *Trichodorus similis*. Z Pflanzenkr Pflanzenschutz 82: 685–694
76. Wyss U (1975b) Feeding of *Trichodorus*, *Longidorus* and *Xiphinema*. In: Lamberti F, Taylor CE, Seinhorst JW (eds) Nematode vectors of plant viruses. Plenum, New York, pp 203–221
77. Wyss U (1977) Feeding phases of *Xiphinema index* and associated processes in the feeding apparatus. Nematologica 23: 463–470
78. Wyss U (1978) Root and cell response to feeding by *Xiphinema index*. Nematologica 24: 159–166
79. Wyss U (1981) Ectoparasitic root nematodes: Feeding behavior and plant cell responses. In: Zuckerman BM, Rohde RA (eds) Plant parasitic nematodes, Vol 3. Academic Press, New York, pp 325–351
80. Wyss U, Zunke U (1985) *In vitro* feeding of root parasitic nematodes. In: Zuckerman BM, Mai WF, Harrison MB (eds) Plant nematology laboratory manual. Univ Mass Agr Exp Sta, Amherst, pp 91–96
81. Wyss U, Zunke U (1986a) The potential of high resolution video-enhanced contrast microscopy in nematological research. Rev Nematol 9: 91–94

82. Wyss U, Zunke U (1986b) Observations on the behaviour of second stage juveniles of *Heterodera schachtii* inside host roots. Rev Nematol 9: 153–165
83. Wyss U, Jank-Ladwig R, Lehmann H (1979) On the formation and ultrastructure of feeding tubes produced by trichodorid nematodes. Nematologica 25: 385–390
84. Wyss U, Lehmann H, Jank-Ladwig R (1980) Ultrastructure of modified root-tip cells in *Ficus carica*, induced by the ectoparasitic nematode *Xiphinema index*. J Cell Sci 41: 193–208
85. Wyss U, Stender C, Lehmann H (1984) Ultrastructure of feeding sites of the cyst nematode *Heterodera schachtii* Schmidt in roots of susceptible and resistant *Raphanus sativus* L. var. *oleiformis* Pers. cultivars. Physiol Plant Pathol 25: 21–37
86. Wyss U (1988) Reaktion von Wurzelzellen (*Ficus carica*) auf die Saugtätigkeit des Nematoden *Xiphinema index*. Film D 1657 des IWF, Göttingen

Chapter 23

M. Dollet

Infection of Plants by Flagellate Protozoa (*Phytomonas* SPP., Trypanosomatidae)

1 Introduction

The existence of trypanosomes in latex plants has been known since 1909 [12]. Immediately after their discovery, Donovan proposed the generic name *Phytomonas* [9], although these organisms had not been characterized in any way. The classification of trypanosomes is currently as follows: Order Kinetoplastida, Suborder Trypanosomatina, Family Trypanosomatidae [15]. Beside this arbitrarily created genus *Phytomonas*, the following genera exist: *Leptomonas, Herpetomonas, Crithidia, Blastocrithidia*, and *Rhynchoidomonas*, which are insect trypanosomatids. *Endotrypanum* and *Trypanosoma* are animal and human trypanosomatids transmitted by insects.

Phytomonas are also transmitted by insects belonging to the Lygaeidae, Coreidae and Pentatomidae families [2, 10, 13]. This field of protozoology has long been ignored by phytopathologists, partly because this problem virtually only concerned noncultivated latex plants and, therefore, had no economic repercussions, and partly because these organisms had never been cultured in vitro and it was not known how to characterize them. The discovery in the 1970's of *Phytomonas* specifically associated with decay in cultivated plants, with considerable economic consequences [6, 7, 16], and the successful *in vitro* culture of *Phytomonas* [3, 14], created new interest in this field [8].

It was Stahel [21] who revealed the first phytopathological problems linked to trypanosomatids, in his study of coffee tree phloem necrosis in Surinam. The possible reasons for the failure to appreciate his work have been discussed elsewhere [2].

Division Virologie IRHO, Laboratoire de Phytovirologie des Régions Chaudes (LPRC), CIRAD, BP 5035, F-34032 Montpellier Cedex 1, France

2 Phloem Restricted Phytomonas

Hartrot of coconut and Marchitez sorpresiva (lethal wilt) of oil palm remained diseases "of unknown origin" for a long time. The hypothesis of a viral or mycoplasmal etiology was the one most considered during the 1970's, which led to diagnostic research with electron microscopy. With both diseases, examination under the electron microscope of phloem of coconut or oil palm inflorescences or young leaves, revealed the presence of microorganisms with a nucleus and a flagellum (Figs. 1.2) [6, 16]. These organisms, measuring about 12 to 20 μm in length and 0.5 to 2 μm in diameter have all the characteristics of Trypanosomatidae (Figs. 3–6). The unit pellicular membrane is lined with a row of microtubules parallel with the longitudinal axis of the organism. A flagellar pocket surrounds the flagellar axoneme at the anterior apex. In cross-section, the distal portion of the flagellum has nine peripheral groups of two microtubules and two single central microtubules. The membrane in the flagellar pocket is devoid of subpellicular microtubules except for a group of four short microtubules [23]. For much of its length, the flagellum carries a paraxial rod. Within the cytoplasm, at the base of the flagellum, there is electron-dense fibrillar material corresponding to the DNA circles of the kinetoplast. Electron-opaque bodies of unknown nature and a single nucleus are also present in the cytoplasm. Sieve elements of diseased palms infected with *Phytomonas* generally contain fewer plastids than those of healthy palms. The diameter of the sieve-plate pores in palms is between 0.3 and 1.0 μm. Despite the presence of callose in fixed material, the Trypanosomatids are apparently able to pass from one sieve tube to another (Figs. 7–9). *Phytomonas* are only present in mature sieve tubes, but not all the sieve tubes of a vascular bundle are infected. The percentage of vascular bundles containing *Phytomonas* varies from 10 to 50%, probably depending on the stage of the disease. The number of *Phytomonas* that can be observed in a cross-section may range from one to about 10–15 on average, and up to completely filled sieve tubes.

No clear cytopathic effects, like in virus or mycoplasma infection, can be seen in diseased tissues. However, apparently collapsed cells with an electron-opaque contents occur sometimes in sieve tubes of oil palm affected by Marchitez (Figs. 7–9).

3 Milkweed Phytomonas

The first work on the ultrastructure of milkweed *Phytomonas* with the electron microscope was carried out on *Asclepias syriaca* [17]. The main ultrastructural characteristics described previously apply also to these *Phytomonas* (Figs. 10–13), whose general morphology can be observed by scanning electron microscopy

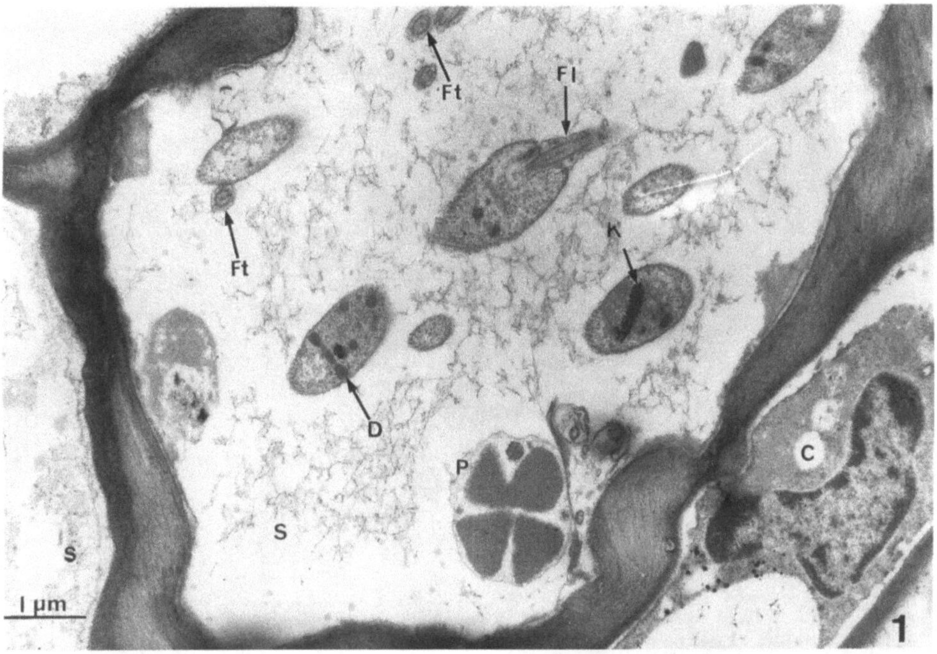

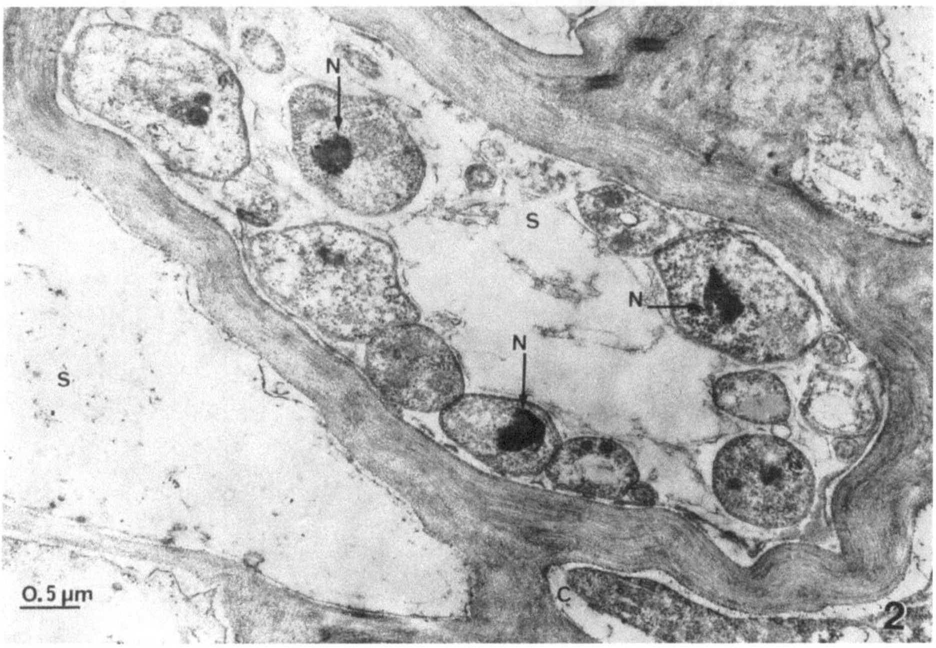

Figs. 1 and **2.** Transmission electron micrographs of *Phytomonas* sp. associated with Marchitez sorpresiva of oil palm in Peru. **1, 2** Transverse section of a vascular bundle in the inflorescence stalk of an oil palm that had early symptoms of Marchitez. *S* sieve element; *C* companion cell; *P* Plastid; *F* flagellum in longitudinal (*l*) or transverse (*t*) section; *N* nucleus; *D* dense bodies; *K* kinetoplast DNA

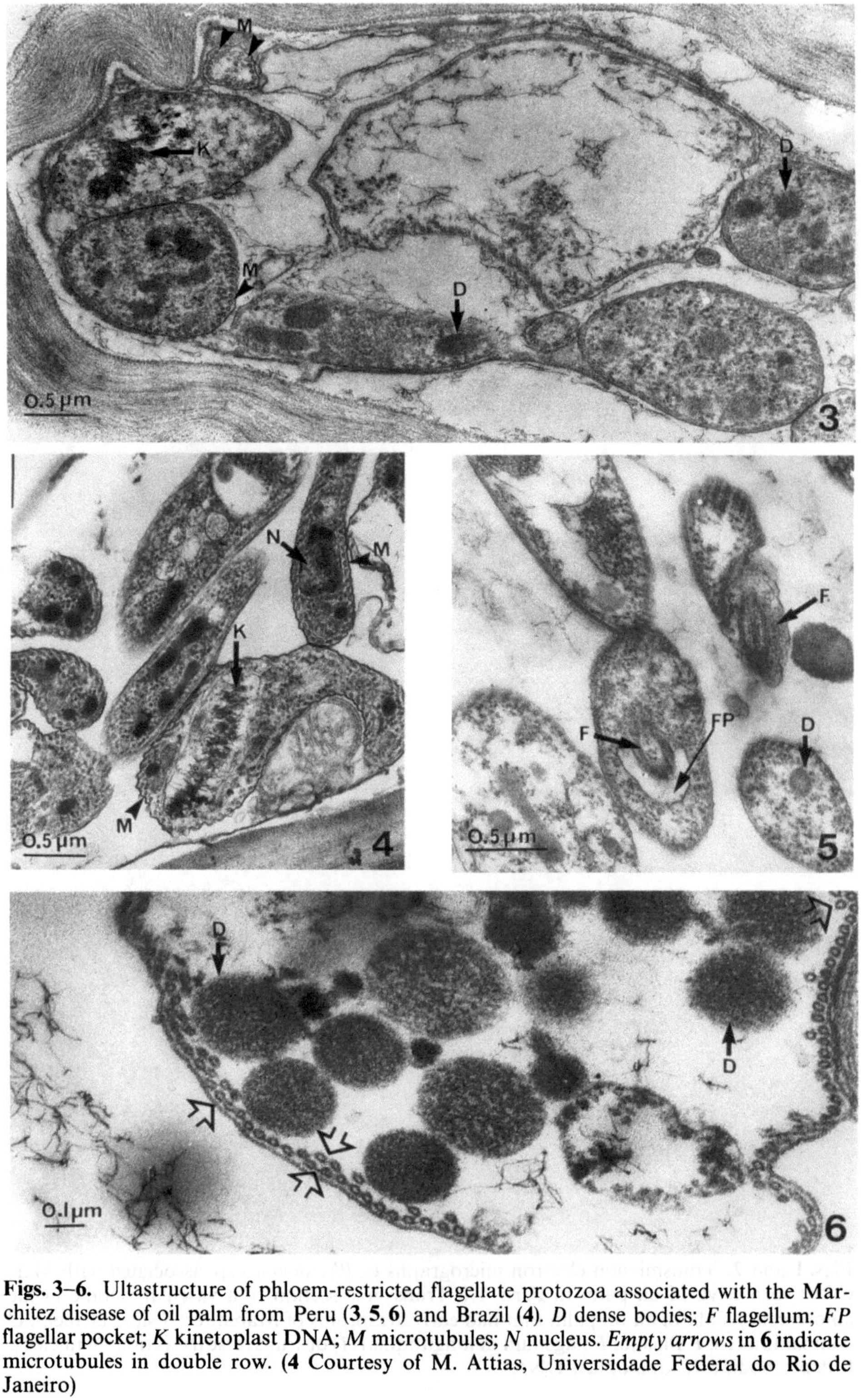

Figs. 3–6. Ultastructure of phloem-restricted flagellate protozoa associated with the Marchitez disease of oil palm from Peru (**3, 5, 6**) and Brazil (**4**). *D* dense bodies; *F* flagellum; *FP* flagellar pocket; *K* kinetoplast DNA; *M* microtubules; *N* nucleus. *Empty arrows* in **6** indicate microtubules in double row. (**4** Courtesy of M. Attias, Universidade Federal do Rio de Janeiro)

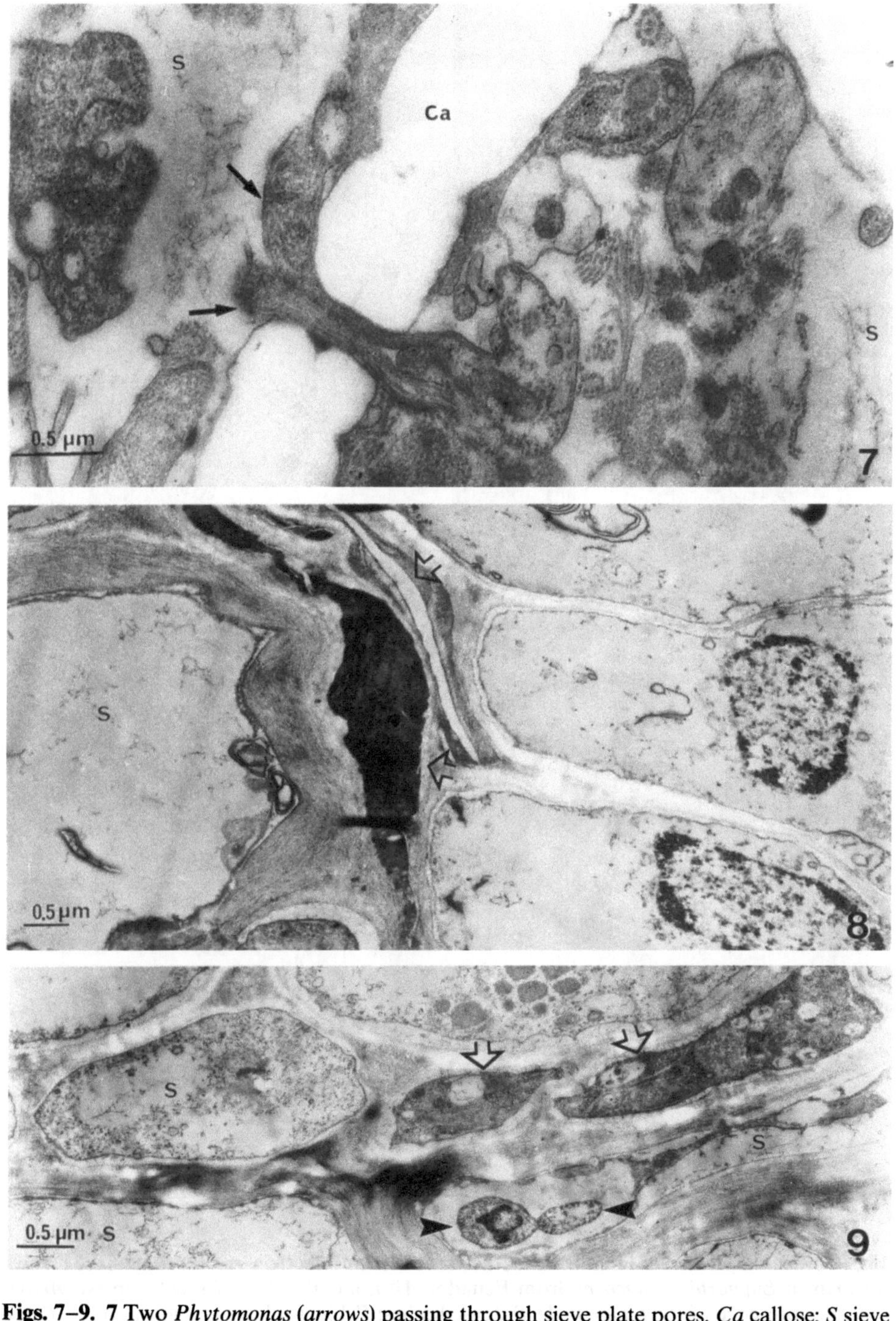

Figs. 7–9. **7** Two *Phytomonas* (*arrows*) passing through sieve plate pores. *Ca* callose; *S* sieve element. **8, 9** Vascular bundle of young inflorescences of oil palm affected by Marchitez showing cells appearing to undergo premature degeneration (*empty arrows*). *Arrow heads* indicate the presence of *Phytomonas* organisms

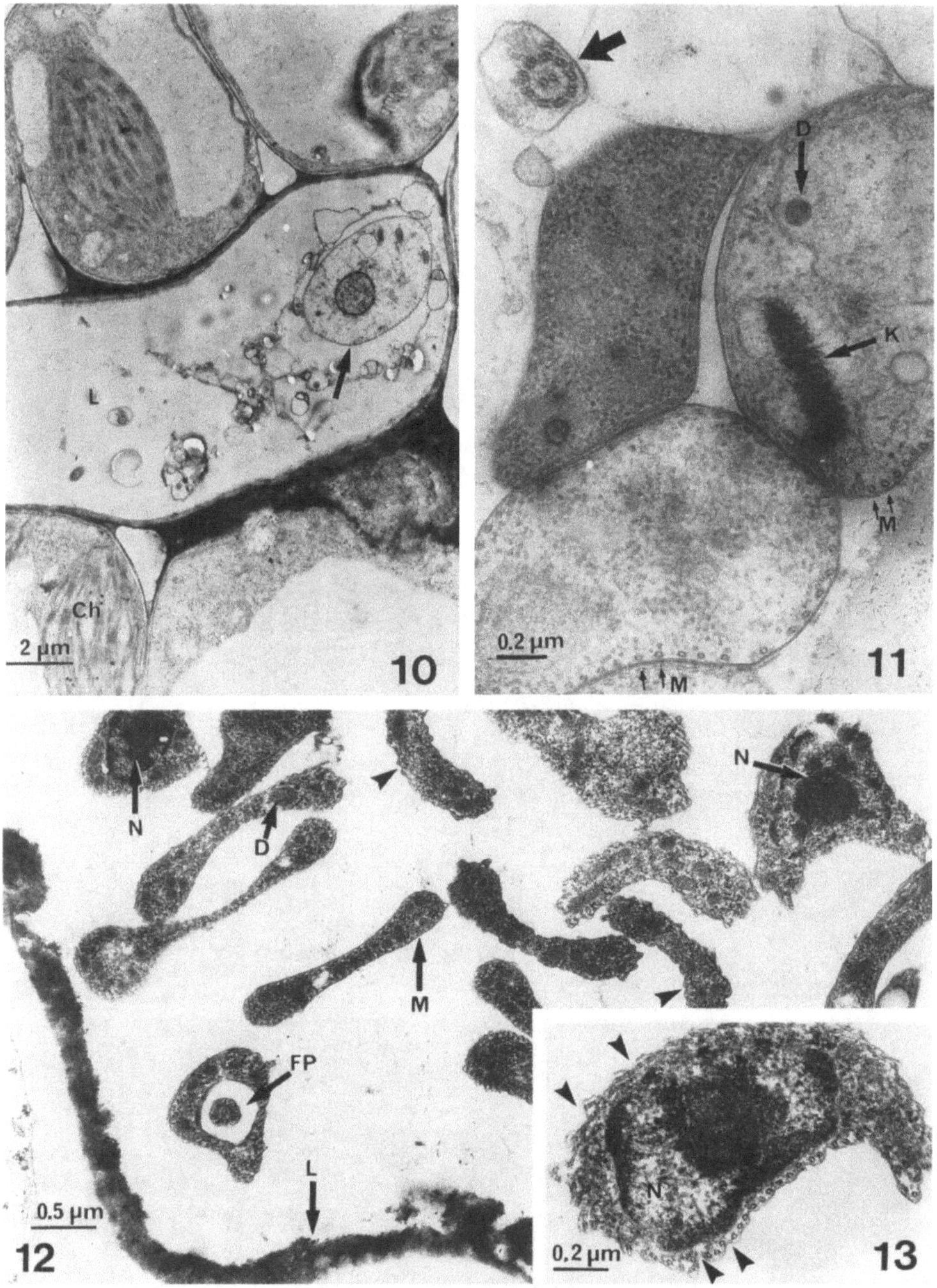

Figs. 10–13. *Phytomonas* in milkweeds in transmission electron microscopy. **10, 11** *Phytomonas* in *Euphorbia lasiocarpa* from Ecuador. **10** *L* laticifer duct; *Ch* chloroplast. *Arrow* indicates presence of *Phytomonas.* **11** *D* dense bodies; *K* kinetoplast; *M* microbutules. *Thick arrow* indicates flagellum. **12, 13** *Phytomonas* in *Asclepias curassavica* from Ecuador. *D* dense bodies; *F P* flagellar pocket; *L* latex; *M* microbutules; *N* nucleus. *Arrow heads* indicate the rippled membrane

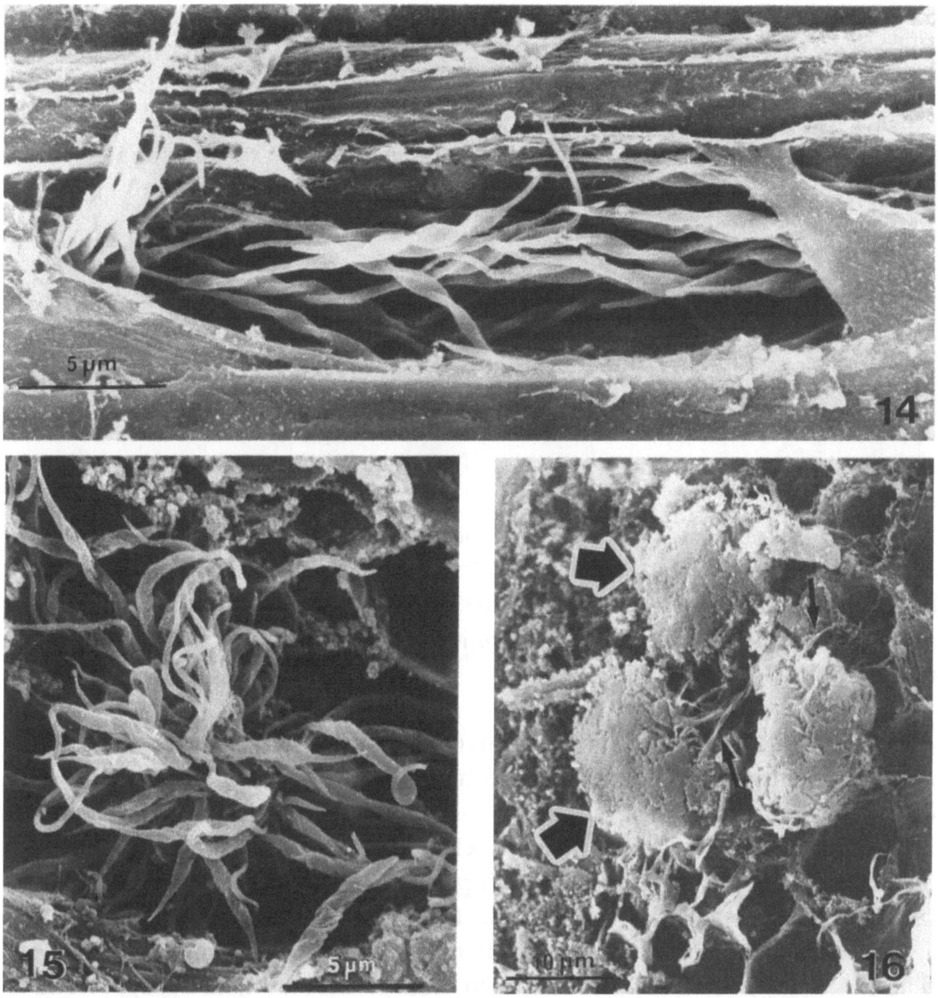

Figs. 14–16. Scanning electron micrographs of *Phytomonas*. **14** Phloem sieve tube of oil palm root, with *Phytomonas*. (Courtesy of M. Attias, Universidade Federal do Rio de Janeiro). **15, 16** Surface of a sectioned leaf petiole of cassava (*Manihot esculenta*) affected with the empty roots disease. Masses of latex (*thick arrows*) intermixed with *Phytomonas* (*thin arrows*). (Courtesy of E.W. Kitajima, Universidade de Brasilia)

(Figs. 14–16). These organisms are elongated and helically twisted, with a single flagellum. Thus, they differ from phloem-inhabiting *Phytomonas*, mainly because of their topological localization. Laticifer tubes are easily recognized by their contents, latex rich in granules and amorphous masses, and by their position. Moreover, the question of *Phytomonas* pathogenicity in latex plants remains an unsolved problem [2]. The only case of specific association between laticifer tube *Phytomonas* and a disease involves the "Chochamento das raizes" (i.e., empty roots), a disorder of cassava in Brazil [11]. However, no cytopathic effects were observed in the tissues and Koch's postulates have not been fulfilled. Certain symptoms seen in latex plants parasitized by *Phytomonas* are probably due to viruses [4]. In addition, the fact that *Phytomonas* can parasitize in latex plants does not mean that cells other than laticifer tubes, such as sieve tubes, should be overlooked. A study of diseased Euphorbia has revealed the existence of *Phytomonas* in the xylem and intercellular spaces [5].

4 Characterization of Phytomonas

Apart from a different cellular location, laticifer tube *Phytomonas* and phloem *Phytomonas* appear to be morphologically "identical" microorganisms. Different approaches have been taken to characterize these organisms since they were first cultured in vitro in 1982 [3].

In Vitro Culture. First and foremost they differ in the degree of difficulty encountered in their primoculture [2]. Whilst certain *Euphorbia* isolates can be grown under axenic conditions on a semi-synthetic medium (with fetal calf serum), others are recalcitrant, so that, for intraphloemic *Phytomonas* an insect cell culture was necessary [14].

Ultrastructural Study. Comparative analysis of the fine structure of four *Phytomonas* isolates from Euphorbiaceae cultured under axenic conditions in the same culture medium revealed variations of some consequences [1]. First of all it was impossible to use the same fixation scheme for the four isolates. Moreover, differences were observed in the endoplasmic reticulum, the peroxysome-like organelle (glycosome) and the kinetoplast DNA structure.

Between the *Phytomonas* from Euphorbiaceae and those from Asclepiadiaceae, differences may exist in the pellicular membrane, which, for instance, appears to be rippled (wavy) in *Phytomonas* from *A. curassavica* (Figs. 10–13), but not in *Phytomonas* from *A. syriaca* [17] or *E. lasiocarpa*. Nonetheless, certain cytological anomalies, such as a double row of subpellicular microtubules, sometimes seen in Marchitez *Phytomonas* (Figs. 3–6), are probably more due to fixation artifacts or to individual abnormality and therefore they may not be used as characterization criteria.

Other Characterization Methods. Different techniques make it possible to differentiate between the various isolates cultured to date, providing complementary data on *Phytomonas* variability—that always lead to the same direction. Techniques involving isoenzyme analysis [22], lectin agglutination [19], serology with monoclonal antibodies [18], and electrophoretic analysis of endonucleases-generated fragments of kinetoplastic-DNA [20] lead to the conclusion that a *Phytomonas* genus is justified, but there are great differences from one isolate to the next. These differences are related to geographical origin as well as to the host plant or the parasitized tissues.

Acknowledgment. Part of this work was supported by the Commission of European Communities.

References

1. Attias M, Roitman I, Camargo EP, Dollet M, De Souza W (1988) Comparative analysis of fine structure of four isolates of trypanosomatids of the genus *Phytomonas*. J Protozool 35: 365–370
2. Dollet M (1984) Plant diseases caused by flagellate protozoa (*Phytomonas*). Annu Rev Phytopathol 22: 115–132
3. Dollet M, Cambrony D, Gargani D (1982) Culture axénique in vitro de *Phytomonas* sp. (Trypanosomatidae) d'Euphorbe, transmis par *Stenocephalus agilis* Scop. (Coreide). CR Acad Sci Ser III 295: 547–550
4. Dollet M, Gargani D (1989) Mixed infection of *Asclepias curassavica* L. in Ecuador: Rhabdovirus-like particles and *Phytomonas* sp. J Phytopathol 125: 269–273
5. Dollet M, Gargani D, Cambrony D, Dzido JL (1983) Observations of flagellate protozoa (*Phytomonas* sp.) in the xylem of *Euphorbia lasiocarpa* in Ecuador. Can J Bot 61: 237–240
6. Dollet M, Giannotti J, Ollagnier M (1977) Observation de protozoaires flagellés dans les tubes criblés de palmiers à huile malades. CR Acad Sci Ser D 284: 643–645
7. Dollet M, Lopez G (1978) Study of the association between flagellate protozoa and Marchitez sorpresiva in South American oil palms. Oléagineux 33: 209–217
8. Dollet M, Wallace FG (1987) Report on the first *Phytomonas* workshop, Cayenne March 1987. Oléagineux 42: 461–468
9. Donovan C (1909) Kala-azar in Madras, especially with regard to its connexion with dog and the bug (*Conorrhinus*). Lancet 177: 1495–1496
10. Kastelein P (1985) Transmission of *Phytomonas* sp. (Trypanosomatidae) by the bug *Edessa loxdali* Westwood (Pentatomidae). De Surin Landb 33: 62–64
11. Kitajima EW, Vainstein MH, Silveira JSM (1986) Flagellate protozoan associated with poor development of the root system of cassava in the Espirito Santo state, Brazil. Phytopathology 76: 638–642
12. Lafont A (1909) Sur la présence d'un parasite de la classe des flagellés dans le latex de *l'Euphorbia pilulifera*. CR Soc Biol 66: 1011–1013
13. Louise C, Dollet M, Mariau D (1986) Recherches sur le Hartrot du cocotier, maladie à *Phytomonas* (Trypanosomatidae) et sur son vecteur *Lincus* sp. (Pentatomidae) en Guyane. Oléagineux 41: 437–449
14. Menara A, Dollet M, Gargani D, Louise C (1988) Culture in vitro sur cellules d'invertébrés des *Phytomonas* sp. (Trypanosomatidae) associés au Hartrot, maladie du cocotier. CR Acad Sci Série III: 597–602
15. Molyneux DH, Ashford RW (1983) The biology of *Trypanosoma* and *Leishmania*, parasites of man and domestic animals. Taylor and Francis, London
16. Parthasarathy MV, Van Slobbe WG, Soudant C (1976) Trypanosomatid flagellate in the phloem of diseased coconut palms. Science 192: 1346–1348
17. Paulin JJ, McGhee RB (1971) An ultrastructural study of the trypanosomatid, *Phytomonas elmassiani* from the milkweed, *Asclepias curassavica*. J Parasitol 57: 1279–1287
18. Petry K, Gargani D, Baltz T, Kastelein P, Dollet M (1989) Use of monoclonal antibodies for differentiation of different isolates of *Phytomonas* (plant trypanosomatids). J Phytopathol 126: 59–68
19. Petry K, Schottelius J, Dollet M (1987) Differentiation of *Phytomonas* sp. lower

trypanosomatids (Herpetomonas, Crithidia) by agglutination tests with lectins. Parasitol Res 74: 1–4

20. Riou JF, Dollet M, Ahomadegbe JC, Coulaud D, Riou G (1987) Characterization of *Phytomonas* sp. kinetoplast DNA, a plant pathogenic trypanosomal species. FEBS Lett 213: 304–308
21. Stahel G (1931) Zur Kenntnis der Siebröhrenkrankheit (Phloemnekrose) des Kaffeebaumes in Surinam I. Phytopathol Z 4: 65–82
22. Vainstein MH, da Silva JBT, Goncalves de Lima VMQ, Roitman I, de Souza W, Dollet M (1987) Electrophoretic analysis of isoenzymes in the identification of trypanosomatids of the genus *Phytomonas*. J Protozool 34: 442–444
23. Waters H (1978) A wilt disease of coconuts from Trinidad associated with *Phytomonas* sp., a sieve tube-restricted protozoan flagellate. Ann Appl Biol 90: 293–302

Chapter 24 SIGRUN HIPPE

Influence of Fungicides on Fungal Fine Structure

1 Introduction

Different strategies of antimicrobial therapy in plant, human, and veterinary medicine have been developed to protect biological tissue from fungal infection and damage. The discovery of systemic fungicides has been an important advance in the development of new chemical therapies. Unlike protective, nonsystemic, and in most cases nonspecific compounds, systemic fungicides are characterized by mobility in plant tissue and selective activity against the pathogen at the infection site. The selectivity is based on structural and physiological differences between host and fungal parasite. For a comprehensive understanding of the mode of action of antifungal agents, biochemical and cellular aspects have to be taken into account. Electron microscopy can provide unique information concerning the influence of fungicides on the fine structure of the pathogen and host. The present review gives a current view of the ultrastructure research in the pharmacology of selectively acting agents.

2 Carboxamides: Inhibition of Respiration Processes

Since the discovery of carboxamide fungicides in 1966, numerous related chemicals have been synthesized and introduced for use in plant protection because of their excellent antifungal activities towards diseases caused by Basidiomycetes. The mechanism of action of carboxamide fungicides (e.g., carboxin, oxycarboxin) has been examined on the molecular, physiological, and cytological level. In vitro studies on isolated mitochondria and on various fungi suggest that carboxin specifically blocks the succinate-ubiquinone oxidoreductase complex, a segment of the mitochondrial electron transport chain (Fig. 1). Decreased respiration and

Christian-Albrechts-Universität Kiel, Botanisches Institut, Biologiezentrum, Olshausenstrasse 40, D-2300 Kiel 1, FRG

altered mitochondrial fine structure are typical features of carboxamide-treated fungal cells. The relationship between the biochemical mode of action and ultrastructural effect of oxycarboxin has been analyzed in haustoria of *Uromyces phaseoli* [27]. Following fungicide application to rust-infected bean plants, fungal mitochondria rapidly became swollen and the cristae disorganized, leading to complete disruption of this organelle. Other changes in the haustorial cytoplasm included the accumulation of lipid bodies, whorled tubular profiles, and increased vacuolization. Dying haustoria subsequently became encased by plant material composed of callose-like substances and diverse electron-opaque extrahaustorial matrices (Fig. 2). Recent data of Heath [11] suggest that the typical response of a bean leaf cell to oxycarboxin-induced fungal death is encasement of the haustorium rather than plant cell death. In this context, selective fungicides have contributed an interesting aspect on the plant-pathogen interaction.

3 Aromatic Hydrocarbons and Dicarboximides: Lipid Peroxidation

The protective and, in some cases, systemic activities of aromatic hydrocarbon (e.g., chloroneb, PCNB) and dicarboximide fungicides (e.g., vinclozolin) against a variety of phytopathogenic fungi have been known for more than 20 years. The mode of action of both chemical groups has recently been described as a lipid peroxidation mechanism based on the generation of activated oxygen products such as superoxide and hydroxyl free radicals and hydrogen peroxide [24]. Free radicals are thought to be produced as a result of a blockage of the mitochondrial or microsomal electron flow after application of aromatic and dicarboximide fungicides (Fig. 1). A cascade process is subsequently induced leading to the peroxidation of unsaturated fatty acid in the cellular membranes of sensitive fungi. The destruction of membranes forming mitochondrial cristae, endoplasmic reticulum (ER) and the nuclear envelope has been observed in cells exposed to this class of fungicides by electron microscopy [24]. In *Mucor mucedo* and *Phytophthora cactorum* the aromatic fungicides chloroneb and PCNB cause lysis of the cristae in swollen mitochondria, damage to the nuclear envelope, pathological thickening of the cell walls and an increased vacuolization of the cytoplasm (Figs. 3, 4) [4, 25]. The increase in cell wall thickening may derive from disturbances in the regulation of plasma membrane-bound chitin and glucan synthetase as a result of the peroxidation of phospholipids adjacent to these enzymes. Similar effects on mitochondria, ER and nuclear membranes were observed in hyphae of *Botrytis cinerea* after treatment with the dicarboximide fungicide vinclozolin [24]. In all cases, the inner mitochondrial cristae appeared to be the most sensitive cellular membrane system, apparently because of its high content of unsaturated lipids and flavin enzymes. In comparison with the ER and nuclear membrane, the plasmalemma appears relatively insensitive

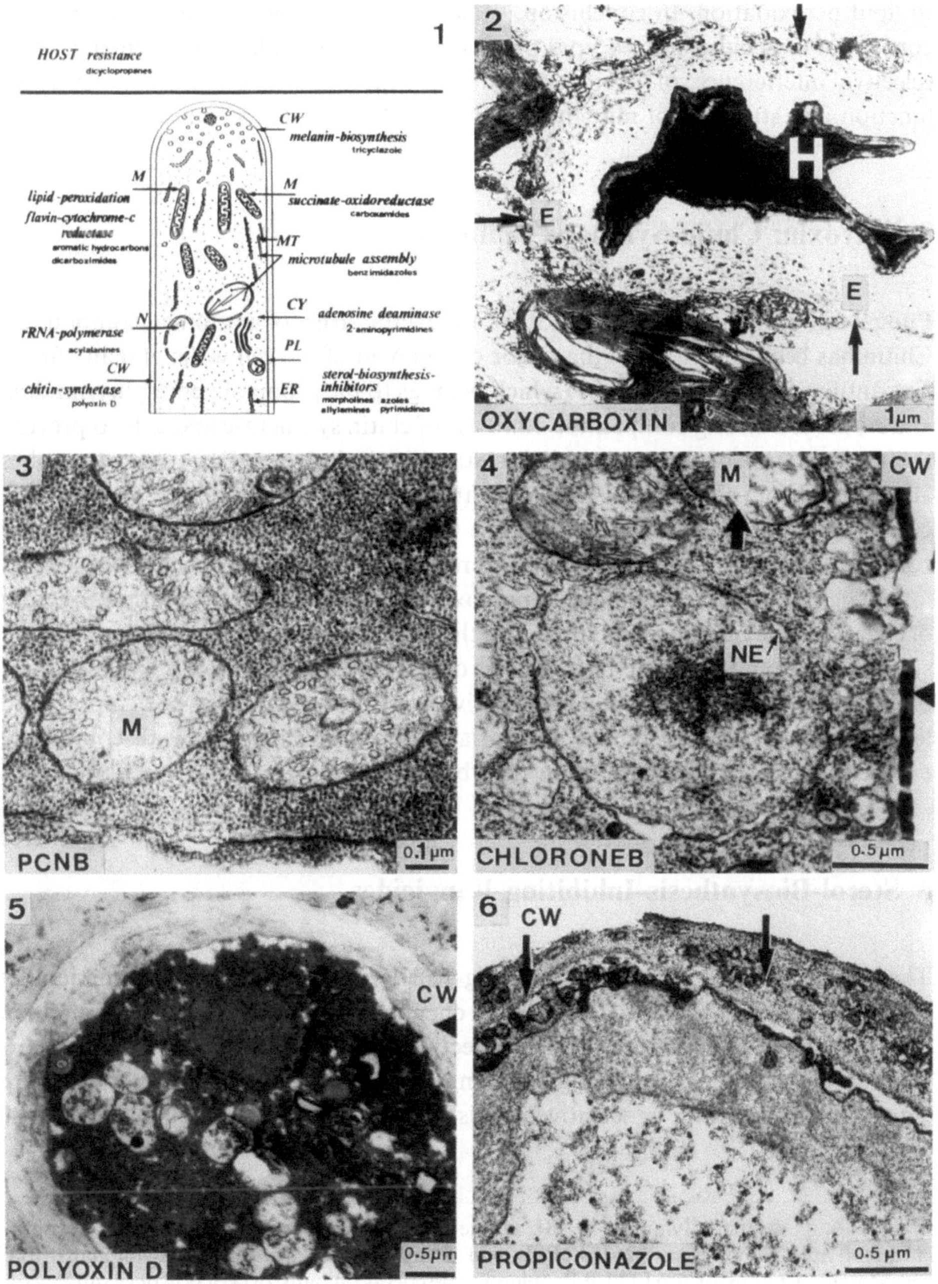

Figs. 1–6. 1 Sites of action of systemic fungicides on fungal fine structure [34]. **2** Oxycarboxin; carboxamide; treatment of *Uromyces phaseoli* on bean leaves [27]. **3** PCNB; aromatic hydrocarbon; applied to *Phytophthora cactorum* [4]. **4** Chloroneb; aromatic hydrocarbon; applied to *Mucor mucedo* [25]. **5** Polyoxin D effects on *Alternaria kikuchiana*; inhomogeneous cell wall thickening [23]. **6** Propiconazole; azole = SBI; treatment of *Taphrina deformans*; thickened cell wall containing membranous vesicles [29]

to lipid peroxidation. In conclusion, these electron microscopic data support the concept of fungicide-induced lipid peroxidation and provide strong evidence for a selective interference of aromatic hydrocarbons and dicarboximides with the function and structure of mitochondrial and other cellular membranes.

4 Polyoxin; Chitin Synthesis Inhibition

Fungal cell wall biosynthesis is another important target of antifungal compounds. Chitin has been shown to be the major component of the cell walls of many fungi. The antifungal agent polyoxin D which belongs to the group of peptidylpyrimidine antibiotics is a strong competitive inhibitor of chitin synthetase and acts to prevent the incorporation of glucosamine into chitin (Fig. 1). Polyoxin D-treated germ tubes and hyphae of *Alternaria kikuchiana* are reported to burst and release their cytoplasmic contents due to a decrease in cell wall rigidity [23]. The cell wall composition and thickness appeared heterogeneous and the cytoplasm degenerated following prolonged treatment with polyoxin D (Fig. 5). Furthermore, the induction of characteristic bulbous structures, and the increase in septation and pigmentation suggest additional effects of the antibiotic on fungal morphogenesis [23]. Treatment of the bean rust fungus infecting bean cultivars with polyoxin D has been reported to lead to degeneration of fungal haustoria [11]. Dead haustoria were completely encased by callose-like material in the absence of plant cell necrosis (Fig. 8).

5 Sterol-Biosynthesis-Inhibiting Fungicides

This class, sterol-biosynthesis-inhibiting (SBI) fungicides, contains functionally related compounds belonging to different chemical groups (e.g., azoles, pyrimidines, morpholines, and allylamines). Their selective fungitoxic action occurs at various steps in the sterol pathway (Fig. 1). Fungi that synthesize their own sterols are generally considered to be sensitive to this class of fungicides, although some of the

Figs. 7–16. Naftifine; allylamine = SBI; applied to *Trichophyton mentagrophytes*; accumulation of lipid depositions [26]. **8** Polyoxin D action on *Uromyces phaseoli* haustoria; encasement of a haustorium [11]. **9** Propiconazole effects on haustoria of powdery mildew on barley; irregular cell wall thickening. **10, 11** Ethirimol; 2-aminopyrimidine; applied to haustoria of powdery mildew; membrane blebbing. **12** Untreated, freeze-substituted haustorium of powdery mildew on barley; improved visualization of the fine structure [17]. **13** PF-face of the plasmalemma of untreated sporidia of *Ustilago avenae*; randomly distributed IMP's [18]. **14** Triadimenol; azole = SBI; influences on the plasmalemma of *Ustilago avenae*; areas of equally sized IMP's in hexagonal arrangement [18]. **15** Fine structure of untreated *Ustilago avenae*. **16** Triadimefon; azole = SBI; treated sporidia of *Ustilago avenae* [13]

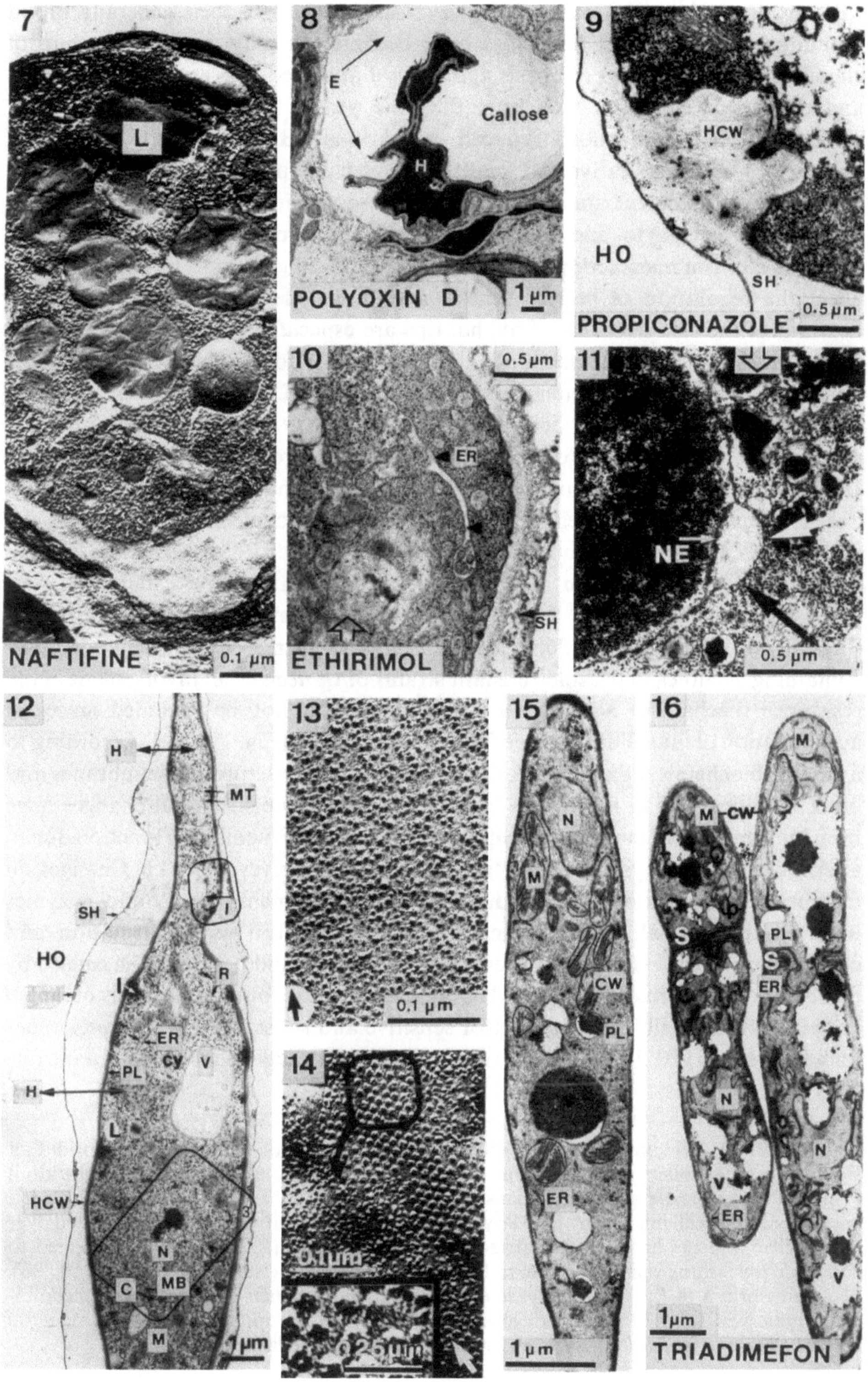
7
L
NAFTIFINE
0.1 µm
8
E
Callose
H
POLYOXIN D
1 µm
9
HCW
HO
SH
0.5 µm
PROPICONAZOLE
10
0.5 µm
ER
SH
ETHIRIMOL
11
NE
0.5 µm
12
H
MT
I
SH
HO
R
ER
CY
V
PL
H
L
HCW
N
C
MB
M
1 µm
13
0.1 µm
14
0.1 µm
0.25 µm
15
N
M
CW
PL
ER
1 µm
16
M
M
CW
S
PL
S
ER
N
N
V
ER
V
1 µm
TRIADIMEFON

SBI's exhibit a high degree of selectivity. On the basis of their generally broad antifungal spectrum, SBI fungicides have become the most important group of chemical agents used at present in agricultural and medicine against filamentous and dimorphic fungi, as well as pathogenic yeasts [34]. Morphological and ultrastructural modifications induced by various SBI compounds have been analyzed in both qualitative and quantitative electron microscopic investigations. Striking morphological abnormalities were observed in cell cultures of *Ustilago avenae* and other phyto- and human pathogenic fungi treated with various SBI's by scanning electron microscopy [1, 14, 26]. These data suggest that SBI's dramatically affect the regulation of morphogenesis and fungal development. Rapid modifications of the fine structure of hyphal tips are especially relevant in this regard. Following short-term treatment with propiconazole, disruption of the Spitzenkörper and cytoskeletal integrity in hyphal tips of germ tubes of *Puccinia graminis* f. sp. *tritici* has been documented [6]. Varying the SBI-fungicide concentration and the time of application, both in vitro and in vivo experiments resulted in a common sequence of ultrastructural changes leading to cellular degeneration and finally cell death. The sequence included undulations of the plasmalemma, increases and destabilization of the ER, disturbances of the mitochondria, accumulation of lipid-bodies, vacuolization and exocytosis, considerable thickening of cell walls, and incomplete septa formation (Figs. 6, 7, 15, 16, 20) [13, 15, 26, 29]. Concerning the increased occurrence of lipid-containing vacuoles in triazole-resistant strains of *U. avenae*, an involvement in the resistance mechanism against SBI compounds can not be excluded since an accumulation of fungicide has been found in these areas (Fig. 17) [19]. According to available biochemical data on the alteration of sterol structures in membranes and other modifications in membrane lipid composition, fine structural changes have been interpreted as a consequence of the deterioration of membrane function due to alterations in biophysical properties of the lipid bilayer [29, 31]. Changes in membrane composition may modify the activity of membrane-bound enzymes involved in cell wall synthesis, leading to irregularites in the formation and composition of the cell walls. However, an increase of lipid peroxidation caused by SBI compounds cannot be excluded as an indirect cause for modifications of fungal growth and cell wall formation in SBI sensitive and resistant fungal strains under long term stress [20, 24]. Further evidence for specific alterations in membrane fine

Figs. 17–22. 17 Triadimenol; azole = SBI; autoradiograph of SBI-resistant sporidia of *Ustilago avenae*; spherical silver grains visible over the vacuole (triadimenol localization) [19]. **18** Propiconazole applied to a haustorium of powdery mildew. **19** Cell wall section of a barley epidermal cell infected by powdery mildew; filamentous silver grains accumulated over the papillae and the host cell wall (triadimenol localization) [21]. **20** Naftifine applied to *Trichophyton mentagrophytes*; globular, lipidic structures on the surface of the cell wall [26]. **21** Appressorium of *Pyricularia oryzae* on *Bryophyllum pinnatum*; electron-dense, melanin-containing wall layer [36]. **22** Tricyclazole treatment of an appressorium of *Pyricularia oryzae*; electron-transparent layer; absence of electron-dense layer [36]

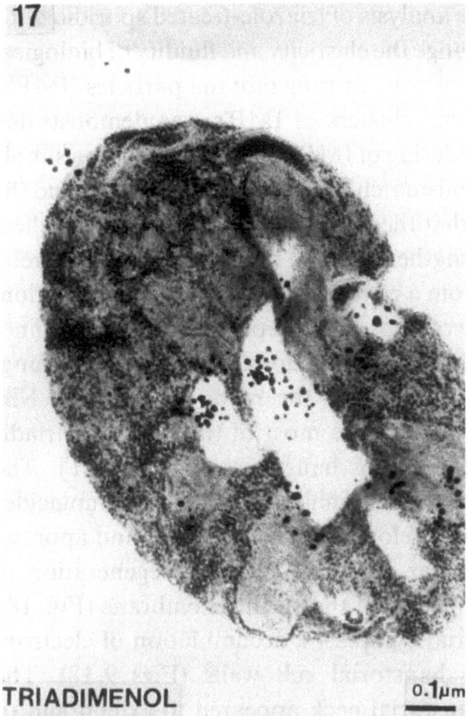
17
TRIADIMENOL
0.1µm

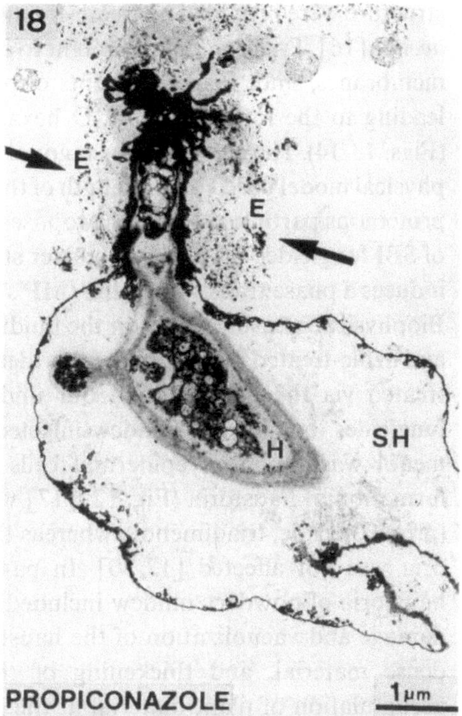
18
E
E
H
SH
PROPICONAZOLE
1µm

20
0.1 µm
CW
L
NAFTIFINE
L
50 nm

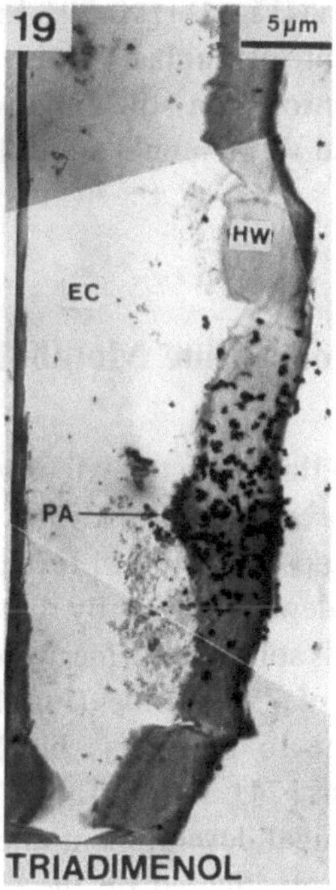
19
5µm
HW
EC
PA
TRIADIMENOL

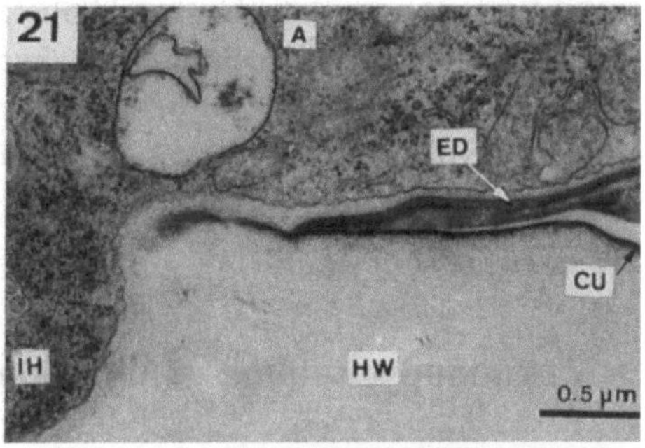
21
A
ED
CU
IH
HW
0.5 µm

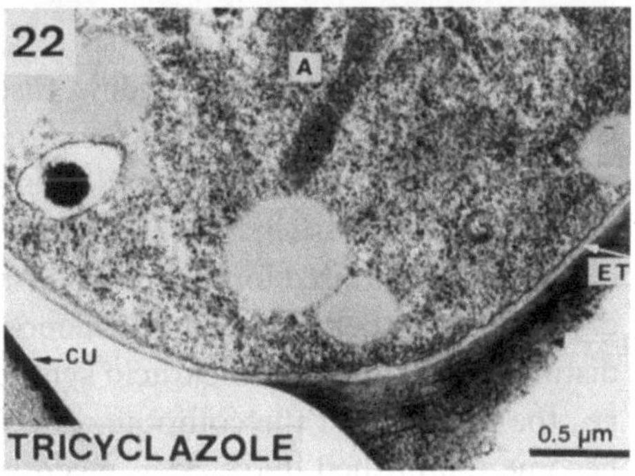
22
A
ET
CU
TRICYCLAZOLE
0.5 µm

structure were provided by freeze-fracture analysis of triazole-treated sporidia of *U. avenae* [16]. Triadimenol is thought to change the elasticity and fluidity of biological membranes, since rearrangements of proteinous intramembrane particles (IMP's) leading to the formation of 2-D hexagonal clusters of IMP's was demonstrated (Figs. 13, 14). The crystalline hexagonal ordering of IMP's was explained by a simple physical model based on the length of the mismatch between the lipid bilayer and the proteinous particles [18]. It is proposed that the liquifying (decreasing order) effect of SBI fungicides on the lipid bilayer strengthens the IMP interaction and thereby induces a phase transition of the IMP's from a random to a hexagonal distribution. Biophysical measurements on the fluidity of liposomes prepared from morpholine- and azole-treated fungi support this idea [31]. Investigations of in vivo-grown fungi treated via the plant add to our understanding of the mode of action of SBI fungicides. In powdery mildew-infected plant tissue most of the fungicide triadimenol was found in epidermal cells containing haustoria (Fig. 19) [21]. The formation of haustoria (Fig. 12) [17] was severely inhibited by triazole fungicides (propiconazole, triadimenol), whereas the development of germ tubes and appressoria was not affected [12, 30]. In particular, fungicide-induced degeneration of haustoria of powdery mildew included swelling of the sheath membrane (Fig. 18), damage and vacuolization of the haustorial cytoplasm, accumulation of electron-dense material, and thickening of the haustorial cell walls (Figs. 9, 18). The accumulation of plant material at the haustorial neck appeared to extend and to completely encapsulate damaged haustoria (Fig. 18) [12, 30]. Effects on the growth, sterol metabolism, and cellular fine structure of plants have also been described, and might partly contribute to antifungal action in a still unknown way. The prime target of SBI agents is, however, believed to be fungal sterol biosynthesis.

6 2-Aminopyrimidines: Inhibition of Purine Metabolism

The 2-aminopyrimidine fungicides (e.g., ethirimol) are extensively used in agriculture to control powdery mildew diseases, and are known for their selective antifungal activities on early infection steps. Ethirimol prevents germination and appressorium formation of powdery mildew on barley by direct interference with adenosine deaminase in the pathogen. This soluble enzyme, which is localized in the cytoplasm, is required for the metabolism of plant purines (Fig. 1). In the presence of ethirimol, the deamination of adenosine to inosine is blocked, resulting in a disturbance of fungal nucleic acid synthesis [24]. The "curative" effects of ethirimol, i.e., the influence of the compound on fungal development after the pathogen has become established, have been recently investigated on the electron microscopic level [12]. In contrast to chemically different powdery mildew inhibitors like SBI agents (propiconazole; Fig. 18), ethirimol did not affect the morphology of fungal haustoria. Rather, it prevented fungal growth on the leaf surface and altered the ultrastructure of haustoria. Hyphal tips became swollen and considerably vacuolized. Incompletely formed septa were common. Ethirimol induced characteristic fine structural changes not observed in SBI-treated mildew samples. Ethirimol-associated changes included blebbing of the nuclear envelope, accumulation of electron-dense material inside the nuclear envelope (Fig. 11) and enlarged cisternae of the endoplasmic reticulum (Fig. 10). Moreover, in contrast to the effects observed with SBI agents, no encapsulation of haustoria by plant material was found after ethirimol treatment, confirming the direct, selective effect of ethirimol on the pathogen [12].

7 Benzimidazoles: Inhibition of Microtubule Assembly

The group of benzimidazole fungicides together with the thiophanates, which are transformed to benzimidazoles, contain a large number of chemically related compounds (e.g., benomyl, carbendazim, nocodazole) which act by binding to β-tubulin and thus disrupting microtubule function (Fig. 1). In agriculture and medicine many benzimidazole derivatives have been developed for practical plant disease control and cancer chemotherapy, respectively [7]. The biochemical basis for their selectivity is due to a differential binding affinity of the tubulin from the various organisms [7]. Microtubules, which are ubiquitous cytoskeleton constituents of the cell, are dynamic structures that continually undergo assembly and disassembly. Inhibition of microtubule assembly in fungi by benzimidazoles disturbs cellular processes dependent on microtubules, such as mitosis and meiosis, the intracellular transport of molecules, and the movement of organelles [10]. Interference of benzimidazoles and thiophanates with nuclear processes in certain plant pathogenic and nonphytopathogenic fungi, as well as in other biological systems, has been convincingly demonstrated by electron microscopy. The main effects of these agents are manifested by a disorganization of the cellular contents of fungal spores and hypha, mammalian and animal cells. Metaphase arrest of mitosis, the aggregation of cytoplasmic vesicles and the ER as well as an irregular thickening of the cell wall have been reported in benzimidazole-treated tissues [10, 22, 28]. The effects of carbendazim, the degradation product of benomyl, on the fine structure of *Fusarium acuminatum* have been documented in detail using advanced low temperature preparation techniques in conjunction with electron microscopy [22]. The ability of carbendazim to disturb specifically the cell structure and influence cell development directly after application was clearly demonstrated by the examination of freeze-substituted hyphal tip cells in a short-term analysis. In untreated cells the Spitzenkörper region contained a large number of apical vesicles and microvesicles, lying within a network of microfilaments (Fig. 25). This accumulation of intracytoplasmic vesicles at the hyphal apex containing material necessary for cell wall

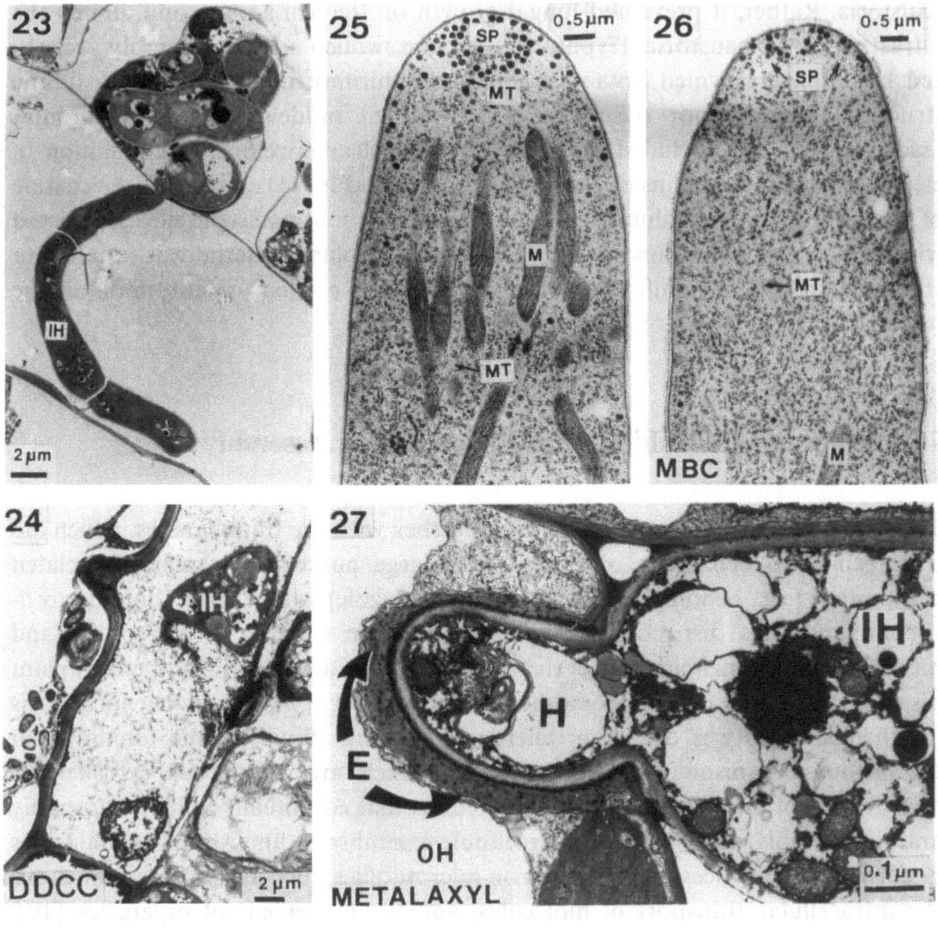

Figs. 23–27. 23 Infection hypha of *Pyricularia oryzae* on *Oryza sativa* [3]. **24** Dichlorocyclopropane effects on the host-parasite interaction; degeneration of the infection hypha in a necrotic epidermal cell; cytoplasmic proliferation in underlying mesophyll cells [3]. **25** Hyphal tip cell ultrastructure of *Fusarium acuminatum* [22]. **26** MBC; benzimidazole; treatment of *Fusarium acuminatum* hypha; reduction of the Spitzenkörper region; displacement of mitochondria from the subapical position [22]. **27** Metalaxyl; acylalanine; treated intracellular hypha and haustorium of *Peronospora tabacina* in *Nicotiana tabacum*; encasement of a haustorium and increased fungal vacuolization [33]

Abbreviations: *A* appressorium; *CW* cell wall; *C* crystal; *CU* cuticle; *CY* cytoplasm; *DDCC* 2,2-dichloro-3,3-dimethylcyclo-propane; *ED*/*ET* electron-dense/transparent layer; *ER* endoplasmic reticulum; *E* encasement; *EC* epidermal cell; *H* haustorium; *HCW* haustorial cell wall; *HO* host; *HW* host cell wall; *I* inclusion; *IH* infection hypha; *IMP* intramembrane particle; *L* lipid body; *MBC* methyl benzimidazole-2-ylcarbamate; *MB* microbody; *MT* microtubule; *M* mitochondrion; *N* nucleus; *NE* nuclear envelope; *PA* papillae; *PCNB* pentachloro-nitrobenzene; *PL* plasmalemma; *R* ribosomes; *S* septum; *SH* sheath; *SBI* sterol biosynthesis inhibitor; *SP* Spitzenkörper; *V* vacuole

synthesis indicates the direction of cell expansion. Microtubules which are generally orientated parallel to the long axis of the cell are believed to maintain the Spitzenkörper integrity [22]. Carbendazim altered the tip cell ultrastructure and the Spitzenkörper integrity within ten minutes of exposure (Fig. 26). Microtubules were no longer observed as ubiquitous components in the apical cytoplasm. The fungicide caused a simultaneous redistribution of apical vesicles, including a gradual disappearance of the Spitzenkörper. Presumably, the reduction of cell expansion is a result of the disruption of microtubule-mediated transport of cell wall material [22]. Prolonged exposure to carbendazim was associated with increased cytoskeleton destruction, an irregular deposition of cell wall layers, and an appearance of autolytic vacuoles and multivesicular bodies in the cytoplasm. These bodies have also been observed in other fungal organisms after treatment with benzimidazoles [28]. The relative insensitivity of both cytoplasmic microtubules in subapical cells and of spindle body-associated microtubules to carbendazim has been explained on the basis of different rates of metabolic turnover of cytoplasmic and nuclear microtubules [22]. In accordance with biochemical results, these electron microscopic data support the hypothesis that microtubules are the primary target for benzimidazoles. However, the disruption of microtubule assembly may also result from alterations to some other target important in microtubule regulation and microtubule functioning [10].

8 Tricylazole: Inhibition of the Melanin Biosynthesis

Tricyclazole is a commercially important compound which affects melanin-producing pathogens by blocking early stages of fungal infection. The inhibition of melanin biosynthesis in *Pyricularia oryzae* and other fungi by tricyclazole is thought to be the primary reason for the inability of treated fungi to penetrate plant tissue (Fig. 1). The melanin affected by tricyclazole is derived via the polyketide pathway and is an integral part of the cell walls of many phyto- and human pathogenic fungi. In many cases melanized cells are directly involved in the pathogenesis processes. Appressoria of some fungi become characteristically melanized prior to penetration of the host epidermal cells [35, 36]. According to electron microscopic investigations, tricyclazole does not prevent spore germination or appressorium development, but rather it blocks melanization of the cell walls of appressoria of *P. oryzae* [36] and *Colletotrichum lindemuthianum* [35] and the cell walls of sclerotia of *Sclerotinia sclerotiorum* [2]. During appressorium development, *P. oryzae* and *C. lindemuthianum* form slight depressions on the host cuticle. At this stage the basal area of the appressorium is characterized by a thin cell wall whereas the remaining part of the appressorium is surrounded by a thicker, three layered cell wall (Fig. 21). The middle, electron-dense layer has been identified as melanin-containing since in tricyclazole-treated fungi this particular layer is absent (Fig. 22). Treated appressoria fail to penetrate the cuticle, probably due to the lack of rigidity of the unmelanized appressoria [35, 36]. In untreated control samples frequent penetration of the plant cuticle and cell wall occurs (Fig. 21). The absence of the melanin-containing cell wall layer appears to be the only ultrastructural change in fungi after tricyclazole treatment. No ultrastructural modifications have been found in plant cells [35, 36]. These observations suggest that tricyclazole acts as an antipenetrant solely by interfering with appressorium functions.

9 Acylalanines: Inhibition of rRNA Synthesis

Acylalanine fungicides (e.g., metalazyl, furalaxyl) are compounds of commercial importance in plant protection because of their excellent properties against a large number of diseases caused by members of the Peronosporales. In studies on isolated nuclei from *Phytophthora* spp., metalaxyl inhibited rRNA-synthesis by specific interference with a chromatin bound RNA-polymerase complex [8]. As a consequence, inhibition of fungal growth, disturbances of the apical dominance of hyphae and degeneration of the fine structure, including thickening of the cell walls, were found after treatment of *Phytophthora infestans* in vitro [9]. The stimulation of cell wall formation following metalaxyl treatment presumably derives from an accumulation of rRNA precursors since nucleoside triphosphates are known to stimulate $\beta(1-3)$glucan synthetase activity [8]. However, early developmental stages of *P. infestans* such as the germination of encysted zoospores are not significantly affected by metalaxyl. No ultrastructural changes were visible in treated sporangia and zoospores [5]. The inability of metalaxyl to interfere with spore germination is probably due to the large number of preformed ribosomes and amount of protein in sporangia which sustain germination and early mycelial growth [8]. The relationship between in vitro and in vivo fungitoxicity of metalaxyl has been critically investigated. In plants, metalaxyl has a higher fungitoxicity than in vitro, suggesting that this fungicide must also induce a plant-mediated antifungal action. However, the increase in phytoalexin production (glyceollin) in infected and metalaxyl-treated plant tissue (soybean) was shown to be a secondary effect, probably due to an increased release of cell wall elicitors from metalaxyl-damaged hyphae [8, 32]. Electron microscopic studies particularly support the direct destructive effect of metalaxyl fungal haustoria of *Peronospora* spp. and *Phytophthora* spp. and help to resolve subsequent hypersensitive modifications at the host-parasite interface [32, 33]. Changes in the ultrastructure of intercellular hyphae and haustoria of downy mildew during systemic infection of tobacco include disintegration of the cytoplasm and organelles, convolution of the plasmalemma, increased exocytosis as well as cell wall thickening (Fig. 27) [33]. Moreover, degenerated fungal haustoria frequently became encased by a single, double, or multiple layered structure (Fig. 27). However, whether or not encasement of haustoria occurs is dependent only on the particular host parasite interaction. Encasement as well as phytoalexin production is triggered by the plant itself and not by metalaxyl treatment.

10 Nonfungicidal Disease Control Agents

The introduction of nonfungicidal chemicals for the control of plant diseases is a new strategy in modern plant protection. These nonfungicidal agents control the development of pathogens in vivo at concentrations which do not prevent growth of the parasite in vitro. They may act as agents, like the dichlorocyclopropanes, by stimulating host defence mechanisms.

11 Dichlorocyclopropane: Action on Host-Resistance Mechanisms

Among the cyclopropanes, DDCC has been extensively investigated because of its interesting biological and chemical properties at concentrations that are nonfungitoxic [3]. Unlike tricyclazole, DDCC controls rice blast disease after penetration by *P. oryzae* has occurred (Fig. 1) [24]. Necrotic reactions in penetrated plant cells of DDCC-treated rice leaves have been detected at the initial stage of the infection, whereas in untreated leaves no visible response to either penetration or the development of the infection hyphae was found [3]. Under control conditions fungal invasion is characterized by a rapid proliferation and branching of the initial intracellular hypha in adjacent epidermal and mesophyll cells (Fig. 23). In contrast, in DDCC-treated rice leaves, fungal development was halted soon after penetration by a hypersensitive response of the host cell resulting in cytoplasmic granulation, accumulation of brown pigments, and fragmentation of degenerated fungal hyphae (Fig. 24). The suppression of fungal growth in DDCC-treated rice plants was apparently associated with the stimulation of the production of fungitoxic phytoalexins, momilactone A and B [3]. Even though the mechanism of action of DDCC, leading to an enhancement of resistance response in the host plant, is not yet known, these data provide a good example of chemical manipulation of host defence mechanisms.

12 Conclusion

Electron microscopic studies can be seen as an important step to a better understanding of the mode of action of fungicides on the fungal pathogen and host parasite interface. Ultrastructural data provide leads for biochemical and biophysical investigations and help to resolve the relationship between the influences of fungicides on the function and structure of treated biological systems.

Acknowledgement. Financial support of the Dr. Christian Dräger Stiftung (Lübeck/FRG) during the preparation of the manuscript is gratefully acknowledged.

References

1. Barug D, Samson RA, Kerkenaar A (1983) Microscopic studies of *Candida albicans* and *Torulopsis glabrata* after in vitro treatment with bifonazole. Light and scanning electron microscopy. Drug Res 33: 528–537
2. Buchenauer H, Zeun R, Schinzer U (1985) Wirkung von Tricyclazol auf das Myzelwachstum sowie die Entwicklung und Pigmentierung der Sklerotien von *Sclerotinia sclerotiorum.* Z Pflanzenkrankh Pflanzenschutz 92: 17–26
3. Cartwright DW, Langcake P, Ride JP (1980) Phytoalexin production in rice and its enhancement by a dichlorocyclopropane fungicide. Physiol Plant Pathol 17: 259–267
4. Casperson G, Lyr H (1982) Wirkung von Pentochlornitrobenzol (PCNB) auf die Ultrastruktur von *Mucor mucedo* und *Phytophthora cactorum.* Z Allg Mikrobiol 22: 219–226
5. Coffey MD, Young LH (1984) Responses to metalaxyl of sensitive and resistant isolates of *Phytophthora infestans.* Phytopathology 74: 615–620
6. Dahmen H, Hoch HC, Staub T (1988) Differential effects of sterol inhibitors on growth, cell membrane permeability, and ultrastructure of two target fungi. Phytopathology 78: 1033–1042
7. Davidse LC (1986) Benzimidazole fungicides: mechanism of action and biological impact. Annu Rev Phytopathol 24: 43–65
8. Davidse LC (1987) Biochemical aspects of phenylamide fungicides—action and resistance. In: Lyr H (ed) Modern selective fungicides—properties, application, mechanisms of action, VEB Gustav Fischer Verlag Jena, Jena, pp 275
9. Grohmann U, Hoffmann GM (1982) Licht- und elektronenoptische Untersuchungen zur Wirkung von Metalaxyl bei *Pythium-* und *Phytophthora* Arten. Z Pflanzenkr Pflanzenschutz, 89: 435–446
10. Heath IB (1982) The effect of nocodazole on the growth and ultrastructure of the fungus, *Saprolegnia ferax*: evidence against a simple mode of action. In: Cappuccinelli P, Ronald Morris N (eds) Microtubules in microorganisms, Marcel Dekker, New York, pp 275
11. Heath MC (1988) Effect of fungal death or inhibition induced by oxycarboxin or polyoxin D on the interaction between resistant or susceptible bean cultivars and the bean rust fungus. Phytopathology 78: 1454–1462
12. Heller A, Frenzel B, Grossmann F, Hippe S (1990) Effects of ethirimol and propiconazole on morphology and ultrastructure of *Erysiphe graminis* f. sp. hordei Can J Botany 1990 (in press)
13. Hippe S, Buchenauer H, Grossmann F (1980) Einfluß von Triadimefon auf die Feinstruktur der Sporidien von *Ustilago avenae.* Z Pflanzenkr Pflanzenschutz 87: 423–426
14. Hippe S (1983) Morphology of *Ustilago avenae* after treatment with systemic fungicides as studied by scanning electron microscopy. Phytopathol Z 106: 321–328
15. Hippe S (1984) Electron microscopic investigations on the cytology of *Ustilago avenae* influenced by the systemic fungicides triadimefon, nuarimol and imazalil-nitrate: a thin-section and freeze-fracture study. Pestic Biochem Physiol 21: 170–183
16. Hippe S (1985) Hexagonal clustering of intramembrane particles in the plasmalemma of *Ustilago avenae* after fungicide treatment. Pestic Biochem Physiol 24: 161–173
17. Hippe S (1985) Ultrastructure of haustoria of *Erysiphe graminis* f.sp. *hordei* preserved by freeze-substitution. Protoplasma 129: 52–61

18. Hippe S Lüth H (1986) A simple physical model for fungicide induced hexagonal clustering of intramembrane particles in the plasmalemma of *Ustilago avenae*. J Theor Biol 21: 351–366
19. Hippe S (1987) Combined application of low temperature preparation and electron microscopic autoradiography for the localization of systemic fungicides. Histochemistry 87: 309–315
20. Hippe S, Giesen U (1988) The effect of triadimenol on the cytology and growth of sensitive and resistant strains of *Ustilago avenae*. Ann Appl Biol 112: 79–90
21. Hippe S Hermanns M (1988) Intracellular localization of the systemic fungicide triadimenol in phytopathogenic fungi. J Phytopathol 122: 13–24
22. Howard RJ, Aist JR (1980) Cytoplasmic microtubules and fungal morphogenesis: ultrastructural effects of methyl benzimidazole-2-ylcarbamate determined by freeze-substitution of hyphal tip cells. J Cell Biol 87: 55–64
23. Ishizaki H, Kohno M, Kunoh H (1977) Effect of polyoxin on fungi VIII. Microscopic observation of bursting of the bulbous germ tubes of polyoxin-treated *Alternaria kikuchiana* Tanaka in pear leaf extract. Ann Phytopathol Soc Jpn 43: 471–474
24. Lyr H (1987) Modern selective fungicides. VEB Gustav Fischer, Jena
25. Lyr H, Werner P (1982) On the mechanism of action of the fungicide chloroneb. Pestic Biochem Physiol 18: 69–76
26. Meingassner JG, Sleytr U, Petranyi G (1981) Morphological changes induced by naftifine, a new antifungal agent, in *Trichophyton mentagrophytes*. J Invest Dermatol 77: 444–451
27. Pring RJ, Richmond DV (1976) An ultrastructural study of the effect of oxycarboxin on *Uromyces phaseoli* infecting leaves of *Phaseolus vulgaris*. Physiol Plant Pathol 8: 155–162
28. Richmond DV, Pring RJ (1971) The effect of benomyl on the fine structure of *Botrytis fabae*. J Gen Microbiol 66: 79–94
29. Sancholle M, Dargent R, Weete JD, Rushing AE, Miller KS, Montant C (1988) Effects of triazoles on fungi. VI. Ultrastructure of *Taphrina deformans*. Mycologia 80: 162–175
30. Smolka S, Ebrahim-Nesbat F, Wolf G (1988) Elektronenmikroskopische Untersuchungen zur Beeinflussung der Wirt-Parasit-Beziehung Gerste-Mehltau durch Triadimenol (Baytan). Z Pflanzenkr Pflanzenschutz 95: 270–284
31. Steel CC, Baloch RI, Mercer EI, Baldwin BC (1989) The intracellular location and physiological effects of abnormal sterols in fungi grown in the presence of morpholine and functionally related fungicides. Pestic Biochem Physiol 33: 101–111
32. Stössel P, Lazarovitis G, Ward EWB (1982) Light and electron microscopy of *Phytophthora* rot in soybeans treated with metalaxyl. Phytopathology 72: 106–111
33. Trigiano RN, Van Dyke CG, Spurr HW (1984) Effects of metalxyl on *Peronospora tabacina* infecting tabacco. Phytopathology 74: 1034–1040
34. Trinci APJ, Ryley JF (1984) Mode of action of antifungal agents. Cambridge University Press, Cambridge
35. Wolkow PM, Sisler HD, Vigil EL (1983) Effect of inhibitors of melanin biosynthesis on structure and function of appressoria of *Colletotrichum lindemuthianum*. Physiol Plant Pathol 22: 55–71
36. Woloshuk, Sisler GD, Vigil EL (1983) Action of the antipenetrant, tricyclazole, on appressoria of *Pyricularia oryzae*. Physiol Plant Pathol 22: 245–259

Subject Index

MIX
Papier aus verantwortungsvollen Quellen
Paper from responsible sources
FSC® C105338

If you have any concerns about our products,
you can contact us on
ProductSafety@springernature.com

In case Publisher is established outside the EU,
the EU authorized representative is:
Springer Nature Customer Service Center GmbH
Europaplatz 3, 69115 Heidelberg, Germany

Printed by Libri Plureos GmbH
in Hamburg, Germany